R

Repair Manual
for British Cars

Robert Bentley's
Repair Manual
for
British Cars

John Organ

2009
BENTLEY PUBLISHERS
1734 Massachusetts Avenue
Cambridge, Massachusettts 02138

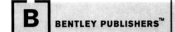

BENTLEY PUBLISHERS™ | Automotive Reference™

Bentley Publishers, a division of Robert Bentley, Inc.
1734 Massachusetts Avenue
Cambridge, MA 02138 USA   Information that makes
800-423-4595 / 617-547-4170   the difference®

BentleyPublishers™
.com

WARNING—Important Safety Notice

Do not use this manual for repairs unless you are familiar with basic automotive repair procedures and safe workshop practices. This manual illustrates the workshop procedures for some maintenance and service work. It is not a substitute for full and up-to-date information from the vehicle manufacturer or for proper training as an automotive technician. Note that it is not possible for us to anticipate all of the ways or conditions under which vehicles may be serviced or to provide cautions as to all of the possible hazards that may result.

We have endeavored to ensure the accuracy of the information in this manual. Please note, however, that considering the vast quantity and the complexity of the service information involved, we cannot warrant the accuracy or completeness of the information contained in this manual.

FOR THESE REASONS, NEITHER THE PUBLISHER NOR THE AUTHOR MAKES ANY WARRANTIES, EXPRESS OR IMPLIED, THAT THE INFORMATION IN THIS MANUAL IS FREE OF ERRORS OR OMISSIONS, AND WE EXPRESSLY DISCLAIM THE IMPLIED WARRANTIES OF MERCHANTABILITY AND OF FITNESS FOR A PARTICULAR PURPOSE, EVEN IF THE PUBLISHER OR AUTHOR HAVE BEEN ADVISED OF A PARTICULAR PURPOSE, AND EVEN IF A PARTICULAR PURPOSE IS INDICATED IN THE MANUAL. THE PUBLISHER AND AUTHOR ALSO DISCLAIM ALL LIABILITY FOR DIRECT, INDIRECT, INCIDENTAL OR CONSEQUENTIAL DAMAGES THAT RESULT FROM ANY USE OF THE EXAMPLES, INSTRUCTIONS OR OTHER INFORMATION IN THIS MANUAL. IN NO EVENT SHALL OUR LIABILITY WHETHER IN TORT, CONTRACT OR OTHERWISE EXCEED THE COST OF THIS MANUAL.

Your common sense and good judgment are crucial to safe and successful service work. Read procedures through before starting them. Think about whether the condition of your car, your level of mechanical skill, or your level of reading comprehension might result in or contribute in some way to an occurrence which might cause you injury, damage your car, or result in an unsafe repair. If you have doubts for these or other reasons about your ability to perform safe repair work on your car, have the work done at an authorized dealers or other qualified shop.

Copies of this book may be purchased from selected booksellers or directly from the publisher. The publisher encourages comments from the readers of this book. These communications have been and will be considered in the preparation of this and other books. Please write to Bentley Publishers at the address listed at the top of this page or e-mail us through our web site.

Library of Congress Catalog Card No 77-118658

ISBN 978-0-8376-0041-3
Bentley Stock No. X041
Mfg. code: X041-P1-0906

First published 1970

The paper used in this publication is acid free and meets the requirements of the National Standard for Information Sciences-Permanence of Paper for Printed Library Materials.

Robert Bentley's Repair Manual for British Cars by John Organ
© 1970, 2009 John Organ and Robert Bentley, Inc.

Manufactured in the United States of America

Contents

overhauling universal joints—Renewing front wheel bearings. Rear axle noises. Drive shafts BMC Mini, 1100, 1300 and 1800—Removing a drive shaft—Overhauling a drive shaft—Fitting a new bell joint. Drive shaft modifications

To Joanne

Acknowledgements

Acknowledgement is gratefully made to the following for supplying information and for their co-operation in allowing the use of illustrations: AC Delco Division of General Motors, Armstrong Patents, Associated Engineering Co. (Glacier Metal Co., Hepworth & Grandage), Automotive Products Group (Lockheed Brake Co.), British Leyland (Triumph Car Division, Austin Car Division, MG Car Division, Morris Car Division, SU Carburetter Co.), Burman Gear Co., Cam Gears Ltd, Champion Sparking Plug Co., Cords Piston Ring Co., Ford Motor Co., Girling Ltd, GMA Reconsets, Mr. R. A. Hall (Author of 'Electrical Equipment of British Cars'–FOULIS), Joseph Lucas (Electrical), Rootes Motors, Timken Bearing Co., Vandervell Products, Vauxhall Motor Co., Weber Carburetter Co. and Zenith Carburetter Co.

Introduction

Poor performance, high oil consumption, difficult starting, an oil warning light constantly flickering, and unfamiliar noises may indicate that an expensive repair bill is in the offing. The symptoms may alternatively indicate the need for a simple repair that can be carried out for the outlay of a few pounds and often only a few shillings. The principal key is accurate diagnosis for with modern material, design, oil and fuel, the amount of wear occurring in an engine is phenomenally small in comparison with that occurring in the engines of 30 years ago, especially where regular servicing has been carried out; indeed engine reconditioning (see page 83) is often carried out unnecessarily early. Engines today, however, still produce problems for, long before any appreciable bore wear occurs or bearings need renewing, the timing chain rattles (see page 41), cam followers become pitted, and numerous troubles occur with component parts.

Although regular decarbonizing is virtually a thing of the past, the cylinder head should nevertheless be removed every 20,000 miles for the valves and seatings will require attention, the oil seals need renewing (see page 56) and the valve springs, though they may last the life of the engine without breaking, may have become weaker to contribute to a 'woolliness' of performance that is not always immediately discernible, or troublesome, but which can lead to more expensive jobs.

Brake deterioration may go undetected until an emergency arises. Regular attention can avoid accidents and expense—working on one's own car can save money; moreover, such repairs need not be confined to minor servicing and adjustment. A brake overhaul (see chapter 9) can be carried out for about a quarter of the cost of a garage job. Carrying out one's own servicing and repairs is not only money saved but, as one becomes more familiar with motor cars, the likelihood of getting stranded due to a simple fault is lessened, especially if a few tools are carried in the car.

The basis of economy is efficiency, but mpg not only depends upon a well-tuned engine and a correctly adjusted fuel system, but also upon using the correct tyre pressures and having brakes that do not bind. Moreover, better consumption figures are more likely with a car that has some 40,000 miles to its credit for the frictional resistance is lower than in a car with a low mileage. The proviso of course is that the car with the higher mileage is kept in good mechanical order.

The requirement of special tools for carrying out some types of repair is often a deterrent to many who would enjoy the fascination and experience of working on their own car but just as previous experience is not essential, neither in many cases are special tools, indeed if they are needed they can often be hired for a very moderate charge.

Torque wrench settings may be considered a problem when doing some jobs, but a torque wrench is not an essential tool. For instance, when tightening a cylinder head if a socket of the correct size and a 14-in.-long bar is used, a fairly firm pull (without using brute strength) is satisfactory, providing the correct tightening sequence is followed (see page 51). Anyone with a torque wrench can obtain the correct settings for cylinder-head nuts and big-end nuts, etc. from the manufacturer's workshop manual, but common sense is of much more value than a list of torque settings and a torque wrench of doubtful accuracy.

Non-starting and misfiring can be troublesome with any motor car and chapter 1 and chapter 11 with their charts, etc. will help in the location and rectification of all non-starting and misfiring problems. This does not call for the use of expensive oscillographs and air/fuel meters, etc.; indeed, a very high percentage of poor-performance complaints prove extremely easy to locate and anyone prepared to spend a little time can reduce the cost of minor adjustments and even major repairs to a few shillings instead of the many pounds which professional labour costs naturally involve.

English/American Equivalents

English	*American*
GENERAL	
bush	bushing (bronze, rubber, etc.)
circlip	snap ring
distance piece	spacer
end float	end clearance
extractor	gear or bearing puller
joint washer	gasket
jointing compound	gasket cement, sealing compound
laden	loaded
methylated spirits	denatured alcohol
paraffin	kerosene
perished	rotted (from oil, etc.)
petrol	gasoline
renew	replace
set screw	bolt
spanner	end wrench
spigot	pilot
split pin	cotter pin
spring washer	lock washer
swarf	chips from cutting, drilling, etc.
ENGINE	
choke tube	venturi
cotters	split valve locks
float chamber	carburetor bowl
gudgeon pin	piston pin, wrist pin
oil sump	oil pan
silencer	muffler
CLUTCH	
clutch housing	bell housing
clutch release bearing	throwout bearing
clutch withdrawal fork	throwout arm
spigot bearing	pilot bearing
GEARBOX	
baulk ring	synchronizing ring, synchro cone
first motion shaft	input shaft
laygear	counter gear, cluster gear

English	American
layshaft	countershaft, cluster gear shaft
propeller shaft	driveshaft
third motion shaft	output shaft

REAR AXLE

crown wheel	ring gear

ELECTRICAL

distributor suction advance	vacuum advance
earth	ground (positive earth = positive ground)
micro adjuster	octane selector

SUSPENSION AND STEERING

hydraulic damper	shock absorber

BODY

bonnet	hood, engine compartment cover
boot	trunk compartment
bulkhead	firewall

Abbreviations

a b d c	after bottom dead center
b t d c	before top dead center
A C	alternating current
A F	'across face'—bolt or spanner measurement
B A	British Association screw threads
B S F	British Standard fine screw threads
B S W	British Standard Whitworth screw threads
b h p	brake horsepower
cc or cm^3	cubic centimeters
D C	direct current
H T	high tension (current)
l h	left hand side (viewed from driver's seat)
L T	low tension (current)
lb ft	pounds foot—torque, force
lb/in^2	pounds per square inch—pressure
mile/h	miles per hour
m p g	miles per gallon
o h v	overhead valves
rev/min	revolutions per minute
r h	right hand side (viewed from driver's seat)
s v	side valves
S W G	standard wire gauge
U N C	Unified National coarse screw threads
U N F	Unified National fine screw threads

CHAPTER 1

Locating and Curing Starting, Misfiring and Operational Troubles

If an engine rotates but will not start when the normal starting procedure is followed (automatic transmission cars, see also page 211), locate the cause of the non-starting by testing:

The ignition system (see chart 1:2)

The fuel system (see chart 11:1)

The compressions (see page 38)

(The charts should be used in conjunction with the text; page references are given on each chart.)

Engine fails to rotate

If the starter motor fails to rotate the engine (see chart 1:1) switch on the headlights and operate the starter switch. If the lights dim (considerably) this may indicate:

An almost flat battery

A high resistance at a battery terminal or earth connection (see fig. 1:1)

The starter motor pinion stuck in the flywheel ring gear (see fig. 1:3)

A short circuit in the starter motor

A mechanical fault in the engine

If the headlights do not light at all and the starter does not operate, an open circuit or completely flat battery is indicated. Test the battery by removing the leads and briefly shorting the negative post to the positive with a piece of heavy duty wire. A strong blue spark indicates the battery is satisfactory and the fault is probably an open circuit in the main feed. On solenoid-operated starter systems the heavy gauge brown wire from the solenoid to the control box may have become detached. If there is no spark when the battery posts are shorted together, or a very weak spark, the battery must be removed for charging or testing, and renewal if faulty.

If the headlights remain bright when the light/starter test is carried out, possibly the starter motor switch, the solenoid or starter motor is faulty, or a main starter lead has become disconnected.

If the starter motor is operated through a solenoid switch (see fig. 1:2) controlled by a second position on the ignition switch (or by a separate starter button) the solenoid should click when the control switch is operated. If a click does not occur, the control switch (ignition switch or button) is probably faulty or there is an open circuit between the ignition switch and solenoid, or a faulty feed to the ignition switch. With a thick piece of wire (whilst the starter switch is operated by an assistant) briefly short the starter terminal to earth. If a blue spark occurs and the starter motor pinion is not jammed in the flywheel ring gear, the starter motor must be removed and the fault located. If a blue spark does not occur, the starter switch is probably not functioning.

If the starter switch is manually operated through the medium of a piano wire or Bowden cable, ascertain that the cable is pulling the switch control lever to its full extent. On this type of starter motor the electric cable to the motor is normally 'alive', being attached to a terminal connected to one of the starter switch 'contacts'. To find out if the feed is 'live', briefly short the terminal to earth. If a blue spark occurs and the starter pinion is not stuck

1

in the ring gear, the starter motor must be removed from the engine for a complete examination.

If the lights dim during the headlight/starter test, examine the battery terminals to ascertain whether they are loose, or if corrosion is causing a high resistance.

If old clamp-type battery cable connectors are fitted, use a screwdriver to bridge the battery post with the cable connector, whilst an assistant operates the starter switch. Try both the positive and negative terminals. If the starter now rotates, remove the connector from the battery post and scrape the connector, and the post, until bright metal appears. Replace the connector and tighten securely.

On the later cap-type battery connectors (see fig. 1:1), if corrosion is forming a high resistance, remove the connector holding-down screw, take off the connector, remove the corrosion deposit from the battery post and connector and refit them together again. Make sure to tighten the securing screw.

Be sure to inspect the battery earth strap, where it attaches to the body, and follow by examining the battery leads to ascertain whether they are loose at the starter switch connections. If the trouble is not located, connect the switch terminals together. If the engine turns, the starter switch needs renewing.

Fig. 1:1 A loose or corroded battery connection can be a constant source of trouble. The connector should be removed, scraped clean, refitted securely and coated with an anti-corrosion grease. If the screw cannot be tightened an over-size screw should be fitted

Next ascertain if the starter pinion gear has jammed in the flywheel ring gear. Use an open-ended spanner (see fig. 1:3) and turn the square shaft which projects from the end of the starter. If the shaft is tight, the starter is undoubtedly jammed but can be freed by turning the shaft, or by engaging top gear and pushing the car to and fro. Make sure the ignition is switched off.

If the starter motor rotates when the starter switch is operated but the pinion fails to engage the flywheel, and the electrical system is functioning satisfactorily, the pinion is probably seized to its shaft due to a combination of oil and clutch dust. If a grating metallic noise occurs immediately the starter rotates and the starter pinion then fails to engage the flywheel, a severely worn flywheel ring gear is indicated. Failure to engage, due to a worn ring gear, is usually preceded by the starter pinion frequently jamming with the ring gear. If a worn ring gear is suspected, remove the starter and inspect the ring gear teeth. A new ring gear must be fitted if they are severely worn.

The starter motor may fail to rotate the engine if the water in the cooling system is frozen, for the water pump impeller will be unable to rotate. If the fan belt is slack, the engine may turn slowly, but if the coolant is frozen, thaw it out before trying to start the engine.

An engine that has suffered mechanical damage may not rotate, but non-rotation due to a mechanical defect is invariably preceded by considerable noise.

Non-starting (starter turns engine)

If the starter turns the engine but the car does not start, the first step is to find out whether

Chart 1:1 (see page 1)

Locating Starting System Faults

Engine fails to rotate when starter switch is operated.

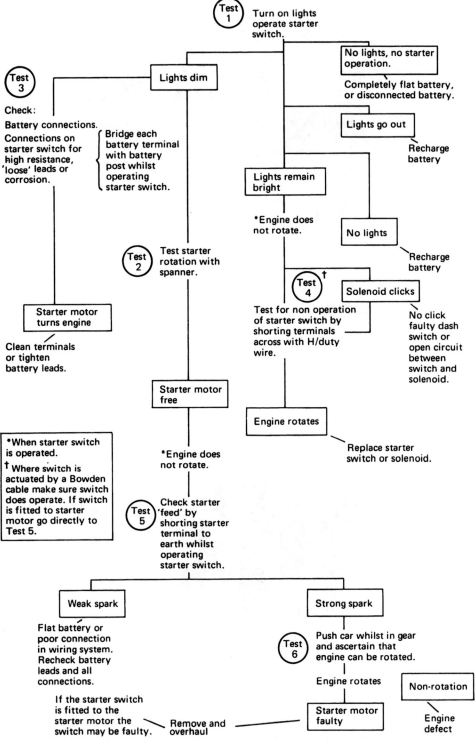

If the engine rotates when the starter switch is operated but the car cannot be started refer to charts 1:2, 1:5 and 11:1—automatic transmission cars see page 212. The above tests assume that the starter motor drive mechanism is functioning correctly. If it is not, the starter will be heard rotating freely without engaging the flywheel. In this event refer to page 2.

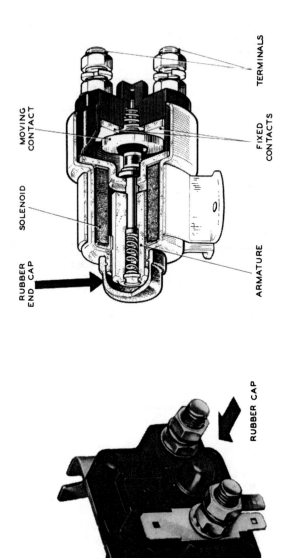

TERMINALS

MOVING
CONTACT

FIXED
CONTACTS

SOLENOID

ARMATURE

RUBBER
END CAP

RUBBER CAP

Fig. 1:2   Most modern cars are fitted with a starter solenoid which is operated by remote control from the ignition/ starter switch. The types illustrated may also be operated manually from the engine compartment by depressing the rubber cap.

Some cars have a safety arrangement so that the solenoid cannot be accidently operated by the switch whilst the engine is running. Where this type of circuit is used the solenoid cannot be manually operated, moreover should the engine be stalled it is necessary to switch off the ignition and switch on again before the starter motor can be operated. Some starter solenoids also incorporate an ignition connection (see page 19) which allows an ignition ballast resistor to be by-passed and so provide a higher coil voltage for starting

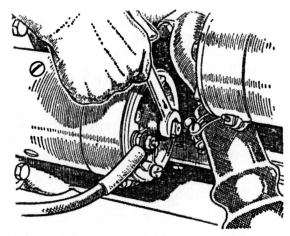

Fig. 1:3 A jammed starter motor can usually be freed by turning the squared shaft-end with a spanner. If the starter motor is of a type without a square end, the car should be placed in gear and pushed so that the flywheel rotates in the reverse direction to normal rotation. This should cause the starter pinion to be thrown out of mesh. Should this fail, the bolts retaining the starter motor to the engine/clutch-housing should be slackened and the process repeated

non-starting is due to an electrical fault, lack of fuel, or to a mechanical reason. Ideally, the compression pressures should be taken to ensure that the trouble is not due to lack of compression. As this is seldom immediately possible it is best to try to locate the cause of the non-starting on the assumption that the mechanical condition of the engine is satisfactory. The compressions can be checked to ensure they are good enough for starting purposes by removing the plugs and rotating the engine a few times (disconnect the LT wire from the distributor) whilst holding a finger over a plug hole. Repeat the procedure on each of the cylinders. Unless a considerable difference in the compressions can be felt, non-starting is unlikely to be due to an engine fault.

If the non-starting has been preceded by misfiring, or overheating, this may be a useful clue. Misfiring usually indicates electrical trouble (plugs, points, etc.) though it can indicate mechanical faults, but spitting through the carburetter is due to a fault in the fuel system, which may either be a mechanical or (with SU pumps) an electrical fault in the petrol pump, a blocked fuel pipe or inadequate fuel availability in the combustion chamber due to carburetter, or valve, trouble.

Usually the easiest thing to check first is the fuel supply (see also chart 11:1) at the carburetter, and the strangler operation. Remove the air cleaner and observe whether the strangler is closing when the control knob is operated and whether fuel is discharging into the carburetter venturi when the engine is rotated. With SU carburetters, if it is difficult to see into the mouth of the carburetter, remove the dashpot and piston and observe whether fuel discharges from the jet and whether the jet drops when the mixture control knob is

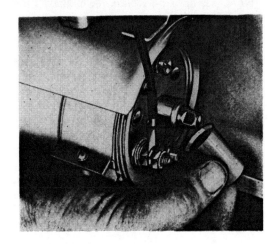

Fig. 1:4 The square end of the starter armature, which can be turned with a spanner if the starter jams, is sometimes covered with a cap. If the screw retaining the inspection cover is slackened, the cover can be pushed back to inspect the brushes. After a starter motor has been overhauled, when refitting the cover, make sure to place it so that the retaining screw will be in an accessible position

pulled. If the control knob operates, and drops the jet satisfactorily and/or the fuel discharge appears satisfactory, the ignition system must be checked.

If an engine is rotated numerous times in an endeavour to start, the plugs may become wet with petrol through excessive use of the mixture control (or throttle if the carburetter is fitted with an accelerator pump). If an ignition fault is corrected and the car still fails to start, the plugs should be taken out, examined and any surplus fuel removed.

When a car has been standing in the open for a long time during winter or has been washed and will not start, the plug leads may be shorting due to an accumulation of moisture around the plug insulators. Wiping the insulators with a dry cloth will cure the trouble but as an added safeguard remove the distributor cap and wipe the inside.

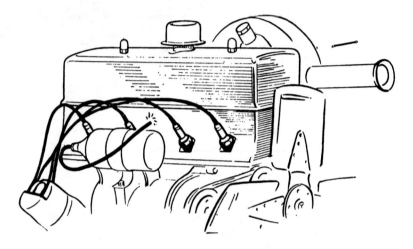

Fig. 1:5   Checking the HT voltage of a coil

A car may fail to start even when the battery, ignition, and fuel system are in order, due to dampness inside the cylinders or poor fuel evaporation. Removing the spark plugs and spinning the engine usually helps to clear the moisture, but if there is poor fuel in the carburetter it may be necessary to warm the plugs.

If the starter rotates the engine at a satisfactory speed, switch on the ignition, disconnect the HT coil lead from the distributor cap and hold the lead $\frac{1}{4}$ in. away from the cylinder block (see fig. 1:5). Rotate the engine. If there is not a bright blue spark at the lead, check the LT circuit.

Defective battery

A 6 volt system that has a battery with a defective cell will not supply enough current to rotate the engine, but a 12 volt system with a defective battery may supply enough current for the starter to rotate the engine but the current absorbed in doing so may prevent the coil from providing an efficient spark.

If a 12 volt battery with a defective cell has been recharged during a journey, it may operate the starter motor and start the car satisfactorily, but if the battery is allowed to stand for a few hours it may lose its charge and the engine will fail to rotate satisfactorily when the starter motor is operated.

If continual starting trouble is experienced and the battery is suspect, carry out the starter/headlight test as suggested on page 3. If the headlights become dim, unless there is a high-resistance connection somewhere in the starter circuit, the battery is at fault. Battery cable connectors with a high resistance should be rectified as suggested on page 2.

If the battery appears faulty, remove it for testing. If a defective cell is discovered, a new battery should be fitted.

Locating a low tension fault

Ascertain whether low tension current is available at the coil SW terminal (see Chart 1:2, test 1). If electricity is not present, the fault is probably in the ignition switch, or the wiring to or from the switch.

Chart 1:2 (see above)

Testing the Low Tension Circuit

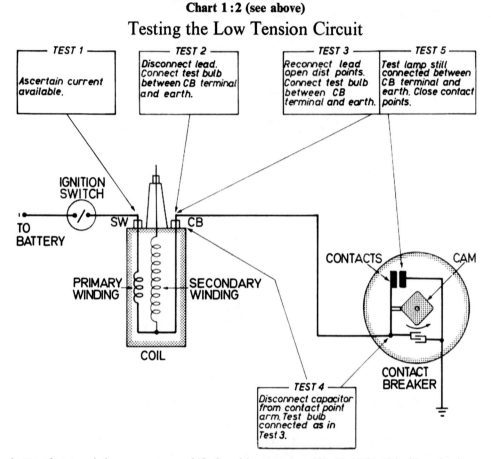

— TEST 1 —

Ascertain current available.

— TEST 2 —

Disconnect lead. Connect test bulb between CB terminal and earth.

— TEST 3 —

Reconnect lead open dist points. Connect test bulb between CB terminal and earth.

— TEST 5 —

Test lamp still connected between CB terminal and earth. Close contact points.

IGNITION SWITCH

TO BATTERY

SW   CB

PRIMARY WINDING   SECONDARY WINDING

COIL

CONTACTS   CAM

CONTACT BREAKER

— TEST 4 —

Disconnect capacitor from contact point arm. Test bulb connected as in Test 3.

Automatic transmission cars see page 212. Overdrives see page 207. 'Cold Start' ignition circuit see fig. 1:18

Next, check the coil primary winding (test 2). Remove the wire which goes to the CB side of the coil and connect a test lamp between the coil terminal and earth and switch on the ignition. If the bulb lights, the coil primary winding is in order. A positive outcome of the test is sufficient only to ensure that the coil is satisfactory for starting purposes. It is not conclusive where misfiring has been occurring, particularly when this has happened at high rev/min, for internal wiring fractures causing an open circuit, or short circuit, may occur only when the coil has been in use and has thoroughly warmed up. In the event of an elusive misfire, the coil should be temporarily replaced by one of known efficiency and the car tested. If the misfire still occurs it may be assumed that the original coil was not at fault.

Continuing the tests, with the wire reconnected to the coil CB terminal, the test lamp still between this terminal and earth and with the ignition switched on, carry out test 3 by opening the contact points; if the bulb does not light, either the capacitor is defective or, on Lucas distributors, the contact-breaker spring-arm-terminal insulation washer is perished, or the assembly of the washer and leads is incorrect (see fig. 1:6).

Check the capacitor by disconnecting the lead from the spring arm and insulating the free end (test 4). If the lamp now lights and the contact-breaker assembly is correct, the capacitor is faulty and should be renewed. Unless special equipment is available, it is only possible to check the capacitor for a possible short circuit. If a complete check of the LT circuit fails to reveal any fault, return to the capacitor and replace it by one of known efficiency. If this eliminates the misfire or non-starting, the original capacitor was faulty.

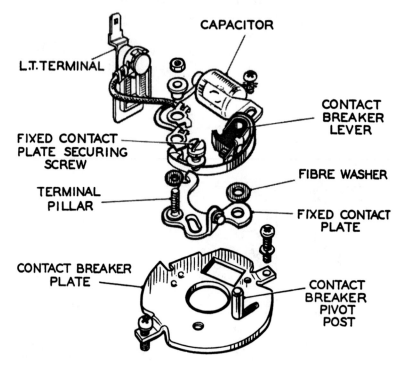

Fig. 1:6   Exploded view of contact breaker used with Lucas (non-vacuum) distributor

If these tests prove that all is in order, reconnect the test bulb between the coil CB terminal and earth and close the distributor contact points (test 5). If the bulb remains alight or glimmers, the contact-breaker moving plate lead is broken (see fig. 1:7) or the contact points are dirty and should be cleaned or renewed and adjusted (see chapter 2, 'Setting the Contact Points').

Non-starting due to the HT lead becoming detached or pulling away from the distributor or coil, are not uncommon, but this apart, faults in the HT circuit, other than those of dampness, seldom result in non-starting, as before a complete breakdown in the HT circuit happens, severe misfiring usually occurs. HT faults that result in hard starting may be traced to a cracked distributor cap, or a faulty or incorrect rotor arm, or faulty spark plugs (see also chart 1:5 and 'Checking the HT Circuit', page 17).

If HT current is available, and fuel discharges into the venturi when the engine is rotated, the timing of the ignition or the valves may be incorrect. If the distributor has just been refitted to the engine, check the ignition timing and make sure that the plug leads have been connected correctly according to the firing order (see Appendix 3 for firing order).

Engine starts but immediately stops

Ensure that an adequate supply of petrol is available, but do not accept petrol gauge readings as indicating there is sufficient petrol in the tank.

If an electrically operated pump is fitted it should be immediately suspect, for a worn pump, before completely ceasing to function, will often operate intermittently.

Mechanical pump failure is often preceded by a bout of difficult starting and the engine spitting back through the carburetter due to inadequate fuel.

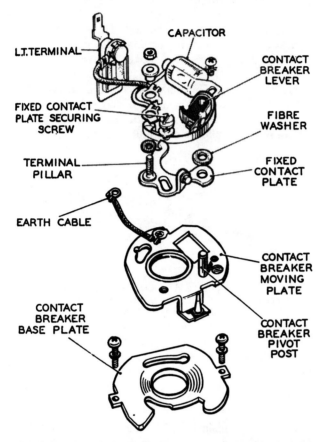

Fig. 1:7   Exploded view of contact breaker used with Lucas (vacuum) distributor

Incorrect mixture control, a sticking carburetter needle valve causing carburetter flooding, water in the carburetter and a blocked exhaust system are a few common troubles that may cause an engine to stop after it has been running for a brief period (see also chart 1:6).

An engine may 'fire' briefly but cut out due to incorrect plugs, *i.e.* too cold (see page 14).

Difficult starting

Difficult starting is usually due to an incorrectly closing strangler flap, or on SU curburetters to faulty operation of the jet mechanism, or to an insufficiently charged battery, dirty battery connector, or a fault in the ignition circuit.

Old cars that have engines with poor compressions due to burnt exhaust valves, or badly seating valves, or engines where the plugs and points need cleaning are often troublesome to start, especially during cold weather. Before carrying out any test procedure, clean the plugs and examine the distributor cap for dampness. Check that the distributor points are opening and closing, clean and adjust them and make sure that a low tension lead has not pulled off.

If a mechanical pump is fitted, hard starting may be due to a faulty petrol-pump outlet valve allowing the petrol to drain from the pipe which connects the petrol pump to the carburetter (see also pages 23 and 231). Although on occasions adequate fuel for starting may remain in the float chamber, should the car be extremely hot, the volatile fractions of the fuel (which assist easy starting) evaporate rapidly and the float chamber is depleted in the quality and quantity of petrol available.

Although some mechanical pumps are fitted with a hand priming lever, a flow of petrol when the primer is actuated is not a conclusive test. The full diaphragm stroke is obtained when using the hand primer (provided it is actuated when the pump rocker arm is on the back of the camshaft eccentric) whereas the rocker arm may not have been operating the diaphragm fully when running.

If all other difficult starting possibilities have been eliminated and the bad starting occurs when the car has been standing overnight and the petrol pump is 'suspect', try injecting a little petrol into the carburetter intake when next trying to start the engine. If the car starts immediately, the petrol pump should be overhauled or a service replacement pump should be fitted.

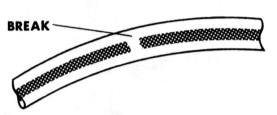

BREAK

Fig. 1:8   Section of a suppressed HT cable. Because the break is inside the lead it cannot be seen and although the spark will jump the break the condition will worsen leading to severe misfiring

Should the car be difficult to start every time, even when the engine is hot, the trouble may possibly be due to a fault in the carburetter. With multi-jet carburetters it may be due to a blockage in the slow-running circuit (see Chapter 12). With SU or Stromberg CD carburetters (see fig. 1:9) a sticking piston may be the trouble.

Locating and curing misfiring troubles

The cause of uneven firing, or misfiring, is sometimes deducible from the type of misfire. Electrical faults, such as indicated in fig. 1:8, tend to produce a definite 'miss'; sometimes it may be very pronounced when the car will feel 'jerky' as the ignition cuts out and in. Alternatively, a slight but definite misfire due to electrical trouble may occur when the car is pulling hard or cruising at high rev/min. Fuel troubles usually result in spitting back through the carburetter or stalling in traffic and erratic performance.

Unless fuel trouble is suspected, an investigation into misfiring should begin by checking the plugs and ignition system.

With the engine idling fairly fast, remove No. 1 plug lead and observe whether the engine speed drops. If the speed drops, the cylinder was firing and the lead may be replaced. Try the same procedure with each cylinder. If a lead is removed and the speed does not drop, hold the lead $\frac{1}{4}$ in. from the cylinder block and ascertain whether a spark is produced.

If the plug lead connector is of the insulated-shroud type, it may be necessary to detach the connector from the plug lead. A strong blue spark indicates that all is well but a weak red spark indicates the HT is faulty (see page 17).

If the plug lead gives a strong spark and the engine speed does not drop when the lead is removed, the plug is faulty or the compression is low, possibly due to poorly seating valves, and a compression pressure test should be carried out (see page 38) if a new plug does not cure the trouble.

If the sparks from all of the leads appear rather weak and the engine has not been starting well, the complete ignition system should be investigated; probably the contact points will be dirty, but the fault may also lie in the coil or capacitor. When a good spark is produced at the plug leads and a drop in engine speed is obtained when each lead is removed, if the misfire occurs only under conditions of load, the HT circuit and spark plugs should be suspected. Severe fouling of a plug due to excess oil or excess fuel might be the cause. In the case of the latter, if two SU carburetters are fitted the trouble may be a carburetter piston stuck in the dashpot*, in such cases the engine may run briefly on all cylinders when started and then 'hunt' and fire spasmodically on two cylinders before stopping. If the engine is accelerated rapidly and driven fast, although the engine may be 'rough' and spit back

Fig. 1:9 Checking to ascertain whether the piston and/or needle is sticking on an SU carburetter. If sticking, see page 259

through the carburetter, the trouble may not be fully apparent until the speed is dropped. If the jet needle of one SU carburetter is loose in the piston (see fig. 1:9) it may result in the needle failing to rise with the piston every time, with the result that the car performs normally for a brief period and then fails to accelerate properly due to only two cylinders firing correctly. On cars fitted with a single SU carburetter much the same symptoms can occur—sometimes the car will accelerate and drive normally while at other times the engine will not pick up (see also chart 1:6).

If the plugs are found to be oily and the engine has a high mileage to its credit, a compression test is essential to ascertain whether there is a likelihood of the cylinders or pistons being severely worn. Temporary relief may be obtained by fitting a 'hotter' plug but the mechanical condition of the engine should be attended to at the earliest opportunity. If the mileage of the engine is under 40,000, the oil may be penetrating the combustion chamber due to worn inlet-valve guides and is unlikely to be due to cylinder or piston wear.

Pre-ignition

Pre-ignition can cause misfiring because it means that something is igniting the mixture in the combustion chamber before it should be ignited. Under prolonged high-speed conditions, such as motorway driving, once pre-ignition has started, severe engine damage can occur very quickly and early recognition of the symptoms can save a large bill.

Constant slow-speed driving or the use of poor-quality fuel may cause a build-up of carbon deposits. If the car is then taken on a high-speed run, the deposits may become hot enough to glow and ignite the fuel.

* Emission controlled cars—see also Chapter 14

Chart 1:3

Plug Condition as an Aid to the Diagnosis of Engine and Carburetter Faults

Plug Condition	Possible Cause	Cure
Sooty deposits (see fig. 1:14)	Engine running too rich and plugs not getting hot enough.	Locate the cause of the rich mixture. Ensure the 'choke' is returning, that the carburetter jets or jet-needle have not been changed, that the air cleaner is not partially blocked, and that the correct fuel-level is maintained. If the car is used for short journeys and in heavy traffic, a 'hotter' plug may be advisable.
Oily deposits (see fig. 1:11)	If the engine has a considerable mileage to its credit, oil may be entering the combustion chamber due to broken piston rings, worn cylinder bores or worn piston-ring lands. The trouble may also be due to worn valve guides or faulty valve-guide oil seals. Oily plugs must not be confused with plugs wet with fuel.	Worn cylinders must be rebored or special rings must be fitted. Worn valve guides can be replaced during a top overhaul. As a temporary measure a plug one grade 'hotter' may help. Overfilling of the sump will sometimes result in plugs oiling up, particularly on a new engine before the rings are fully bedded in. Plugs fouled with petrol are often found when the engine refuses to start from cold. In this case warm the plugs to evaporate the fuel, refit them and try starting again without using the 'choke' and with the throttle fully open.
Insulator, etc. covered with whitish deposit.	Combustion temperature too high, possibly due to over-advanced ignition timing, or a weak mixture. Possible that the compression ratio has been raised.	If a weak mixture is suspected, check the carburetters for blockage or incorrect jets. Check for air leaks on the inlet manifold and carburetter flange. Check the tappets. If a compression pressure test indicates a higher compression pressure than the standard, the cr has probably been raised and plugs one grade cooler may be tried. Check the ignition timing.
Sooty and oily.	Plugs too cool. Too much oil penetrating combustion chamber. Mixture too rich.	Unless the carburetter jets have been changed from standard or the engine specification has been changed, a 'hotter' plug may be tried.

Plug Condition	Possible Cause	Cure
Burned electrodes, cracked, broken or blistered insulator (see fig. 1:12).	An over-advanced ignition timing may cause plug 'break-up'. Over-advance results in the combustion chamber temperature being raised and detonation. The electrodes and insulator break up due to a combination of high temperature and thermal shock waves.	Although advancing the ignition may give an increase in power it also raises the temperature of the combustion chamber. This is counteracted on racing engines by using 'cold' plugs and high-octane fuel. Advancing the ignition for road use may ruin standard plugs. In extreme cases, overheating and breaking-up of the pistons may occur. The correct ignition timing, plugs and fuel should always be used. Fit new plugs. One grade cooler could be tried.
Eroded earth electrode. Plug misfires after 2000–3000 miles.	This could be caused by reverse polarity of the ignition system resulting in the spark jumping from the earth electrode to the centre electrode and not *vice-versa*. After a few thousand miles the voltage required is too high for the ignition system to cope with and misfiring occurs under hard acceleration.	As a check, remove a plug lead and hold about ½ in. from the plug terminal. Put the point of a soft pencil in the middle of the gap and start the engine. If, as the spark jumps the gap, it shows red towards the spark plug the polarity is correct but if towards the plug lead the polarity is wrong and the LT leads on the coil require changing over (see fig. 1:10). Fit new plugs.
Clean insulator, straw- to coffee-coloured with clean body (see fig. 1:13).	Plug condition correct.	(Engine and carburation satisfactory).
Fluffy grey deposit (see fig. 1:13).	Plug condition correct. (Modern fuels have the tendency to form this type of deposit.)	(Engine and carburation satisfactory).
Clean insulator, straw- to coffee-coloured with hard carbon deposits on body.	Undue oil penetration to combustion chamber or too much upper cylinder lubricant.	Make sure to keep oil to correct dipstick mark, engine may require decarbonizing if not carried out recently and 'running-on', is experienced. Reduce the amount of upper cylinder lubricant.

Poor seating of the exhaust valve also causes pre-ignition by allowing an escape of gases at blow-torch temperatures, and making the thin outer edge of the valve glow. Grinding the valves down to a sharp edge when decarbonizing can also result in a hot spot which will pre-ignite the mixture. Using plugs which are too 'hot' for the engine or which have been overheated by detonation, or over-advanced ignition, is another cause.

Often the first sign of pre-ignition is misfiring or 'holding-back' at high speeds, similar to fuel starvation. Investigation may show up burning of the spark plug electrodes (see fig. 1:15), but if it is assumed that the plugs are to blame and they are replaced and high-speed driving is continued, a holed piston may well result (see fig. 1:16). The best procedure in the case of severe pre-ignition is to drive at reduced speed and at the earliest opportunity to remove the cylinder head and carry out a top overhaul (see page 50).

Inspection of spark plugs

If an engine has been 'pepped up', a grade of plug colder than standard may be used (see charts 1:3 and 1:4), otherwise it is advisable, unless oiling-up troubles occur, to adhere to the car manufacturer's recommendation (see Appendix I) regarding the type and grade. The make is really unimportant and principally a matter of individual preference.

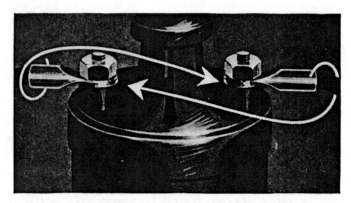

Fig. 1:10   If a new HT coil is fitted, or if the leads to a coil are disconnected it is essential that the leads are correctly re-connected for HT current flow should be negative at the spark plug terminal. If polarity is reversed, voltage requirements will greatly increase and rough idle or misfiring under load may result. Polarity may be checked as suggested on page 13.

If the polarity is incorrect reverse the leads at the coil. In the positive earth system, the coil primary terminals are marked SW and CB. The latter going to earth through the contact breaker is thus the positive terminal of the coil. Terminal SW connected to the insulated side through the ignition switch is the negative terminal. The direction of current flow through the primary winding is thus from CB to SW. To use the same coil on a negative earth system the connections to the primary winding must be reversed, the cable to the contact breaker and earth being connected to the SW (or negative) terminal, and the insulated feed from the ignition switch to the CB (or positive) terminal. Most coils are now marked according to polarity. The contact breaker lead is always the 'earth' lead and so must be connected to terminal − on negative earth systems and + on positive earth systems

Sometimes it is possible to detect a faulty spark plug by looking at it, but not always, for although cracked insulators, burned electrodes and excessive gaps are easily discernible, other faults are not. Taking the plug from the engine, reconnecting the lead and laying the plug on the engine whilst the engine is switched on and rotated, is inconclusive, for a spark plug has to operate efficiently when hot under pressure of some 80–90 lb/in^2. Although the plug may spark well when lying on the engine, it may cease to function completely under pressure.

The condition of a spark plug can indicate where to look for engine and carburation troubles by revealing, if the plug has been firing, whether too rich a mixture is prevalent or too weak a mixture, and whether excess oil is reaching the combustion chamber (see chart 1:3).

Retarded ignition, excessive oil consumption, overheating or an incorrect air/fuel ratio are among the principal factors that can affect the performance of plugs and which are revealed by the condition of the plug. If overheating, pre-ignition and rapid burning of plug

Fig. 1:11 Wet oily deposits may be caused by oil leaking past worn piston rings. 'Running in' of a new or overhauled engine before rings are fully seating may also produce this condition. A porous vacuum booster diaphragm or excessive valve guide clearances can also cause oil fouling. Usually these plugs can be degreased, cleaned and reinstalled. While hotter type spark plugs will reduce oil-fouling, an engine overhaul may be necessary to correct this condition

Fig. 1:12 Badly burned or eroded electrodes or burned or blistered insulator are indications of spark plug overheating. Improper spark timing or low octane fuel can cause detonation and overheating. Weak fuel/air mixtures, cooling system stoppages or sticking valves may also produce this condition. Sustained high-speed, heavy-load service can lead to high temperatures which require use of colder spark plugs

Fig. 1:13 Brown to greyish tan deposits, and slight electrode wear, indicate correct spark plug heat range and mixed periods of high- and low-speed driving. Spark plugs having this appearance may be cleaned, re-gapped and reinstalled

Fig. 1:14 Dry fluffy black carbon deposits may result from over-rich carburation, over-choking, or clogged air cleaner. Faulty breaker points, weak coil or condenser, worn ignition cables can reduce voltage and cause misfiring

Fig. 1:15 Pre-ignition damage to spark plug. If plugs are found in this condition they must be discarded and the ignition timing checked at once

Chart 1 : 4

Plug Recommendations

(Use in conjunction with Appendix 1)

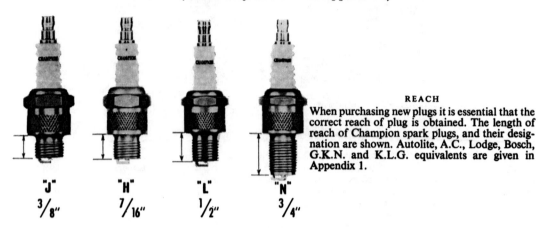

"J" 3/8"　　"H" 7/16"　　"L" 1/2"　　"N" 3/4"

REACH

When purchasing new plugs it is essential that the correct reach of plug is obtained. The length of reach of Champion spark plugs, and their designation are shown. Autolite, A.C., Lodge, Bosch, G.K.N. and K.L.G. equivalents are given in Appendix 1.

HEAT RANGE

It is essential also that the correct heat range is used. Cold plugs having short heat-flow paths are used for high-temperature applications—racing, heavy duty service—to avoid overheating of the plug and possible pre-ignition. Hotter plugs with longer heat-flow paths can be substituted for low-temperature conditions like slow-speed driving, light-duty service. This helps minimize cold-fouling deposits such as carbon and oil.

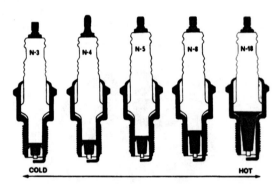

COLD　　　　　　　　　　　　HOT

PROJECTED CORE NOSE

Many overhead-valve passenger-car engines of suitable design also use projected core nose plugs. (There isn't sufficient clearance for these plugs in the older side-valve engines and certain overhead-valve engines do not give the cooling effect required.) In this design, the plug tip is cooled by the incoming fuel charge at higher engine speeds. This means that the same plug can run hotter at low speeds— in effect extending its heat range. Recommendations for use are given in Appendix 1.

Any Champion plug with the letter 'Y' on the end of its symbol has the projected core nose.

AUXILIARY GAP

On some Champion plugs the letter 'U' is sometimes used in the prefix designation. This indicates a 'series' air gap or auxiliary gap which is built into the centre electrode on certain Champion plugs. The gap isolates the ignition coil from the carbon deposits on the insulator nose which can drain away voltage, this gives the coil time to build up enough voltage to break down the series air gap and spark across the electrodes. This produces additional cold-fouling protection and is particularly useful on some high-powered o h v engines which tend to foul up in traffic conditions. Recommendations for use will be found in Appendix 1.

electrodes are occurring, a colder plug may be used to give temporary relief until the true fault can be found and rectified.

When fitting new plugs or refitting plugs which have been serviced, avoid using too much force to tighten them and ensure that the correct type is used and that a gasket washer is fitted.

The cost of having a set of plugs sandblasted and tested is so inexpensive that in cases of misfiring and poor performance this might well be one of the first jobs to do.

Plug gaps increase at the rate of approximately 0·001 in. per 1000 miles and if the gaps are not reset every 5000–6000 miles there is a likelihood of misfiring at high speeds. Plug manufacturers suggest renewing plugs every 10,000–12,000 miles. On high-performance engines, or where considerable mileage is normal and absolute reliability of prime importance, this recommendation must be adhered to, but if the plugs are cleaned properly and the gaps reset regularly, plugs often operate satisfactorily for up to twice this mileage.

In cases of misfiring, if the spark plugs are found to be satisfactory, the HT circuit should be examined.

Chart 1 : 5

Common High Tension Defects

Insufficient LT current due to coil-robbing effect of operating starter motor when battery almost flat, or the occurrence of a high resistance in the circuit

Distributor cap 'tracked'

Distributor cap cracked

Faulty plug lead

Plug leads incorrectly refitted

Incorrect ignition timing

Defective carbon brush in distributor cap

Broken or faulty plugs

Segments in distributor cap burnt

Incorrect rotor arm

HT 'leak' due to moisture

Faulty coil

'Crossfire' due to HT leads running parallel

Reversed polarity, due to incorrect refitment of coil LT leads

Checking the HT system

In cases of misfiring or poor performance, if a compression test has been carried out, the spark plugs tested and the gaps reset, the HT circuit should be thoroughly checked (see also chart 1:5).

Examine the HT leads between the distributor and spark plugs. Many modern engines have suppressed-type plug leads that have a flexible carbon conductor running through the centre. The carbon acts as a suppressor and cuts down TV and radio interference. If the leads are mishandled, a break in the carbon may occur. The break cannot be seen (see fig. 1:8) and while the spark may jump the break, the gap gradually becomes larger until a misfire occurs when accelerating. New plugs, because of their lower voltage requirement (to produce a good spark), may cure the misfire and the old plugs will get the blame for the misfire. Unless the faulty lead is renewed, however, as soon as the voltage requirement of the new plugs rises to the level of the voltage available (lower than standard because of the gap in the plug lead) the misfire will return. If a faulty lead is suspected, the continuity of the carbon conductor should be tested.

Faulty HT lead insulation is not particularly common but where HT cables run through a ring or clip to keep the cables in position and the leads run parallel and close together 'crossfire' may occur, resulting in rough idling, misfire under load, and even severe detonation.

Fig. 1:16   Detonation caused by over-advanced ignition, low octane fuel, combustion chamber deposits, etc. can result in a holed piston (see page 11). Exhaust valve burning see fig. 4:10

Continuing the examination of the HT circuit, remove the distributor cap and wipe the inside of the cap with a clean dry cloth. Ascertain that the carbon brush is in place and moves freely in its holder. Give each HT lead a gentle pull to ensure that each is retained securely, and examine the inside of the cap to ascertain whether it has become tracked. Tracking is a conducting path (due to the deposit of nitrous powder formed as the result of the constant sparking and ionization of the air in the spark gap) seen as a thin black line between two electrodes, or between an electrode and part of the distributor body which is in contact with the cap. A tracked cap must be renewed, but temporary 'get you home' relief may be obtained by scraping the track and polishing the area with emery tape.

If the rotor and the segments in the distributor cap are corroded due to the continuous sparking between the rotor and segments, scrape the segments with a penknife and wipe away any loose powder from the cap with a clean rag.

Examine the rotor arm for cracks and tracking. If a new rotor arm has to be fitted, compare it with the rotor removed, thereby ensuring that it is of the correct type.

As moisture is a frequent cause of HT trouble, all electrical ignition components should be kept in a damp-proof condition. Ideally, spray the distributor cap and leads with a proprietary damp-start lacquer.

The ignition coil

Ignition coils require no maintenance but all terminals and connections should be kept clean and tight.

An ignition coil has a primary winding consisting of a few hundred turns of relatively

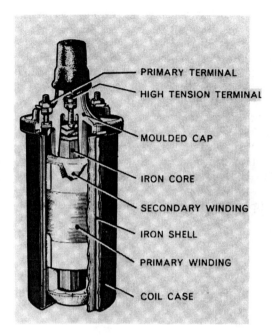

Fig. 1:17   Section view of a typical ignition coil

heavy wire, and a secondary winding with many thousand turns of fine gauge wire. These
windings (see fig. 1:17) surround a soft iron core and are enclosed by a soft iron shell. This
assembly is inserted into a one-piece aluminium case which is then filled with oil, for permanent insulation and better heat dissipation, and hermetically sealed by a moulded cap.

   If the performance of the HT coil is suspect, it is best substituted by a coil known to be in
a satisfactory condition.

Cold-start coils

Since September 1964, Vauxhall cars have been equipped with a special 12 volt Delco Remy
resistor coil, indeed this type of coil can also be fitted to other vehicles to improve cold

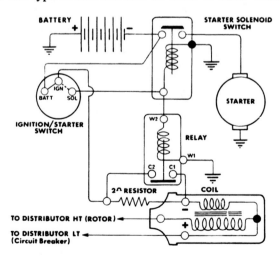

Fig. 1:18   A typical 'cold start' ignition coil circuit using a relay

starting. Resistor coils are designed to operate at a lower-than-system voltage (about $7\frac{1}{2}$ volts) and a 2 ohm ballast resistor is connected in the LT circuit (see fig. 1:18) to reduce the battery voltage to this level for normal running. At the moment of starting, however, a separate relay or special starter solenoid connection causes the resistor to be by-passed so that the full battery voltage is momentarily applied to the coil. Starting is greatly assisted by the higher HT voltage thus obtained while battery drain and starter motor wear are correspondingly reduced. This is particularly beneficial during cold weather when the available battery voltage is lower, and starting voltage requirements are higher than normal.

It should be noted that cold start (resistor) coils must not be used without the resistor in circuit during normal running. If it should be necessary to disconnect the coil, special care must be taken when refitting to reconnect the terminals to the solenoid and relay (when fitted) exactly as before. In positive-earth systems the resistor is attached to the coil negative terminal. The ignition switch should not be left 'on' with the engine stopped as this will cause overheating and possibly damage to the resistor.

Chart 1:6

Common Engine Troubles*

EXCESSIVE PETROL CONSUMPTION

(Any condition where an engine is lacking in power will also affect fuel consumption)

Fault	*Remedy*
Excessive fuel pump pressure.	Reduce pressure. Possibility of incorrect pump fitted or faulty fitting of a mechanical pump.
Fuel leakage.	Locate and rectify. Possibly loose float bowl or on SU carburetters float-chamber mounting rubbers perished.
'Choke' not fully returned.	Examine for possible faulty cable. Automatic choke not functioning properly—repair.
Blocked air cleaner.	Clean or fit new element (see fig. 11:16).
High float level or faulty or incorrect float.	Adjust level or replace float if faulty or incorrect.
Broken jet return spring (SU carburetter).	Fit new spring.
Sticking or dirty needle valve.	Fit new needle valve assembly.
Worn jet.	Renew.
Idling mixture over rich or slow-running set too fast.	Adjust.
A stuck accelerator pump valve.	Free off.
Faulty ignition causing misfiring and lack of power.	Check coil, condenser, plugs, points.
Loss of compression.	Test compressions and rectify trouble.
High rolling resistance due to low tyre pressures, brakes binding or wheel misalignment.	Check and correct.
Clutch slipping.	Adjust or overhaul clutch.
Faulty economy device (Zenith).	Remove and clean. Fit new diaphragm.

ENGINE LACKS POWER

Incorrect ignition timing.	Check.
Accelerator pump faulty.	Locate reason and rectify.
Low float level.	Set level.
Insufficient fuel supply.	Locate reason:
	Fuel line blockage.
	Fuel pump not operating satisfactorily.
	Blocked carburetter jets.
	Fuel tank vent blocked.
	Air leaks on suction side of pump.
	Fuel leakage.
'Choke' not operating satisfactorily.	Adjust or repair.
Manifold heat control valve stuck.	Free off.
Air leak at carburetter flange joint or inlet manifold.	Fit new gasket, tighten nuts. Reface fitting flange if necessary.
Throttle not opening fully.	Adjust.
Rich mixture.	Clean and repair and fit new parts as necessary.
Worn jets.	
High float level	
'Choke' stuck on.	
Choked air cleaner.	See fig. 11:16.
A blocked inlet manifold drain.	
Vapour lock.	Badly placed fuel lines, high temperature in engine compartment.

* Emission controlled cars, see also Chapter 14

Fault	*Remedy*
Fuel pump defective.	Overhaul
Blocked exhaust.	Clean out.
Defective ignition.	Check systematically (see Chapter 1).
Loss of compression	Check compression pressures, rectify as necessary.
Excess carbon deposits in engine.	Decarbonize.
Defective valve operation.	Check compression pressures, check valve timing.
Cooling system defective, engine overheating.	Flush system, check thermostat (see Chapter 13).
Excessive rolling resistance.	Locate and rectify cause.
Clutch slipping.	Overhaul clutch.
Plugs defective.	Renew.
Moisture around plugs or distributor cap.	Dry off.
Contact breaker gap too small.	Adjust.

ERRATIC IDLING

Idling mixture, or idling speed incorrectly adjusted.	Readjust.
Air leaks.	Locate and rectify.
Slow running jet blocked.	Clean.
Most faults listed under 'Engine Lacks Power'.	
Incorrect ignition timing.	Check and reset ignition timing.
HT leads wrongly fitted.	Refit HT leads according to firing order.
Damaged volume control screw (Solex and Zenith carburetters).	Replace screw.
Float bowl overflow tubes blocked.	Remove and clean.
Inlet manifold drain pipe blocked.	Remove and clean.
Jet from drain pipe not refitted.	Remove pipe and fit jet (Hillman engine).

SMOKY EXHAUST

Rich mixture (black smoke).	See 'Excessive Fuel Consumption'.
Excess oil in combustion chamber (blue smoke).	Test compression pressures to ascertain the cause of oil penetrating the combustion chamber.

ENGINE STALLS

Idling mixture or idling speed incorrectly adjusted.	Adjust.
'Choke' control stuck.	Free and adjust.
Manifold heat control valve stuck.	Free off.
Engine overheats.	See Chapter 13.
Slow running jet blocked.	Clean.

ENGINE STALLS AFTER DRIVING

Defective fuel pump.	Repair and replace.
Engine overheats.	See Chapter 13.
Fuel percolation occurring.	Check manifold/carburetter insulation.
Vapour lock.	Reposition petrol pipe if necessary so that it is not subjected to excessive heat.
Stalls after brake application.	Fuel surge in carburetter due to incorrectly mounted carburetter, or faulty design.

Fault   *Remedy*

BACKFIRING

Fault	Remedy
Exhaust system blowing.	Cure exhaust system for leaks.
Faulty plugs or plug of wrong heat range.	Test, renew as necessary.
Sticking valves.	Top overhaul required.
Ignition or valve timing incorrect.	Check.

MISFIRING

Fault	Remedy
Fuel pump erratic in operation.	See page 233.
Carburation fault.	Clean, examine and overhaul fuel system if necessary.
Fuel level incorrect.	Adjust float level and fit new needle valve.
Ignition system defective.	Test ignition circuit.
Blocked exhaust.	Check and clean if necessary.
Engine overheating.	See chapter 13.
Mechanical fault.	Test compression pressures.
Carburation flat spot.	Compensating jet incorrect. Constant vacuum carburetters: wrong needle, incorrect damper or dashpot spring. Incorrect oil in damper. Faulty economy device diaphragm.
CV carburetters (SU and Stromberg CD), sticking piston.	Clean piston. Re-centre jet.
Blocked nylon float-bowl/jet tube.	Remove dashpot and piston, float bowl lid and float and blow through the tube with air pressure.

ENGINE WILL NOT START UNLESS PRIMED

Fault	Remedy
Defective fuel pump.	Overhaul.
Carburetter jets, or fuel line blocked.	Clean as necessary.
Air leaking at carburetter or inlet manifold.	Replace gasket, tighten nuts, reface carburetter flanges or manifold as necessary.

HESITANT PICK-UP

Fault	Remedy
Piston damper not functioning (CV carburetters).	Fill damper orifice with SAE 20 oil.
Blocked jets.	Remove and clean.
Pump diaphragm faulty.	Overhaul pump.
Wrongly positioned accelerator pump linkage.	Reposition pump linkage according to season.
Most faults listed under 'Erratic idling'.	

DIFFICULT STARTING (NON-ELECTRICAL TROUBLE)
(Engine warm and rotates)

Fault	Remedy
'Choke' sticking on.	Inspect cable. Adjust, or repair, mechanism as necessary.
Manifold heat control valve stuck on.	Free-off valve.
Vapour lock.	Reposition petrol pipes so that they are not subjected to excessive heat.
Fuel percolation.	Check manifold/carburetter insulation.
Mechanical fault.	Examine engine.

ENGINE SLOW TO WARM UP AND WILL NOT IDLE

Fault	Remedy
Manifold heat control valve not operating.	Free off valve.
Wrong thermostat or thermostat stuck.	Fit new.

CHAPTER 2

Overhauling Generators, Distributors and Starter Motors

When a car has covered some 35,000 miles, engine auxiliary equipment has reached a stage where its reliability must be questioned. The DC generator may need new brushes, the starter motor pinion will probably be partially seized to its shaft, and the amount of brush dust around the starter commutator and rear bearing will contribute to a high wear factor. On older cars, the distributor spindle will probably have excessive clearance, the distributor weights and springs may be worn, and the suction advance diaphragm perished. If a fault develops in an auxiliary unit about the 35,000 mileage mark, it is worth stripping and over-hauling it, for unless the generator or starter motor field coils have to be taken out for re-insulation (this is seldom needed) any auxiliary unit can be overhauled in less than three hours.

Overhauling the generator

Although the following instructions apply to Lucas DC generators, the principles are applicable to most popular types.

When the generator has been removed from the car, undo the pulley-securing nut and withdraw the pulley from the armature shaft. Remove the cover band (if fitted) pull back the brush springs and remove the brushes. On generators with screw-type terminals, remove the nut and washers from the field terminal on the commutator-end bracket.

Take out the generator 'through' bolts (see fig. 2:1) and take off the commutator-end bracket. If a fibre washer is fitted on the shaft between the commutator and the end bracket, remove it from the shaft but make certain that it is refitted before the end bracket is replaced on the yoke.

Remove the drive-end bracket and armature from the yoke as a complete unit. If the armature appears in good condition do not separate it from the bracket unless the ball-bearing is worn. If the bearing is worn, or the armature has to be renewed, remove the bracket.

Remove the brushes from the commutator-end bracket by undoing the retaining screws. Check the new brushes in the brush holders to ensure that they can slide freely. If a brush is tight in the holder, lightly file the side of the brush to remove any high spots.

If the bearing (bush type) in the commutator-end bracket is worn, press the bush from the bracket. Insert the new bush into the bracket and, using a shouldered mandrel (see fig. 2:2), the spigot of which is the same diameter as the bearing, press the bush into the bracket. (Prior to fitting the bush, allow it to stand completely immersed in an SAE 30 engine oil for 24 hours to enable it to absorb lubricant.)

Fit new brush springs with the new brushes but temporarily position each spring at one side of the brush (see fig. 2:3) to prevent it protruding from the base of the holder. This greatly facilitates positioning the bracket over the commutator when the bracket is refitted to the yoke. The springs are correctly repositioned with a thin screwdriver before the bracket is right home—or through the ventilator holes.

If the generator has not been charging and the armature windings appear blackened or smell burned, the armature must be given a 'growl' test. Most of the larger garages can do this in less than five minutes. If the commutator is faulty, a service replacement armature must be obtained.

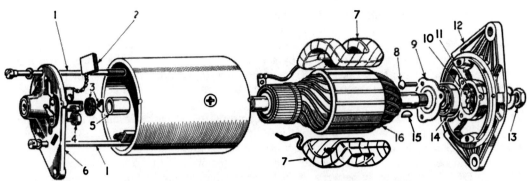

Fig. 2:1 General assembly view of a Lucas C40-1 two brush, two-pole DC generator

1. Bolts
2. Brush
3. Felt ring and aluminium sealing disc
4. Brush spring
5. Bearing bush
6. Commutator end bracket
7. Field coils
8. Rivet
9. Bearing retainer plate
10. Corrugated washer
11. Felt washer
12. Driving end bracket
13. Pulley retainer nut
14. Bearing
15. Woodruff key
16. Armature

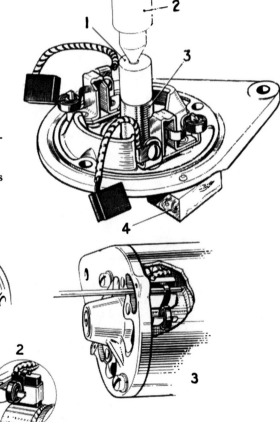

Fig. 2:2 Fitting a new bearing to the commutator end bracket

1. Mandrel
2. Press
3. Bush
4. Wood blocks

Fig. 2:3 Fitting commutator-end bracket to 'windowless' yoke generator

1. Method of trapping brush in raised position with spring
2. Normal working position
3. Method of releasing brush on to commutator

Common armature faults are short- or open-circuited windings. Visual indication of the latter is burnt commutator segments, whereas short-circuited armature windings are identifiable by burnt commutator segments and discoloured windings caused by overheating.

If the commutator is worn, denoted by a deep ridge at the side of the brush path, the commutator will have to be skimmed in a lathe, or a replacement armature fitted. If the ridge is shallow it can be removed by polishing the commutator with coarse emery tape whilst rotating the armature.

When the ridge has been removed from the commutator, the mica insulator between the commutator segments must be undercut (see fig. 2:4) by approximately $\frac{1}{32}$ in. The ideal tool is a hacksaw blade ground to the thickness of the mica insulator.

The field coils may be tested, without removing them from the yoke, by connecting a 12 volt DC electrical supply between the field terminal and yoke with an ammeter in series with one of the leads. A reading of approximately 2–$2\frac{1}{2}$ amps should be obtained. If a reading cannot be obtained, the field coils are open circuited. If the readings are much higher than $2\frac{1}{2}$ amps the fields are short circuited due to a breakdown in the insulation. In either case, the field coils must be removed from the yoke and should be renewed.

If a new armature is to be fitted or if the armature support bearing requires renewing, remove the woodruff key from the armature shaft (and distance collar if fitted), support the bearing-retaining plate and press the armature from the bracket. *On some generators the distance collar is fitted behind the bearing but on others it is fitted to the front of the bearing.*

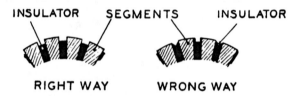

Fig. 2:4 The correct method of undercutting the dynamo commutator

If a press is not available remove the bracket from the armature in the following manner. Place two pieces of metal (approximately $\frac{1}{4}$ in. thick and 2 in. × 6 in.) across the jaws of a vice to support the bearing-retainer plate. Hold the armature vertical (an assistant will be required) and with a piece of wood resting on the shaft give the piece of wood a few sharp hammer-blows.

To remove the ball-bearing in the drive-end bracket, knock out the rivets which secure the retaining plate and remove the plate. Press the bearing from the end bracket, remove the corrugated washer, felt washer and oil-retaining washer.

Before fitting the new bearing, pack it with a high-melting-point grease. Place the oil-retaining washer, felt washer and corrugated washer in the bearing housing in the end bracket (see fig. 2:1). Locate the bearing in the housing and press it home, using a tube of a suitable diameter to fit over the inner journal of the bearing. Fit the bearing-retaining plate. Insert the new rivets from the inner side of the bracket, and spread the rivets to secure the plate in position.

Place the distance collar in position (see previous note in italics), refit the bearing and bracket to the armature shaft.

When the generator is reassembled, position the brushes and springs correctly (see fig. 2:3), but before fitting the generator pulley, spin the armature and give the end bracket a few light taps with a rubber mallet, so as to align the end brackets and settle them into position. This done, retighten the 'through' bolts, replace the pulley spacer collar (see earlier note) fit the woodruff key, and refit the pulley, the spring washer and nut and tighten securely. Finally replace the brush cover and refit the generator to the engine.

Checking the distributor

Distributors vary according to make, type and design but their overhaul does not present any insurmountable problems.

For a complete overhaul, the following parts are necessary, but the list may be reduced according to the condition of the distributor.

2 Distributor shaft bushes	1 Distributor cap (if necessary)
1 Distributor shaft	1 Driving dog (or skew gear)
1 Vacuum advance diaphragm (if fitted)	Lubricator pads (as necessary)
1 Pair distributor weight springs	1 Carbon brush for distributor cap
1 Capacitor	1 Retaining pin for gear
1 Set contact points	1 Thrust washer and collar (as necessary)
1 Rotor arm	

Before removing the distributor unit from the car, examine the unit to ascertain what parts are likely to be required and make sure they will be readily available. The distributor cap and rotor arm can readily be inspected for evidence of tracking (see page 18) and corroded segments. Try the centrifugal advance mechanism by twisting the rotor arm in the direction of rotation. The arm should turn slightly and, when released, return to its original position. If it fails to return, or if the spring feels slack, the springs require renewing.

Check the cam by trying it for side movement. Unless excessive side movement can be felt the cam may not require renewing, but carry out the following check: Set the points fully open and with feeler gauges measure the gap. Next, rotate the engine so the rotor arm is turned through exactly 180°. Make sure the points are fully open and carefully measure the gap. If the measurements differ more than 0·003 in., the cam is unduly worn on one lobe, or the cam is not running 'true'. If excess side movement (more than 0·0015 in.) of the cam is detected or if the gap measurements showed undue variation, the cam and distributor shaft need renewing.

Checking the suction advance mechanism

Most distributors are fitted with a suction advance mechanism. The unit, usually housed as a separate part of the distributor, operates by means of the depression in the inlet manifold and is connected by a pipe to a drilling in the carburetter on the atmospheric side of the throttle. If the engine is accelerated a few times, it is a simple matter to ascertain whether the mechanism is functioning. On some distributors, where the mechanism is fitted internally, its action cannot be observed but the diaphragm, which tends to become perished and is usually the cause of non-operation, may be tested by removing the pipe from the diaphragm chamber and pushing on the diaphragm link rod (it is necessary to remove the distributor cap to gain access). With the link rod pushed to the full extent of its travel place a finger over the union to which the pipe was previously fitted. On release of the link rod, suction should be retained briefly in the diaphragm chamber. When the finger is released from the union, the diaphragm rod should return to its original position. Failure to retain the suction indicates that the diaphragm is faulty and a new one is necessary. If the diaphragm is satisfactory, before refitting the diaphragm-chamber/carburetter link pipe, push the link rod a few times to expel fuel that may have accumulated in the diaphragm chamber.

Dismantling the distributor

To strip Lucas distributors (see fig. 2:5), remove the distributor cap, take off the rotor, remove the nut and washer from the moving contact anchor pin. Withdraw the insulating sleeve from the capacitor lead and low tension lead connectors (see fig. 2:6 and note the order in which they are fitted). Lift the moving contact from the pivot pin, and remove the large insulating washer from the pivot pin and the small one from the anchor pin.

On Delco Remy (see fig. 2:7) and other distributors, depending upon the make and type, removal of the contact points is slightly different, but in practice little difficulty is experienced

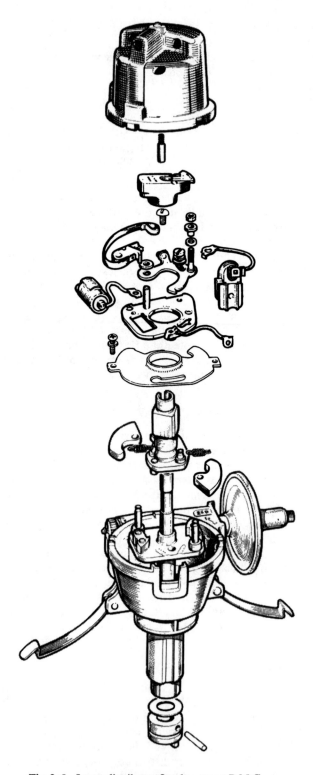

Fig. 2:5   Lucas distributor fitted to many B.M.C. cars

so long as it is noted where the various insulating washers and electrical leads fit. The Ford Cortina distributor is illustrated in fig. 2:8.

On Lucas distributors, take out the two screws (each has a spring and flat washer) which secure the fixed-contact plate. Remove the plate.

Take out the capacitor by removing the securing screw (note the earthing lead for correct replacement). On Lucas 25D distributors, remove the spring clip retaining the advance unit to the contact-breaker base plate. Take out the two screws securing the base plate to the distributor body (one also secures the earthing lead) and remove the base plate.

Mark the position of the rotor arm drive slot (in the cam spindle) in relation to the offset drive dog on the end of the spindle. This will facilitate correct reassembly and avoid the timing being 180° out when the cam spindle is re-engaged with the centrifugal weights.

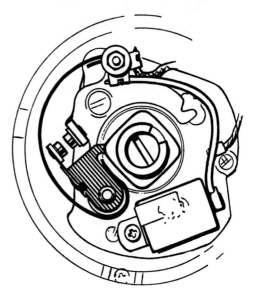

Fig. 2:6 Contact breaker plate showing position of capacitor and connecting leads. If 'Quikafit' contacts are used, the insulating washer must be discarded. The terminal pillar is of a plastic material and is therefore completely insulated

Remove the cam-retaining screw and cam. Take out the centrifugal weights. On some distributors they may be lifted as individual assemblies, each complete with a spring and toggle.

Release the suction advance unit by removing the circlip, adjusting nut and spring.

If the drive shaft or shaft bushes are worn, release the shaft from the body, by driving out the parallel driving pin which passes through the driving dog, or gear. Take off the dog, or gear, noting any thrust washers or collar between the gear and body.

Reassembly of the distributor is basically a reversal of the dismantling procedure using new parts as necessary, but if new bushes in the distributor body are necessary, prior to fitting them it is important to stand the bushes in an SAE 30 oil for 24 hours. (The bushes are porous and absorb a certain amount of the oil.) When the bushes have to be used at once, warm the oil to about 180° F, immerse the bushes and allow the oil to cool before using them.

When the distributor shaft bushes have been fitted it is not necessary or desirable to use a reamer.

Important points are: When refitting the gear, prior to drilling the shaft and inserting the retaining pin, check that 0·002 in. to 0·003 in. end-float exists between the gear and body. Replace the cam correctly, consulting the previous marks. Do not stretch the small springs that control the centrifugal advance mechanism.

On some distributors the springs are barely in tension; on some, one spring is of thicker

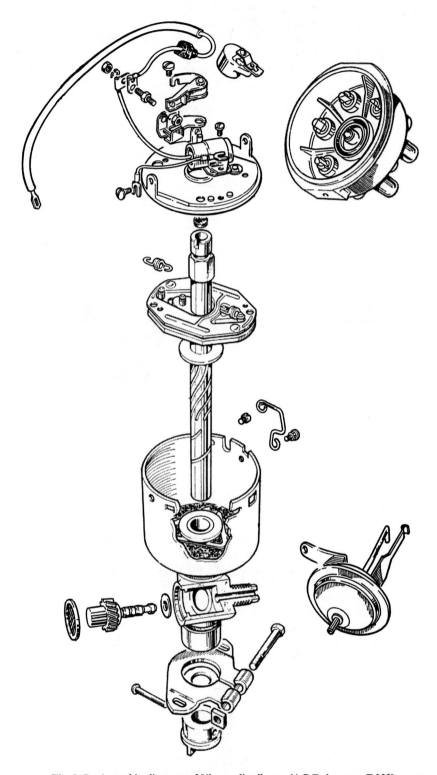

Fig. 2:7   Assembly diagram of Vitesse distributor (AC Delco type D202)

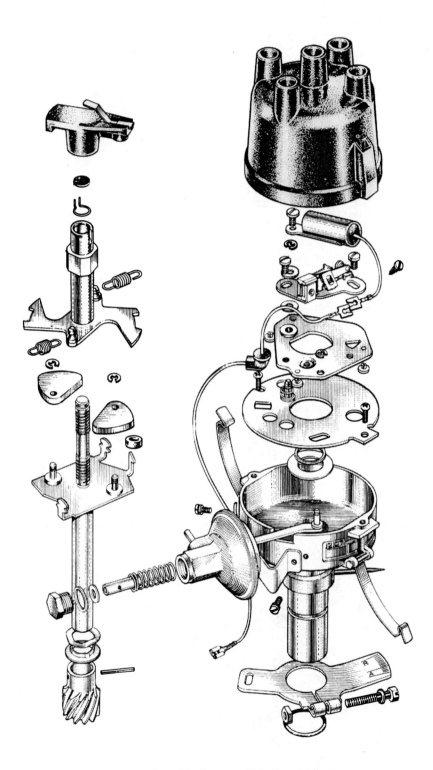

Fig. 2:8 Assembly diagram of Cortina distributor

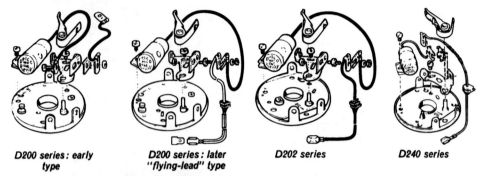

D200 series: early
type

D200 series: later
"flying-lead" type

D202 series

D240 series

Fig. 2:9 Assembly details of Delco Remy distributor points. It is possible to service D200 and D202 distributors with the D240 contact set 7953383 simply by discarding the old nuts and lockwashers together with the contact arm centre screw or stud. If it is necessary to remove the distributor from the engine, first note the position of the rotor and then release the stud securing the clamp to the engine. Do not loosen the housing clamp bolt or turn the engine whilst the distributor is removed otherwise the ignition will need retiming

gauge than the other, and one of the springs may also be retained in more tension than its partner.

Make sure that all electrical wires are repositioned correctly, all terminals are tight and, when fitting the contact points (see figs. 2:6 and 2:9) ensure the insulating washers are refitted. If a new distributor cap is fitted make sure to position the leads correctly and ensure that a carbon brush is fitted to the cap (see fig. 2:10).

Setting the contact points

When setting the points, make certain the heel of the points is on the lobe of the cam (see fig. 2:6). The gaps of the distributor points vary slightly according to the model of the distributor.

On Lucas distributors prior to 1962, the recommended gap is 0·012 in. but on distributors produced since 1962 the gap is 0·014–0·016 in. Vauxhall models require a gap of 0·019–0·021 in. (Continental cars are usually 0·016–0·020 in. but the Volkswagen 0·017 in. and most Fiats 0·017–0·020 in.)

On cars where the recommended distributor points gap cannot be ascertained 0·015 in. will be found satisfactory.

If a new capacitor is fitted, it must be of the correct capacity. Each capacitor is marked

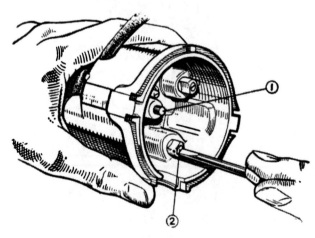

Fig. 2:10 The correct method of connecting HT leads. 1. Carbon brush. 2. Cable securing screw

Fig. 2:11 The breaker plate and vacuum assembly on some Ford cars can be removed as a complete unit

and the colour of the lead on the capacitor usually signifies the capacity. The wrong capacitor is likely to cause excessive pitting at the contact points.

Sluggish starter-motor rotation

Sluggish starter rotation is not necessarily the result of an internal starter defect (see chart 1:1) for it may be due to poor electrical connections between the engine and body, or a high resistance somewhere else in the starting circuit.

A break in the field coils results in complete starter failure, but an earthed field coil is more likely to result in loss of starter power, resulting in the starter motor not rotating the engine sufficiently fast for the engine to start, especially when cold. At a later stage, complete failure of the starter may occur.

Fig. 2:12 Fitting new points to the AC Delco D200 distributor on a Spitfire. Some manufacturers now fit ventricated-type contact points to the distributor as standard equipment. On ventricated contact points the fixed point has a hole through the centre. Ventricated types are available for most distributors and though they cost more they last longer

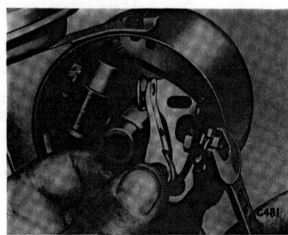

Overhauling the starter motor

When overhauling a starter motor, unless the drive assembly does not operate satisfactorily or is otherwise faulty it is not usually necessary to dismantle it, a thorough wash in petrol often being all that the drive assembly will need.

Dismantling the starter motor

Remove the cover band (see fig. 2:13), pull back the brush springs and release the brushes from their holders. Undo the two 'through' bolts and remove the armature, drive-end bracket and drive assembly as one unit. Remove the terminal nuts and washers from the terminal post on the commutator-end bracket and carefully detach the bracket from the yoke.

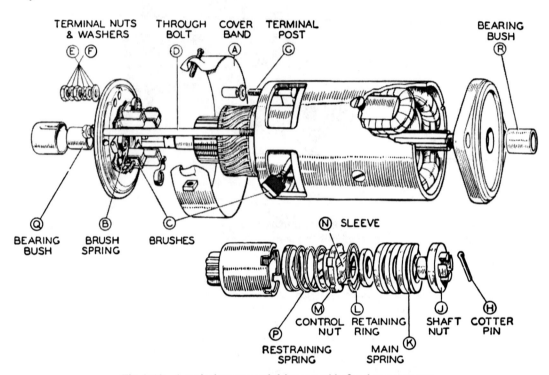

Fig. 2:13   A typical starter and drive assembly fitted to many cars

Dismantling the drive assembly

Although starter motor drive assemblies differ in several respects, provided care is taken to observe the correct order in which the various parts are removed, little difficulty is encountered.

Before dismantling the drive gear on starters such as depicted in fig. 2:13, check the main spring. If it can be rotated on the shaft with the fingers the spring is weak and should be renewed. To remove the drive assembly from the armature shaft, compress the main spring slightly and extract the circlip.

On drive assemblies fitted with split pins instead of circlips, the pin must be taken out and the shaft nut removed to release the main spring, collar and the rest of the assembly.

To strip the drive assembly, hold it in a vice and compress the return spring with the hand to enable the retaining spring clip to be removed.

On drive assemblies fitted with the larger diameter type main spring (see fig. 2:14), remove the circlip from the outer end of the drive-head sleeve, take off the front spring-anchor plate,

Fig. 2:14 Another commonly used type of drive mechanism

main spring and rear spring-anchor plate. Extract the pin securing the drive-head sleeve to the armature shaft, push the assembly down the shaft, remove the woodruff key and take the complete drive assembly from the shaft.

Remove the barrel-retaining ring from the inside of the barrel-pinion assembly and withdraw the barrel and anti-drive spring from the screwed sleeve. From the inner end of the drive-head sleeve remove the circlip, locating collar, control-nut thrust washer, cushioning spring, control nut, screwed sleeve and drive-head thrust washer.

Reassembling the drive assembly is a reversal of the dismantling procedure. Starter drive assemblies should not be lubricated otherwise they will collect clutch dust and grit and cease to function.

Renovating the armature

Inspect the armature, giving particular attention to where the conductors attach to the commutator, if the conductors have 'lifted', the armature should be renewed.

An armature can become severely distorted and it is therefore important that it is checked by placing it between centres and slowly rotating the shaft. If the shaft is bent, a new or reconditioned armature should be fitted.

If the armature shaft runs 'true', clean the commutator with fine glass paper whilst rotating the shaft. Do *not* undercut the mica on a starter-motor commutator. If the commutator is severely worn, denoted by a ridge at the side of the brush path, the armature should be placed in a lathe and rotated at high speed and the commutator lightly skimmed, with a sharp tool, sufficiently to remove the ridge. A final polish should be obtained by using fine glass paper.

Field coils

If the insulating band is not fixed in position, remove it from the 'yoke' and thoroughly wash the field coils in petrol and allow to dry.

Test the coils for open circuit by using a 12 volt battery and connecting a lead to one of

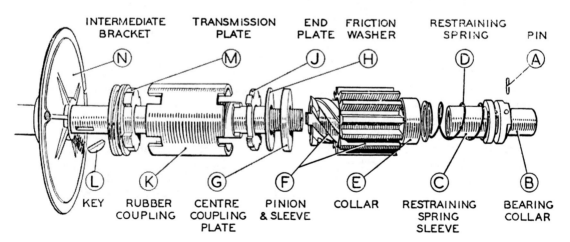

Fig. 2:15 Assembly of the 'RS' type drive mechanism

the field brushes and another lead, with a test bulb in circuit, to the field terminal. If the bulb lights, remove the lead connected to the brush and connect it to the other brush. If the bulb fails to light, there is an open circuit in the field winding.

To check that the field coils are not short circuited, continue the test but make sure the brushes are not fouling the yoke, and remove the lead from a brush and hold it onto a clean part of the starter yoke. If the bulb does not light carry out the test again, but this time connect the lead to the other brush. If the bulb lights, the field coils are earthed and they must be renewed.

Starter-motor bearings

If the bearing in the drive-end bracket, or the bearing in the commutator-end bracket, is worn, a new bearing must be fitted. The bearings, however, are of a porous, phosphor-bronze type, and must be completely immersed in lubricant for 24 hours before fitment, or, alternatively, immersed in oil heated to 180° F. for two hours and allowed to cool (in the oil).

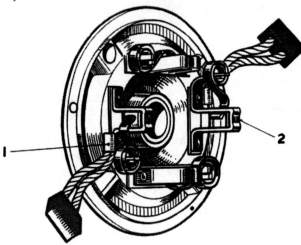

Fig. 2:16 A commutator-end bracket showing the brushes which may be renewed by unsoldering at the terminal 1. The brushes must be a sliding fit to the brush box 2

Commutator-end bracket

Thoroughly wash the bracket in petrol and allow it to dry. If the commutator is pitted or burned, the brush holder springs are probably weak and should be renewed. If the bearing which previously supported the armature is worn, the bearing should be pressed from the bracket and a new bush fitted. If the old bush is in good order, lightly oil it when reassembling.

Fitting new brushes

The brushes fitted to the end bracket can easily be unsoldered (see fig. 2:16) and unless the brushes are fairly new they should be renewed.

The brushes that connect by means of leads to the field coil 'tappings' (see fig. 2:17) are also soldered in position and may be removed by unsoldering them unless the field coils are made of aluminium when the process of resoldering, even if the correct solder is available, can be difficult. Rather than unsoldering the brushes from the field coil it is better in this case to cut the original flexible copper leads so as to leave short lengths to which the new brush leads can be soldered. It is advisable however, to trim the new leads for if they are too long they will chafe against the cover band.

Reassembling the starter motor

Reassembly of the starter motor is a reversal of the dismantling process. Points to note:

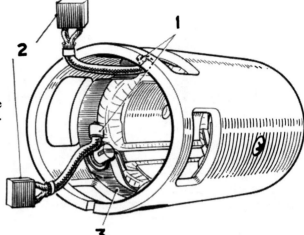

Fig. 2:17 Brush connections to the field coils. 1. Field coil connections. 2. Brushes. 3. Yoke

Refit the insulating band between the field coils and yoke at the commutator end

Refit all thrust and insulating washers

Position the field brush leads so that they cannot 'chafe' against the cover band or 'through' bolts

Refit the pinion assembly in a dry condition.

When the starter is assembled, rotate the armature and give the end brackets a light tap with a hide-faced hammer and then retighten the 'through' bolts

Test the starter motor before refitting it to the car

CHAPTER 3

The Power Unit—Is an Overhaul Necessary?

If an engine is consuming an excessive amount of oil, making strange noises and the oil light comes on, a thorough investigation of the unit should be carried out to ascertain:

The cause of the trouble

What repair is necessary

If the required repair is an economical proposition

A compression test is indispensable for judging the condition of valves, pistons and cylinders. All cylinders should register within 5 or 6 lb/in² of each other.

If a compression pressure gauge is not available most garages will do the test for a very moderate charge.

Run the engine until normal operating temperature is obtained, switch off and adjust the tappets. This done, run the engine for a few more minutes and switch off again. Remove any loose deposits from around the base of the spark plugs and take them out.

Next, prop the throttle wide open and if the starter is operated from the ignition switch, disconnect the distributor LT wire. Insert the compression gauge connector into No. 1 hole and press the starter button. With every compression stroke of the cylinder under test, the gauge should rise, until after 4 or 5 compression strokes the optimum pressure of each cylinder is reached. Count the number of compression 'pulses' and when 6 'pulses' have been felt, remove the gauge. Record the readings on a note pad, zero the gauge, and take the readings of the other cylinders.

Whilst taking the compression pressures, observe the pattern of pressure increase, for the rapidity with which the maximum is reached is important. If the readings do not rise steadily with each pulse, reaching a peak after 5 or 6 revolutions of the engine, the valves may be sticking slightly.

Any cylinder reading 10 lb/in.² less than the highest reading obtained from the other cylinders indicates leakage past the piston rings or valves. To decide which, squirt a little oil on to the piston in the cylinder which indicates a low pressure, and recheck the compression pressure. On some cars, to avoid the oil getting on to the valves, it is advisable to inject the oil on to the piston with a syringe. The idea is to get the oil to form a temporary seal to prevent compression leakage past the piston rings. If the pressure on the cylinder being tested is increased by 8–10 lb/in² or more by the oil treatment, piston ring leakage is indicated. If this does not increase the pressure it denotes burnt or badly seating valves.

When adjacent cylinders are low in pressure, this usually indicates gasket failure between the cylinders. If all of the pressures are different and the pressure does not increase steadily with each pulse and the oil test has little effect the valves are probably not seating correctly or there may be a great deal of piston or cylinder bore wear.

Oil consumption problems

If oil consumption is thought to be on the high side, before deciding on any form of repair it is advisable to check the engine for oil leaks and to ascertain the exact consumption by carrying out an oil consumption test.

To check if an engine is burning oil, warm the engine to operating temperature and, with the car standing still, snap the throttle open quickly a few times and watch the exhaust. Clouds of blue smoke denote excessive oil burning. On high-mileage engines, oil burning will either be due to wear of the piston rings and piston lands, and slack bearings causing

Chart 3:1

Common Engine Troubles (1)

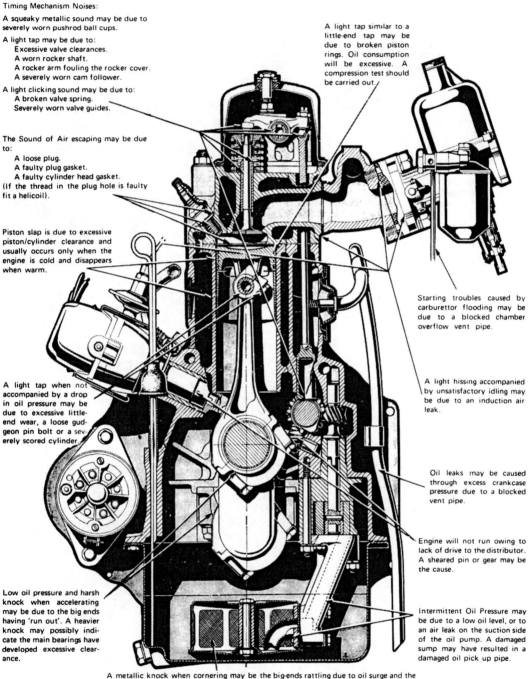

Timing Mechanism Noises:

A squeaky metallic sound may be due to severely worn pushrod ball cups.

A light tap may be due to:
Excessive valve clearances.
A worn rocker shaft.
A rocker arm fouling the rocker cover.
A severely worn cam follower.

A light clicking sound may be due to:
A broken valve spring.
Severely worn valve guides.

The Sound of Air escaping may be due to:
A loose plug.
A faulty plug gasket.
A faulty cylinder head gasket.
(If the thread in the plug hole is faulty fit a helicoil).

Piston slap is due to excessive piston/cylinder clearance and usually occurs only when the engine is cold and disappears when warm.

A light tap when not accompanied by a drop in oil pressure may be due to excessive little-end wear, a loose gudgeon pin bolt or a severely scored cylinder.

Low oil pressure and harsh knock when accelerating may be due to the big ends having 'run out'. A heavier knock may possibly indicate the main bearings have developed excessive clearance.

A light tap similar to a little-end tap may be due to broken piston rings. Oil consumption will be excessive. A compression test should be carried out.

Starting troubles caused by carburettor flooding may be due to a blocked chamber overflow vent pipe.

A light hissing accompanied by unsatisfactory idling may be due to an induction air leak.

Oil leaks may be caused through excess crankcase pressure due to a blocked vent pipe.

Engine will not run owing to lack of drive to the distributor. A sheared pin or gear may be the cause.

Intermittent Oil Pressure may be due to a low oil level, or to an air leak on the suction side of the oil pump. A damaged sump may have resulted in a damaged oil pick up pipe.

A metallic knock when cornering may be the big-ends rattling due to oil surge and the oil strainer sucking in air. It is more likely to happen if the oil level is low. A *blocked* strainer may result in a low oil pressure. A blocked external filter (not visible) can result in a drop of oil pressure of about 15-20 p.s.i. It is essential to renew filter elements at the stipulated periods. A blocked filter causes unfiltered oil to be fed to the bearings.

Chart 3:2

Common Engine Troubles (2)

Inadequate number of threads for adjusting rocker clearance due to pocketed valves or too much metal removed when re-cutting valve seats. Incorrect valves may have been fitted.

Burned exhaust valve and seats may have been caused by valves sticking and/or detonation owing to heavy carbon deposits and/or distortion and overheating. Usual symptoms are difficult starting and poor performance. A compression test will help reveal the trouble.

Coolant seepage may be due to a distorted cylinder head, unsatisfactory pulling down of cylinder head after overhaul or a faulty head gasket or foreign matter becoming trapped between head and gasket when refitting.

Lack of compression on one cylinder may be due to a hole in the piston crown resulting from detonation or valve head becoming detached, or to foreign matter such as a steel nut entering the carburetter intake or through a plug hole. Detonation may be caused by poor fuel, incorrect timing etc.

A heavy rumble from the rear of engine may be due to a loose flywheel. A lighter clicking sound possibly a clutch release plate driving strap broken or a loose ring-gear.

Oil pressure slow to appear may be due to blocked strainer and/or worn oil pump, excessive bearing clearance and blocked external filter see chart 4·3.

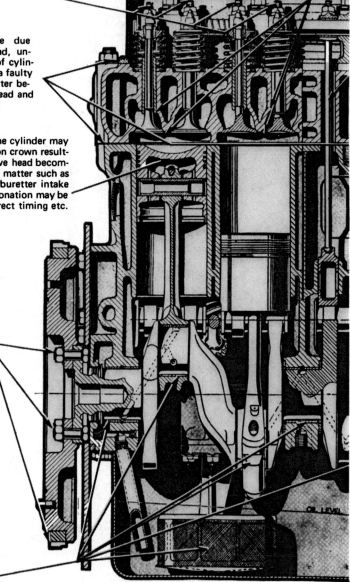

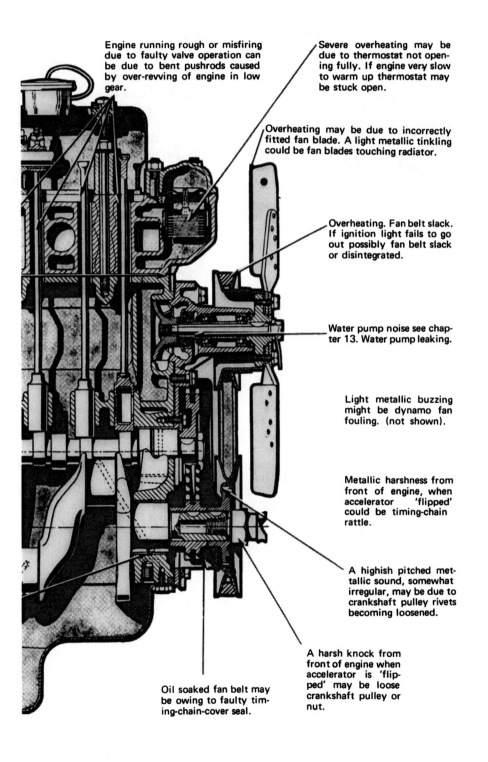

Engine running rough or misfiring due to faulty valve operation can be due to bent pushrods caused by over-revving of engine in low gear.

Severe overheating may be due to thermostat not opening fully. If engine very slow to warm up thermostat may be stuck open.

Overheating may be due to incorrectly fitted fan blade. A light metallic tinkling could be fan blades touching radiator.

Overheating. Fan belt slack. If ignition light fails to go out possibly fan belt slack or disintegrated.

Water pump noise see chapter 13. Water pump leaking.

Light metallic buzzing might be dynamo fan fouling. (not shown).

Metallic harshness from front of engine, when accelerator 'flipped' could be timing-chain rattle.

A highish pitched mettallic sound, somewhat irregular, may be due to crankshaft pulley rivets becoming loosened.

A harsh knock from front of engine when accelerator is 'flipped' may be loose crankshaft pulley or nut.

Oil soaked fan belt may be owing to faulty timing-chain-cover seal.

excess oil to be thrown into the cylinder bore, or to severely worn valve guides or faulty valve-guide oil seals.

To carry out a consumption test, place the car on level ground and allow it to stand for an hour, so that the oil in circulation drains back into the oil pan (sump). Top up the sump to the full mark on the dipstick. When 100 miles have been completed, place the car on level ground and again allow it to stand for an hour and measure the amount of oil required to bring the oil level to the 'full' mark on the dipstick.

Oil leaks

To find how much oil is being lost through leaks, etc. place a clean piece of paper under the car and warm the engine. When the normal operating temperature is reached run the engine fairly fast for a couple of minutes and switch off. When an hour has passed, examine the paper and note the approximate position of any oil patches. This will help to locate the leak. Most engines lose a little oil from leakage of some kind, usually it has a marginal effect on oil consumption but it is surprising how a few drips from a loose oil pipe or a badly seated oil filter bowl mount up and it is advisable to correct the fault.

Fig. 3:1   If an oil light switch is renewed it is essential that the replacement fitted is of the correct type for the engine (see below)

A common cause of oil leaks, particularly with older cars, is the building-up of excess crankcase pressure. This forces oil through the back main bearing and causes leaks from sump gaskets and numerous other places. Usually the pressure build-up is the result of nothing more than a blocked crankcase breather pipe or rocker breather pipe, which is easily cured.

Loss of oil pressure

Oil warning lights are notoriously unreliable and the actual pressure should be tested by temporarily fitting an oil pressure gauge. On a car which is to be kept for a long time it may be worth fitting a permanent oil gauge. This is simple to do and allows faults to be corrected before they lead to more serious trouble. Moreover, oil light switches (see fig. 3:1) on different cars operate at different pressures, and a pressure switch can be changed to give the appearance of a satisfactory oil pressure. When purchasing a second-hand car this is a point to guard against.

Oil pumps are designed to deliver far more lubricant than the maximum requirements of the engine. The surplus oil, under pressure from the pump, is dealt with by the oil-pressure relief valve (see fig. 3:2) which, like a safety valve, comes into operation at a predetermined pressure. The output from the pump is fed to all pressure-lubricated points and to the oil-pressure relief valve. A proportion of the pump output is absorbed by escaping through the sides of bearings, etc. but the surplus oil, when the top pressure is reached, escapes through the relief valve.

An increase in bearing clearance at a pressure-lubricated point can reach the stage where it causes a drop in pressure due to the oil escaping and not building up pressure. On a cold engine, the oil cannot escape so readily through the wear points and the pressure may be reasonable until the oil is warm.

If a good oil is used, engines not driven hard will operate satisfactorily at quite low pressures, but a low pressure accompanied by a knock must be given immediate attention for excessive clearance at a pressure-lubricated point is indicated and may result in ruination of the entire engine.

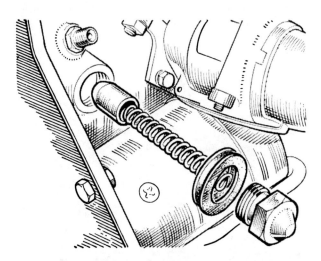

Fig. 3:2 A typical oil pressure relief valve and spring. On a worn engine, renewal of the oil relief valve and spring, and/or stretching the spring, though it may increase oil pressure when the engine is cold, is unlikely to affect the running pressure. Low oil pressure is usually a result of bearing/crankshaft/oil-pump wear

Big-end and main-bearing knocks

Severe big-end trouble is audible as a hard metallic knock accompanied by a severe drop in, or complete failure of, oil pressure. In less severe cases it can be detected when the car is standing, by opening the throttle quickly a few times.

Slight big-end wear can be detected by test-driving the car on a level road and constantly varying the amount of throttle so that the car passes through light-drive to over-run. If a heavy rapid 'purring' metallic noise is generated and the oil pressure is low, the big ends are either worn severely enough to have developed extra clearance, or one of them is practically 'knocked' out.

A light metallic 'purring' noise can be due simply to a blocked oil filter causing oil starvation at the big ends. If the filter is not changed immediately bearing life will be considerably shortened.

With modern engines, bearing wear is less discernible than it was in the past when white metal bearings were used, and on high-mileage engines the noise may denote a combination of crankshaft ovality and bearing wear. Whenever a distinct metallic purring sound or harsh knock occurs, the condition is getting serious and even if the drop in oil pressure is not significant, the sump should be removed, the bearings inspected and if necessary renewed. During the test drive, light rumbling, when the accelerator is depressed quickly and the engine is picking up speed, indicates that the main bearings have excessive clearance. On some cars this may involve engine removal, but if the sump can be taken off it may be possible to remove the main-bearing caps and renew the bearings with the engine *in situ*.

Piston slap—timing-gear noise—gudgeon-pin knock

Some noises are not accompanied by a loss of oil pressure. Piston slap, timing-gear noises and little-end knocks being the most notable. Little-end trouble is identifiable by shorting the spark plugs of each cylinder in turn when the engine is hot and running fairly fast. The tap will disappear when the faulty cylinder is located. If the tap does not disappear it is unlikely to be little-end trouble.

Piston slap is audible when the engine is cold and pulling hard from low speeds up a gradient, but the noise disappears as soon as the accelerator is released and may disappear completely when the engine is hot. Although piston slap is undesirable it is not serious with high-mileage cars and is unlikely to give rise to immediate trouble.

Valve mechanism noises

The valve mechanism can produce light taps due to excessive wear on the rocker shaft and push-rod ball caps, and excessive valve clearance. Such noise is not changed in volume by engine load and its rhythm is at half crankshaft speed.

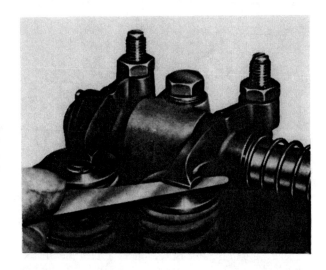

Fig. 3:3 Measuring the valve clearance. See appendix for correct clearance. Sequence tables will be found on page 63. If adjustment of the valve clearances fails to reduce tappet clatter, the rocker arms are probably pitted and/or the rocker shaft and rocker arm bushes worn

Tappet noises may be detected by removing the oil filler cap on the rocker box when the noise will be more discernible. Adjustment of the valve clearances (see fig. 3:3) may not cure it, for the rocker arms can become severely pitted, and it may be necessary to remove the rockers and grind the pit marks out, or renew the rockers. Undue clearance between the rocker arm and shaft also gives rise to excessive tappet noise while lack of lubrication on the ball end cups may give rise to some queer squeaking effects. Excessive clearance between the valve guides and valve stems may result in a clicking metallic sound like a broken valve spring. It will probably have caused slight misfiring, or poor performance, but whereas undue clearance between the valve and guide results in a rise in oil consumption and misfiring due to the oiling up of a spark plug, a broken valve spring will usually cause fairly serious misfiring and if neglected can result in severe damage to the piston.

Timing-chain rattle is easily discernible by listening at the frond end of the engine whilst the throttle is 'flipped' a few times. Where generator or water pump noise conflicts the issue, disconnect the fan belt and 'flip' the throttle again. In cases where a distinct, heavy rattle occurs at the front end, check the crankshaft pulley. A loose pulley can result in the crankshaft boss becoming severely worn and a relatively inexpensive job becomes expensive.

On some Vauxhall engines, if the timing chain becomes noisy, the chain tensioner can be adjusted from outside the engine by turning the square-headed adjuster screw (which is situated on the timing cover, on the generator side) anticlockwise whilst the engine is running until the noise from the chain is eliminated. Care must be taken, however, not to over-tension the chain otherwise excessive wear will take place.

Air leaks

Induction air leaks may be audible as a slight hiss; they can be confirmed by applying oil around the suspected area. If the engine is then accelerated the oil will be visibly sucked into the induction.

Generator noise

Internal generator noise can vary from a 'buzzing' to a distinct knock when the engine is accelerated. Broken generator brackets, and loosened generator bolts too, can produce an assortment of noises varying from a rapid vibratory type to a distinct medium/heavy 'clonk'.

To ascertain from where a noise is emanating, place the blade of a screwdriver against the suspected area and rest the handle of the screwdriver against the ear. If an assistant then accelerates the engine, the noise will be much more audible and the precise location of it made much easier.

Engine vibration

An engine vibration can occur for a number of reasons, but usually a vibration occurs due to the imbalance of a rotating part of the engine. Common faults, when not accompanied by lack of performance, or severe noise are:

A broken fanblade
A distorted or broken clutch-pressure-plate drive strap
A disintegrating driven plate
A loosened flywheel (indicated by a deep rumbling when accelerating)

By locating the cause of an engine noise when it first occurs and taking the necessary steps to cure it at an early stage, a considerable amount of expense and work can sometimes be avoided.

CHAPTER 4

Overhauling the Engine

When the car has been road-tested, compression tests have been carried out and the general condition of the engine has been assessed, it should be possible to decide upon the best course of action. Engine overhaul procedures may be classified as follows:

1 Top overhaul (removing cylinder head, decarbonizing, attending to valves, etc.)
2 Fitting oil-control rings, including category 1
3 Partial overhaul (possible with engine *in situ* if the sump can be removed)
4 Rebore
5 Complete reconditioning of the entire power unit (reground crankshaft, rebore, etc.)

Severe detonation (pinking) when the correct grade fuel for the type and model of car is used, or low or unequal compression, due to burnt or badly seating valves, indicates a top overhaul is required. The extent of the overhaul is dependent upon the condition of the valves, the valve seatings and springs, but on moderate-mileage cars a set of exhaust valves, valve springs and a decoke set of gaskets is probably all that will be necessary.

When carrying out a top overhaul, it is advisable to clean and reset the distributor points and to service the carburetter. After the overhaul, when 100 miles have been completed, the oil filter and engine oil should be changed.

Cylinder wear and piston ring wear affect the performance of an engine in many ways. Common complaints are: excessive oil consumption; lack of power; high petrol consumption; smoky exhaust; blow-by fumes from the engine breather vent; oil leakage from the sump joint and crankshaft oil-sealing owing to excessive crankcase pressure.

Special oil-control rings are available for use when it is not convenient to rebore the cylinders (see fig. 4:1); they are especially useful when oil-fouled plugs keep occurring and it is wished to avoid the expense of having the cylinders rebored. Moreover, as the rings can be fitted with the engine in situ provided the cylinder head and sump can be removed they are particularly useful for fitment where cylinder reboring would entail removing the engine from the car. Several makes and types of ring are available; Cord, Wellworthy, Simplex and Apex (see fig. 4:2) are a few of the better known.

When wear takes place in a cylinder, it is not distributed evenly from the top to the bottom but is greatest at the top of the cylinder just below the non-wear ridge where combustion takes place. As very little wear occurs at the bottom of the cylinder, it also becomes rather tapered. The severest wear usually occurs in the cylinder at right angles to the axis of the crankshaft and worn cylinders are therefore slightly oval.

The conventional compression and one-piece oil-control ring is perfectly circular but as only the bottom edge of a one-piece ring can make regular contact with the cylinder wall, the rings cannot form an effective seal in a worn bore (see fig. 4:3); compression is lost, gases blow past the rings contaminating the lubricating oil and forming carbon deposits in the oil-ring grooves, and oil passes above the rings.

Cord rings are designed to function effectively in worn bores and consist of a number of separate segments (fig. 4:3); each is capable of independent movement and will seal even a severely tapered cylinder. The segments are 'cupped' or 'dished' giving them an up and down tension that allows them to seal completely against the ring groove, so preventing the passage of gases and the occurrence of wasteful 'ring pumping'; the lack of ring movement also effectively reduces groove wear. Because of the constant flexing of the segments, carbon does

Fig. 4:1 The Cords 'Oilguard'—a ventilated type of ring, which is also available with the top and bottom segments chromium-plated. This type of ring is now being adopted as standard practice on a wide scale throughout the motor industry

not get a chance to form in the grooves and, owing to their 'concertina' formation, oil is held in the ring.

The Cord company also manufactures an Oilguard ring (see fig. 4:1) similar to the cupped segment ring but with a spacer dividing two segments so providing a ventilated ring which is extremely successful in worn bores, especially on pistons where no oil-relief is provided below the groove. The Oilguard ring is also ideal for use in rebored engines, and may be used in the groove below the gudgeon pin if desired. There is also another type of ring which has a stainless steel expander spacer fitted between two chrome rails, known as the Cordflex, which the Cord company recommends for some types of engines.

Fig. 4:2 Designed to function independently of piston groove depth, Apex oil rings made by Hepworth and Grandage are three-piece rings comprising two chromium-plated steel rails and a special steel spacer/expander

Cord rings are supplied in sets and, with few exceptions, include one cupped-segment ring for compression and oil control, and one other oil-control ring of one of the types described; in addition there is a cast-iron compression ring to complete the ring installation. When ordering Cord rings, it is not necessary to specify any particular type of ring; sets are made up to include the rings found to produce the best results in the type of engine for which they are supplied.

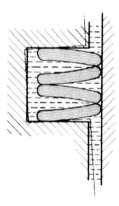 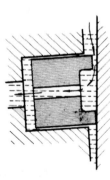

Fig. 4:3 Cords cupped-segment rings. (Left) The lateral tension, provided by the 'cupping', seals the groove and prevents ring pumping, yet at the same time oil is held between the segments to provide instant lubrication at the crucial period when the engine is started from cold. Even if the piston moves out of parallel with the bore (as when piston topple occurs) the whole of the ring face of a Cords cupped-segment ring maintains contact with the bore. With a one-piece ring (right) only one corner of the ring does so

The rings supplied for the top ring grooves will, in the case of popular car sets, normally be stepped rings, so as to avoid contact with the non-wear ridge.

Cord's multi-segment rings cannot be fitted in two-stroke machines because the piston grooves are not usually pegged across the full width of the groove, and one or more of the segments might move round in the groove resulting in the ends fouling the induction ports.

Re-ringing the pistons of modern engines is not difficult and does not require the use of special tools or equipment. Local stockists are always glad to advise and will usually be able to supply the parts needed.

Full instructions for fitting Cord rings and refitting the pistons to the cylinders are supplied with every set.

It is important when fitting 'overhaul' type pistons or Cord rings, to deglaze the bores, otherwise the rings take a long period to 'bed in'.

By the time an engine begins to use oil, other parts of the car are usually 'suspect', particularly big-end bearings, timing chain and cam followers. An overhaul, renewing such items, and using oil-control rings on the pistons, can be surprisingly inexpensive, but where the engine has to be removed for the purpose it may be worth the slightly greater expense of going a stage further.

Fig. 4:4   An 'overhaul type' piston by Hepolite. These pistons are fitted with the latest Apex or Hepocrom oil-control rings and will restore worn engines to peak performance without the necessity of an expensive overhaul. They may be fitted into engines without reboring when cylinder wear measures between 0·002 in. and 0·008 in. and will restore normal oil consumption, fuel economy, good starting, original performance and quietness with the minimum of trouble

If the piston lands are cracked, or the thrust face of the pistons scored or considerably worn, the pistons are not satisfactory to take the rings and, in such cases, pistons such as the Hepolite 'Pep' overhaul type pistons (see fig. 4:4) which are satisfactory for cylinders with up to 0·008 in. wear, and have oil-control rings fitted to the skirt, might well be used. On the other hand, if new pistons have to be obtained, a reconditioning set which includes pistons, rings and bearings, etc. is worth a thought.

If the pistons are not cracked or the skirt scored, piston life may be considerably extended by fitting inserts to the ring grooves and expanding the piston skirts (see page 69).

If it is not possible to remove the sump with the engine *in situ*, it may pay to run the car a few more miles and then do a complete overhaul on the engine; for instance the sump cannot be dropped on the MGA though it can on the MGB; on the Morris Minor the sump is easy to remove; on the Vauxhall Victor the suspension has to be lowered and the engine raised so that the sump can clear the sub-frame. The sump can also be removed on the Austin A40, some Ford Consuls and the Sunbeam Rapier but not of course on BMC transverse engines.

Overhaul kits

Engine reconditioning sets such as those supplied by Associated Engineering (Sales) Ltd. and A. S. Recon Ltd., are boxed kits designed specifically for our category 3 overhaul, and are exceedingly good, but separate parts can be purchased from engine specialists if desired.

A partial overhaul however is not suitable for engines with more than 0·008 in. bore wear, or if the crankshaft has more than 0·003 in. ovality. The mileage at which our category 3 overhaul is advisable can vary tremendously, but about 45,000 miles is probably the ideal, for around this mileage the wear rate accelerates fairly sharply, leading to ovality of the crankshaft and deterioration of the piston lands. By carrying out a category 3 overhaul in time, another 35,000–40,000 miles should be possible without any major repair becoming necessary.

Apart from fitting the basic overhaul kit, the oil pump should also be given attention where possible. Most kits, which are boxed for individual makes of car, contain pistons, connecting-rod bearings, exhaust valves, valve springs, timing chain, decoke set of gaskets, timing cover oil seal, oil filter, piston ring clamp, valve grinding stick, gasket cement, coarse and fine grinding paste, emery cloth, and comprehensive overhaul instructions.

It is important, before purchasing a kit, to ensure that the engine has not previously been overhauled for the cylinder block may have been rebored and/or the crankshaft reground. Piston sizes are usually stamped on the piston crowns and the big-end bearing shells are usually marked on the back.

On the Triumph Spitfire for instance, and on many popular car engines, excessive oil consumption resulting from excess cylinder wear can be overcome by reboring the cylinders whilst the engine is *in situ*, and many smaller garages are tending to specialize in this type of work. After the owner has removed the cylinder head and sump, the garage will tow the car away and return it with the cylinders rebored, and new pistons fitted to the cylinders with connecting rods realigned and new big-end bearings for a surprisingly small fee. If the crankshaft has more than 0·002 in. ovality, however, it may be advisable to carry out a complete overhaul. The cost of a reground crankshaft with bearings and thrust washers for a typical 4-cylinder engine is about the same as for a set of pistons.

If the car is to be kept for another 40,000 miles, although oil-control rings can take care of oil consumption problems on cylinders worn by as much as 0·010 in., if bore wear exceeds 0·008 in. and/or the crankshaft has developed ovality in excess of 0·002 in., a complete overhaul of the entire power unit is advisable. The difference in the cost of overhauling the engine oneself and the price of a reconditioned engine can be so small that, if the guarantee given with a Works unit is taken into account, the Works-reconditioned unit may be the best way, especially as there is always the risk that the cylinder head of one's own engine may have a cracked valve seating, or the cylinders have already been rebored to the maximum permissible, or the crankshaft previously reground.

Several manufacturers supply a 'short' reconditioned engine consisting of a cylinder block and pistons with a reground crankshaft and bearings. This can be an excellent starting point when building a reconditioned power unit.

Tools for the job

When overhauling an engine or, indeed, carrying out other work of quite a minor nature on the car, a set of mechanic's tools is essential. The basic needs are: a large and small screwdriver, a pair of pliers, a small cold chisel, hammer and a set of open-ended spanners, ring spanners and socket spanners.

A comprehensive range of special tools for specific purposes is available for most cars, but while they cut down the time spent on a job and tend to make it easier, few of the tools are actually indispensable. A little initiative and mechanical aptitude allow most jobs to be tackled satisfactorily, but the minimum requirements as regards special tools for an engine overhaul are circlip pliers, a valve-spring compressor, a piston-ring compressor if possible and a clutch-alignment mandrel. On B.M.C. transverse engines, a flywheel puller is also necessary, but it is usually possible to hire one.

Stripped threads

One of the difficulties which occur when carrying out repairs is that caused by stripping the threads; often an offending nut or bolt can be replaced easily but sometimes a stripped thread can put one to a fair amount of trouble. Helicoil inserts, however, can be used to restore threads permanently in many awkward situations. The inserts are formed of stainless-steel diamond-shaped wire and when installed in a properly prepared hole they form standard threads of original diameter. Helicoils are particularly useful for restoring threads in cylinder-head plug holes and for restoring threads for studs in cylinder blocks and allow one to avoid using oversized studs or bolts. Helicoil threads are stronger than the original and resist corrosion, stripping and cross-threading.

Helicoils are available in kits, covering a wide range of sizes. Each kit contains full instructions, a tap and a thread insertion tool. The suppliers are Armstrong Patents Co. Ltd., East Gate, Beverley, Yorks.

Self-locking nuts

Modern self-locking nuts, of such types as the Armalock nut, make invaluable replacements for nuts that are subjected to the severe effects of vibration and will lock without the addition of lockwashers, locknuts, split pins, tabwashers, etc. They are shock-, vibration- and corrosion-resistant and function satisfactorily at any temperature between $-70°$ C and $200°$ C. The self-locking properties are provided by the integration of a nylon 'collar' into the nut. When the nut is screwed onto a bolt, it turns freely until the 'collar' is reached. At this point the first resistance is felt. This is because the bolt threads are impressing themselves into the nylon collar. The bolt threads do not cut into the collar—they impress their thread form into the nylon as a result of further tightening. The end-result is that the assembly is locked firmly by friction resulting from flank loading of the mating threads of the bolt and nut body.

Locktite liquid sealant

This is another product which is useful where nuts are subject to vibration. A little of the sealant when applied to any nut virtually turns it into a locknut by seeping into the threads and setting. Apart from its use on nuts, bolts and screws, however 'Locktite' is particularly useful for putting onto fixed splines and prevents noise, oil seeping down the splines, and wear, by stopping movement between the surfaces. Locktite can also prevent bearing races turning in their housings provided that the housing wear is not more than 0·005 in. The sealant is resistant to heat, fuel, water and oil but, before it is applied, the parts must be completely free from grease and should be cleaned in a solvent. When Locktite has been applied the parts must be allowed to stand for 4 hours before being put into service.

Nuts and bolts treated with 'Locktite' can be released with ordinary spanners but a little extra force will be necessary due to the prevailing bond on the threads.

TOP OVERHAUL

If a top overhaul is to be carried out it will pay, a few hours before starting, to douse the exhaust manifold nuts with a corrosion dispersal solvent. In the case of side-valve engines (see page 63) the head nuts, too, should be treated.

On the majority of cars, unless it is a warm bright day and possible to work in the open, it is advisable to remove the bonnet so as to allow plenty of natural working light. Make sure, however, to mark the bonnet/hinge attachment points with a pencil or scribe mark, so that the bonnet can be replaced in precisely the same position, otherwise considerable re-adjustment may be necessary when it is refitted.

If damage to the paintwork is to be avoided, removing the bonnet is a task for two people.

When the bonnet is off, drain the cooling system, disconnect the battery, remove the air cleaners, the throttle and choke connections, and disconnect any heater hoses attached to the cylinder head. Remove any engine tie rods connected to the cylinder head and take off the valve rocker cover.

On side-valve engines, where the distributor is located on the cylinder head, the distributor must be removed.

Disconnect the plugs and take off the distributor cap. On engines such as the Vauxhall Victor, the complete distributor has to be removed to allow the push rod cover to be taken off.

Fig. 4:5   Rocker assembly (Triumph Herald 13/60 and Spitfire Mk 3)

The exhaust and inlet manifolds must be disconnected from the cylinder head but it may be possible to avoid disconnecting the exhaust pipe from the manifold, by removing the retaining nuts from the manifold flange and pulling the manifold to one side and wiring it into a suitable position.

On some cars, the exhaust and inlet manifolds and carburetter all have to be removed individually, but in many cases the carburetter, or carburetters, can be left attached to the inlet manifold and removed complete as one unit. With twin carburetter installations this avoids disturbing the carburetter interconnecting shaft, etc. The inlet manifold may have circular dowel-locating inserts fitted in the ports. Sometimes they drop out when the manifold is removed and, if this happens, retrieve them and retain them until the manifold is refitted.

As spark plug insulators are easily broken, remove the spark plugs.

Next take off the water hose between the cylinder head and radiator. If the water pump is attached to the cylinder head, disconnect the water hoses. Remove the retaining bolts and take off the pump.

If a by-pass hose is fitted between the front of the cylinder head and water pump, release the clips.

Before removing the rocker assembly (see fig. 4:5), it may be necessary to withdraw an oil-pipe-retaining clip from the rocker shaft and disengage the oil pipe.

Vauxhall Viva, see fig. 4:19.

When undoing nuts which retain the rocker shaft assembly, if any of them also help hold the cylinder head down take the nuts off gradually, and slacken the nuts in the sequence depicted in fig. 4:6 until the load is released. Note any locking plates on the rocker assembly for refitting in the same position when the assembly is refitted.

Keep the washers which fit under the head-nuts or bolts separate from other washers. Head-nut washers are of a slightly harder material and it is important that other washers are not refitted in place of them.

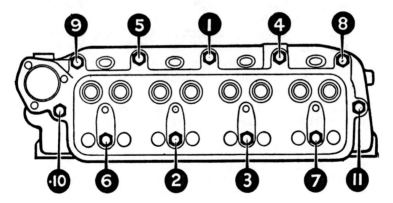

Fig. 4:6   When tightening or releasing a cylinder head, the nuts must be tightened or released in the correct sequence. Alloy heads must not be slackened or tightened when hot

When working on a car with an alloy cylinder head, make sure the head is quite cold before slackening the nuts otherwise the head is likely to distort.

Lift off the rocker assembly (see fig. 4:7) and withdraw the push rods giving them a sharp twist as they are pulled up, otherwise they will pull the cam followers from their bores. If, however, the unit has a high mileage to its credit this does not matter for the cam followers should also be removed and inspected.

Retain the push rods and cam followers in the same sequence in which they were fitted so that they can be replaced in precisely the same positions.

Check that everything necessary is disconnected and, if the back of the cylinder head is close to the bulkhead, make sure there are no brackets or fitments which will foul the head as it is lifted.

The next step should be to lift the cylinder head clear of the retaining studs. If the joint is tight, a light tap on the side of the head with a hide-faced mallet, or a hammer with a piece of wood interposed, will probably do the trick.

Fig. 4:7   Removing the rocker shaft assembly from a Cortina engine. Some engines, such as the Vauxhall Viva, have rocker arms fitted to individual studs (see page 64)

Some cylinder heads prove very difficult but sometimes there are lugs on the side of the head where it is possible to apply a lever. Never drive a screwdriver or metal implement between the head and block face. If difficulty occurs, use a good corrosion solvent on the studs. If this does not do the trick, try screwing the nuts halfway onto the thread of the studs, and give them a few gentle taps on each side with a hide-faced mallet.

If trouble persists, and lifting equipment is not available, refit the spark plugs, make sure that electrical wires are not dangling about, and reconnect the battery, ascertain that nothing can foul the radiator, and rotate the engine using the starter motor. The compression will usually break the joint. If it does not, be persistent with levers and corrosion solvent.

An assistant can be a great help when removing a stubborn head for it is important to lever squarely upon the lugs or other projection.

Cleaning the cylinder block and piston tops

When the head is off, remove the gasket and, preparatory to removing all carbon and extraneous matter from the block face, plug all water ports, etc. with a piece of rag to prevent entry of loose carbon particles. Next, turn the engine so that none of the pistons is at the top of a cylinder and smear a little grease around the top of each bore. When each piston reaches the top of its cylinder, the grease will form a seal and prevent any carbon particles falling between the cylinder wall and the piston.

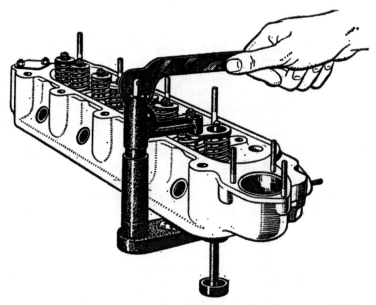

Fig. 4:8   Compressing the valve springs, to remove the valve cotters and release
the valves and springs

Turn the engine so that each piston crown can be cleaned, and scrape the carbon from the crowns. Next, inspect the cylinders and the block face around the base of each cylinder-head retaining stud to ensure that a stud is not pulling from the block and that all of the studs are screwed fully home.

The cylinder head

A valve-spring compressor is required to remove the valve cotters (see Fig. 4:8). This will

Fig. 4:9   Triumph Herald 1200, 12/50 and 13/60. Valves and springs which are removed by compressing the valve spring and sliding the cap sideways

EXHAUST   INLET

release the springs and valves which should then be kept in their relative order until they can be washed off and inspected.

Cone-type split cotters, although varying in size, are common to most cars but not all (see fig. 4:9). Sometimes cone type cotters are exceedingly tight in the spring cap, and difficulty occurs in compressing the spring. If this happens, place the valve-spring compressor in position and then put a short tube (a ¾ in. socket will do) over the valve cap, and keeping a little pressure on the spring compressor, give the tube a sharp blow with a hammer.

As the cotter and spring caps are being removed, note where the valve stem oil-seals are fitted. The seals may be small neoprene rings which should be fitted at the base of the cotter groove, or felt washers fitted to caps which sit over the valve guides, or rubbers in a separate cap beneath the spring cap. If an excessive amount of oil has been penetrating the combustion chamber the oil seals may be fitted incorrectly; this applies especially to the small neoprene type on B.M.C. cars, for when the oil seals are fitted by someone inexperienced, they are often incorrectly placed around the main stem of the valve near the top of the valve guide.

Whatever type of sealing washers are used they must be renewed when the valves are re-fitted.

Before inspecting the head and the valves, decarbonize them. A rotating wire brush operated by an electric drill is ideal for the purpose, but if this is not available a screwdriver or scraper can be used, provided care is taken to avoid damaging the valve seating. The carbon must be removed from the inlet ports as well as the exhaust ports. If the exhaust valve guides have a relief machined in the inner end, it is essential that all of the carbon is removed from it.

Although a wire type valve-guide cleaner can be very useful to help remove the varnish in the valve guides, if it is not available, a piece of rag given a good soaking with a proprietary carbon remover and pushed through the guides is a great help.

Clean the valves thoroughly, taking care to keep them in their proper order, giving particular attention to the neck of the valves, and to the removal of varnish deposits from the stems. If the stems are ridged, or there is more than 0·001 in. wear, the valves should be renewed.

When the head and components are thoroughly clean, check the valve guide clearance. Measurement is difficult without special equipment, and the amount of wear may have to be estimated by trying a new valve in the guide. The clearance between the exhaust guide and valve stem should be 0·002 to 0·003 in. and the inlet clearance from 0·0015 to 0·0025 in.

If possible, retain the old valve guides in the head for, when new guides are fitted, the valve seats invariably have to be recut to realign them with the new guides. If the valve seats are badly pitted, or the seats have sunk into the head, or if they are too wide, they will have to be recut. Usually, when valve guides reach a condition where they need renewing, the seats also want recutting and *vice-versa*. In such an event, it is often better to take the head along to one of the specialist engineering firms, or the local distributor for the make of car.

If the exhaust valves are burnt (see fig. 4:10), inspect the seatings in the head, as a cracked seat may have been a contributory cause of the valve burning. If there is any likelihood of a cracked seat, briefly lap a valve onto the seat so as to obtain a clean surface. The seat can then be examined through a magnifying glass, and the crack, should there be one, is visible as a hair line across the seating. A dye-penetrant type crack test will help by making any crack more immediately discernible. If a crack is found, a valve seat insert will have to be fitted.

Ford valve guides

Ford Consul Mk II, Zephyr Mk II and Cortina engines, and a few other models, do not have replaceable valve guides, the valves operating directly in guides machined in the cylinder head. Valves with oversize stems are available but if they will not adequately compensate the wear, the head will have to be taken to a Ford agent or an engine specialist for drilling and the fitment of valve guides.

Cortina inlet valves

Ford Cortina inlet valves are especially coated to increase the valve's resistance to high temperature and oxidization, and to form a hard wear-resistant surface. These valves must on no account be lapped to the head seats. If the valve faces are pitted, or worn, new valves should be fitted. The head seatings, however, can be refaced (see page 56) and should be lapped until smooth and free from pits using an old valve. Exhaust valves may be lapped in the usual manner. Cortina valves can be obtained with 0·003 in. or 0·015 in. oversize stems.

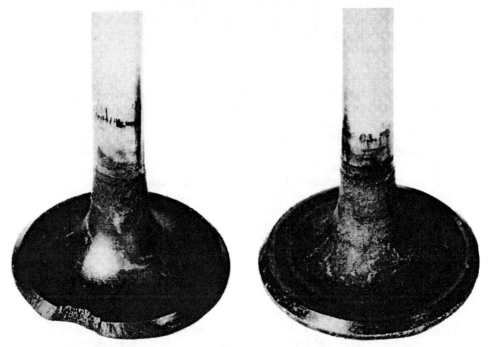

Fig. 4:10   Failure of exhaust valves caused by seat-burning as a result of pre-ignition or weak mixture. Weak mixture is frequently caused by unsuspected air leaks from loose carburetters, distorted flanges (see page 245), blown manifold gaskets etc.

Removing the valve guides

To remove valve guides, rest the cylinder head face-downwards on a piece of hardwood and drift out the guides using a steel shaft with a locating spigot, about 1 in. long, that will fit into the top of the guide. The outer diameter of the shaft should be slightly smaller than the outside diameter of the valve guide.

Before removing the guides, measure the height of the part which stands proud of the head, so that the new guides can be correctly positioned.

Make sure to fit the new inlet guides and exhaust guides into their respective positions, and ensure that they are in the right way up.

Valve seatings

Valve seatings may be refaced with special cutters, or with carborundum seat grinders; the latter being operated through the medium of a portable electric drill. Various sized stones are available, but as they are operated at a moderately high speed, great care must be exercised so as not to remove too much metal from the seating. Valve seat cutters, however, are hand-operated; the equipment includes a 'pilot' which fits into the valve guide, and a glaze breaker. The glaze breaker must be used on the seating first, followed by the cutters which give a smooth finish.

The idea, with valve seatings, is to obtain a smooth, narrow seat, but different manufacturers have slightly different ideas on the precise width of the seats but provided they are 50% less in width than the valve faces, the seatings will be satisfactory.

If the valve seats have sunk into the head a shallow cutter 10°–15° will be necessary to reduce the pocket which has formed. If the seat is wide, a 75° cutter to remove metal from the throat of the port may be required. The valve seat must be cut to conform to the original angle specification which is usually 30° or 45° according to the make of car.

If inspection of the cylinder-head face reveals that there has been 'blowing' between the combustion spaces, or if any distortion is suspected, have the head refaced. This will also help to counteract the decrease in compression ratio which would otherwise result from the increased combustion space caused by recutting the valve seatings.

Refacing the valves

Although valve faces must be cleaned up, should any of the faces show signs of excessive pitting, or if the valves appear bent, distorted or unlikely to clean up satisfactorily, or if the valve heads require an undue amount of metal removed, the valves are best renewed.

If the valve faces and seatings in the head are not severely worn they may be cleaned up by using a suitably shaped piece of emery cloth. Cut a sheet of coarse emery cloth into circles (a little larger than the valve heads) make a hole in the centre of each circle and then cut a wedge-shaped piece of cloth from the circle. To reface the valve, place a piece of emery cloth onto a valve, with the abrasive surface facing towards the valve face. The valve must now be rotated to and fro whilst maintaining a constant pressure on the valve. Valves with screwdriver slots are easy to rotate, but plain-headed valves prove rather more difficult, and the valve stem may have to be gripped with a hand brace and the pressure exerted by pushing lightly on the valve head with a hammer shaft. An assistant will be required, and though a somewhat rough-and-ready method, it nevertheless works.

Lapping in the valves

When the valves and seats have been refaced, use a coarse or medium grade carborundum paste to lap the valve faces onto their seatings. If the valve head does not have a screwdriver slot, a suction valve-grinder tool (see fig. 4:11) must be used to rotate the valves.

When lapping, use a minimum of grinding paste otherwise the paste splashes down the valve guides. Adopt a quick, semi-rotary action lifting each valve occasionally to spread the paste and help avoid 'grooves'.

If the valves have slots in their heads a screwdriver may be used to rotate the valves and a light coil spring may be placed under the valve head to help lift the valve from its seating. This will allow both hands to operate the grinding tool.

When a valve has been lapped, the valve face and the seating should have a dull matt, even finish, free from pitting.

If new valve guides have been fitted, make sure to check each seat with engineer's blue. Lightly rub the blue onto the valve face and, pressing the valve onto its seat, rotate the valve. Afterwards, examine each seat for the complete circle of blue that ensures all is well.

On completion of the valve lapping, thoroughly wash the valves and cylinder head in petrol, or other cleaning fluid, to ensure the complete removal of grinding paste, giving particular attention to the valve ports and guides. Apply oil to the valve seats and guides and refit the valves in their respective locations.

If the engine has covered a high mileage, or if the valve seats have been recut, it is advisable to fit new valve springs. Make sure to replace the bottom collars or caps, and follow with the inner springs (if fitted) and the outer springs (see fig. 4:13), the shrouds, the oil seals and the spring caps.

Compress each valve spring and, on B.M.C. cars, make sure to fit the valve-stem oil-packing ring into the bottom of the cotter groove or cap (see fig. 4:13). Fit the cotters and remove the compressor.

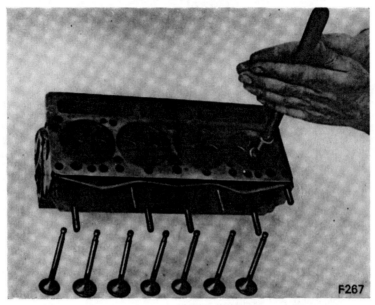

Fig. 4:11   Grinding in valves

On some cylinder heads, the valve-stem oil-seals consist of a felt washer that fits under the bottom cup. Before fitting the felts, it is important to soak them in oil otherwise the seals may become scorched when the car is first started.

Reconditioning the rocker assembly

Before inspecting the rocker assembly, dismantle it, but make sure to place all of the parts on the bench in the order in which they are removed (see fig. 4:14).

Fig. 4:12   Some valve springs are of the variable rate type as on this Triumph 2 litre engine. It is essential that the springs are fitted with the closed coil end against the cylinder head

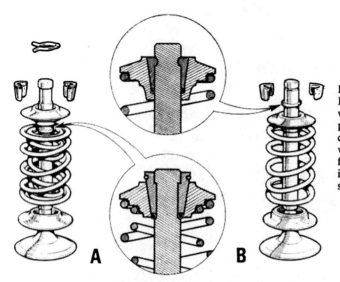

Fig. 4:13 Oil seal arrangements on B.M.C. valves: **A.** Early type valves with wide cotter grooves: inset shows packing ring fitted at the bottom of cotter groove. **B.** Later type valve with narrow cotter grooves and different type cotters: inset shows packing ring below cotters and above spring cap retainer

Wash the parts thoroughly and inspect for: cracked locknuts, pitted rocker arm faces, wear on the rocker shaft and rocker arm bushes, and make sure the rocker shaft pedestals are not cracked.

If the rocker arm bushes are worn, it is advisable to fit new arms (fig. 4:15) for although on some rocker arms it is possible to press the old bushes from the arms and fit new bushes, the new bushes, when fitted, will require reaming to the correct size, and may also require holes drilling in them for lubrication. In this event the holes may have to be drilled before the new bushes are pressed into position.

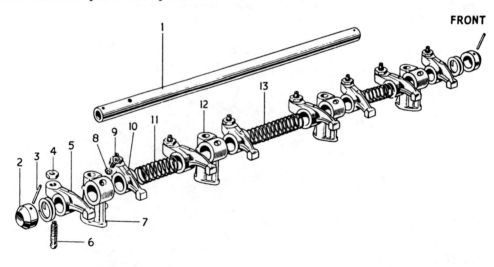

1. Rocker shaft	8. Shakeproof washer
2. End cap	9. Phillips head screw
3. Mills pin	10. Rocker l h
4. Locknut	11. Distance spring
5. Rocker r h	12. Pedestal
6. Adjusting screw	13. Centre distance spring
7. Pedestal rear	

Fig. 4:14   Rocker gear 1200, 12/50 Herald, Spitfire 4 and Mk 2. The Vitesse rocker assembly is similar but has 12 rockers, 6 pedestals and 5 distance springs

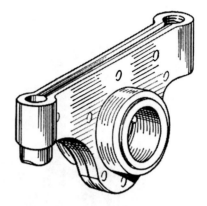

Fig. 4:15 The pressed-steel type of valve rocker fitted to some B.M.C. vehicles must not be rebushed

Worn rocker shafts, tappet screws and pedestals must be renewed. Pitted rocker arms can be ground to remove the pit marks, but care must be taken to ensure that the face radius remains the same. Examine the locknut on each rocker arm carefully, for if the corners are rounded, or the nut is cracked it should be renewed.

Reassemble the rocker mechanism, lubricating the shaft and bushes taking care that the parts are refitted in the correct order.

Fitting the cylinder head

Always use a new gasket. Sealing compound should not be used and is not desirable. Ensure that the cylinder head and the cylinder-block faces, and all of the cylinder-head studs and stud holes are completely free from rust and foreign matter. Ensure that the faces of the

Fig. 4:16 If the cylinder head is retained by bolts, as on this Cortina engine, and not by studs and nuts, two studs should be made (from old bolts) to align the gasket and facilitate fitting the head

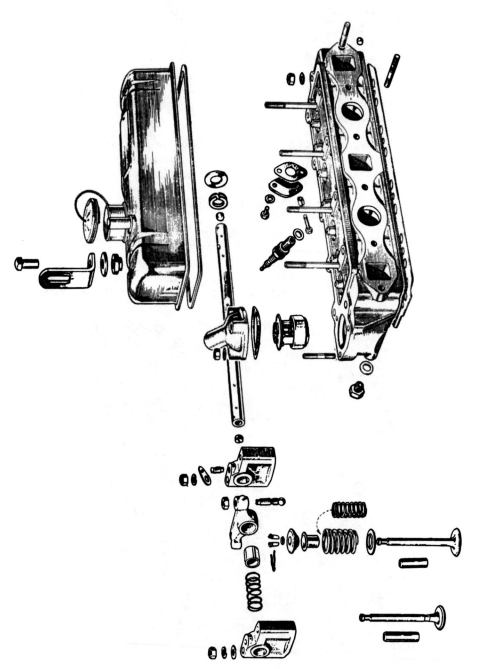

Fig. 4:17   B.M.C. 'B' type cylinder head components

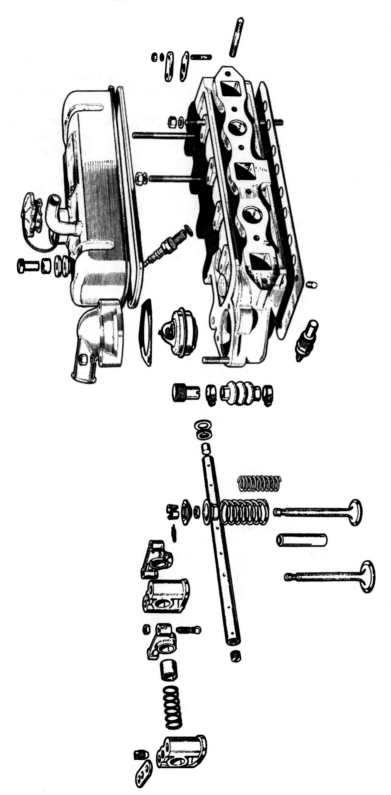

Fig. 4:18   The B.M.C. 'A' type cylinder head and valve gear.

new gaskets are free from dirt, they are specially pre-coated during manufacture and do not need further treatment.

If the old cylinder-head gasket showed any evidence of blowing, thoroughly examine the cylinder head and block faces for any signs of distortion. All traces of burning or distortion should be removed by resurfacing. Check the cylinder-head studs, or bolts, for stretching, or damaged threads, and the nuts for stripped threads. Failure to rectify any of these points may lead to gasket failure.

When positioning the gasket, ensure that any markings are observed for some gaskets are marked 'top' and/or 'front'. Copper/asbestos gaskets are usually fitted with the double edge upwards. Place a liberal amount of oil on the piston crowns and carefully lower the head over the retaining studs.

If bolts are used to retain the head, two dummy studs (see fig. 4:16) will be required to keep the gasket in line as the head is lowered into position. They may be made from two long cylinder-head bolts by cutting the head from the bolts and making a screwdriver slot, with a hacksaw, in the top of the stud. With the cylinder-head gasket in position, screw one stud into position at each end of the cylinder block. If the head is being fitted whilst the engine is in the car, make certain the dummy studs are positioned where it is possible to remove them with a screwdriver, particularly the rear stud which, due to a projection on the bulkhead or equipment fitted to it, may be difficult to take out.

Refit any brackets that were removed from the head.

When tightening the head, in the absence of any nut-tightening data recommended by the vehicle manufacturer, it is vitally important that the correct method of tightening is followed (see fig. 4:6), starting at the centre of the head and working outwards. The same degree of tightening must be applied to all nuts, and after the engine has run for 100 miles it is advisable that all cylinder-head retaining nuts, or bolts, are retightened. It is advisable to retighten cast-iron cylinder heads when the engine is hot, and aluminium heads when cold, but it is essential not to overtighten, as this may lead to undue stretching of the retaining studs.

Refitting the rocker assembly

Before placing the rocker assembly on the engine, refit the cam followers, renewing any that are pitted, scored, cracked or show signs of wear, and refit the pushrods making sure that the base of each rod fits into the cam follower.

Slacken the rocker arm locknuts and 'back off' the adjusting screws. Place the rocker assembly in position and tighten it down, making sure that the rocker arms do not trap a pushrod, and that each rocker-arm adjusting screw ball-end fits into its cup. When the assembly has been placed on the engine, make sure to refit any oil pipes that were removed.

On B.M.C. engines make sure to fit the rocker-shaft dowel-lockplates (see fig 4:17 and 4:18) and tighten the rocker assembly down evenly.

Adjusting valve clearances

The valve clearances on side-valve engines should be adjusted before the head is fitted (see page 65).

It is essential that each valve clearance is set when the cam follower is on the base of its respective cam, but as the cam cannot be observed, the adjustment is best carried out by following the accompanying sequence table. The table also allows engine rotation to be kept to a minimum.

To obtain the clearance, a feeler gauge of the appropriate thickness (see Appendix 4) is required (see also fig. 3:3).

To adjust the clearance, release the adjusting screw by slackening the hexagonal locknut with a spanner whilst holding the screw against rotation with a screwdriver. Set the clearance to the correct measurement by carefully rotating the rocker-arm screw, whilst checking the clearance with the gauge. When the clearance is correct, the feeler gauge should be lightly 'pinched' but not tight. The locknut may then be retightened whilst holding the screw against rotation.

Sequence table for the adjustment of valve clearances

FOUR-CYLINDER ENGINES	SIX-CYLINDER ENGINES
Adjust Valve No. 1 with No. 8 fully open	Adjust Valve No. 1 with No. 12 fully open
— — — 3 — — 6 — —	— — — 6 — — 7 — —
— — — 5 — — 4 — —	— — — 9 — — 4 — —
— — — 2 — — 7 — —	— — —11 — — 2 — —
— — — 8 — — 1 — —	— — — 5 — — 8 — —
— — — 6 — — 3 — —	— — — 3 — — 10 — —
— — — 4 — — 5 — —	— — —12 — — 1 — —
— — — 7 — — 2 — —	— — — 7 — — 6 — —
	— — — 4 — — 9 — —
	— — — 2 — —11 — —
	— — — 8 — — 5 — —
	— — —10 — — 3 — —

If the adjustment is carried out when the engine is cold an additional 0·001 in. should be allowed on clearances specified for adjustment when hot.

Refitting the rocker cover, manifolds and carburetters

When refitting the rocker cover and gasket, make sure to position the cover correctly and not to overtighten the retaining nuts. On some engines there is very little clearance between the rocker arms and the cover and if a rocker arm fouls it will make a noise.

Cork gaskets must not be fitted using oil or grease to hold them in position; use an adhesive (on the rocker-cover side only).

Refit the exhaust manifold, inlet manifold, carburetter and the 'choke' and throttle cables, etc.

Make sure to reposition the manifold gasket correctly, checking that it cannot obstruct the flow of gases. Refit any locating dowel rings that were removed, and tighten the manifold evenly and firmly so as to avoid air or gas leaks.

When refitting the throttle cable, align it with the retaining bracket and the carburetter throttle spindle attachment (see fig. 4:20), avoiding twists or kinks in the cable. Make certain that the carburetter flanges are not bowed (see page 245) and replace any cracked insulating washers. Induction air leaks can cause burned valves, spitting through the carburetter and poor performance, whilst air leaking into the exhaust manifold may cause severe 'bangs' in the silencer.

Side-valve engines

On many side-valve engines the distributor is fitted to the cylinder head and in this case, before the head is taken off, the distributor should be removed. If the distributor is retained by a clamping plate and setscrews, do not slacken the clamping bolt on the plate as this will allow the distributor body to rotate in relation to the plate, and the ignition timing will be lost. In this instance, remove the setscrew that secures the plate to the cylinder head.

The procedure for decarbonizing and fitting new rings is virtually the same as for ohv engines, but with sv engines a special valve-lifter suited to the engine in question is essential.

Before removing the valve cotters make sure to plug with rag all drain holes which connect to the lower crankcase, otherwise a dropped cotter may fall through into the sump. When removing or refitting cotters on an sv engine it is usually best to place the claw of the valve-spring compressor between the valve cap and spring, rather than on the valve cap. Make sure to turn the engine so that there is as much clearance as possible between the tappet and valve stem.

When the valves and cotters have been refitted, the tappet clearance must be reset; this must be done by slackening off a locknut on each tappet (see fig. 4:21). Apart from later type Ford sv engines that have self-locking tappets (see fig. 4:22) this will probably call for

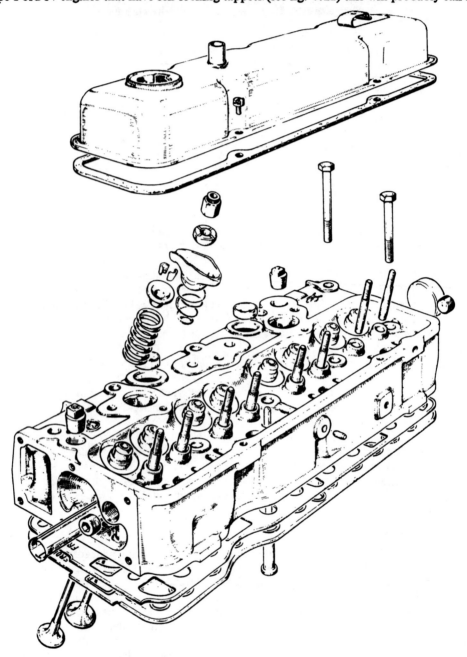

Fig. 4:19   The Vauxhall Viva series HB cylinder head assembly

the use of two, or sometimes three, thin-type tappet spanners. Ensure that each tappet is on the base of the cam before adjustment is carried out (see sequence table, page 63) and set the clearance to the manufacturer's recommendation.

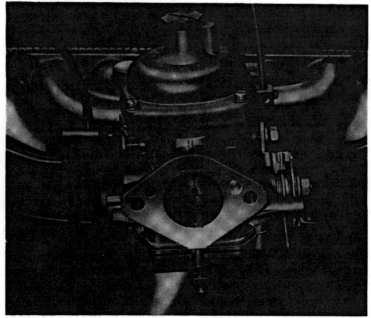

Fig. 4:20 It is essential to
align the throttle cable and
bracket correctly with the
throttle spindle attachment
avoiding twists and kinks
to the cable. Illustrated is
the Stromberg CD 150
single carburetter arrangement fitted to the Herald
13/60

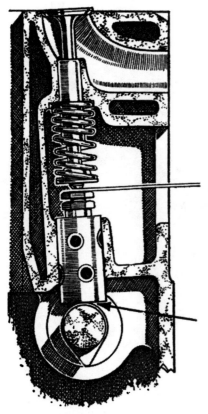

Hot .017 in.
(.43 mm.)
Cold .018 in.
46 mm.)

Tappet at
back of cam

Fig. 4:21 Position of cam and tappet for
maximum clearance on a side valve engine

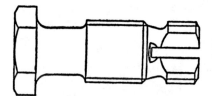

Fig. 4:22   A self-locking tappet as fitted to later type Ford side valve engines

Early Ford sv engines (8 hp and 10 hp 93A models)

Early engines are fitted with valves that have a mushroom-shaped tip to the valve stem, to locate a slotted spring retainer. When the valve spring has been compressed so that the retainer is raised to clear the mushroom foot of the valve, the retainer can be slid sideways from the valve stem. The expanded end of the valve will not pass through the valve guide; each guide is made in two halves and can be driven downwards using a special tool (one is supplied in overhaul kits) after the valve springs have been removed.

Overhead camshaft engines

The removal and refitment of cylinder heads on overhead camshaft engines is slightly more complicated than with push rod engines and unless one is familiar with the type of work it is not advisable to attempt it—certainly not unless one obtains the workshop manual appertaining to the vehicle in question. The adjustment of valve clearances on some overhead camshaft engines is also rather more involved than with a push rod engine although the clearances can be checked easily enough (see fig. 4:23).

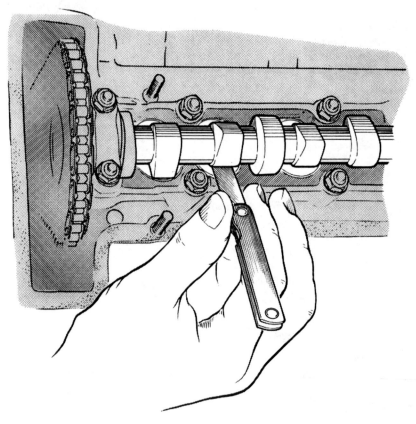

Fig. 4:23   Checking the valve clearances on a Hillman Imp engine

CHAPTER 5

Fitting Oil-control Rings

If the sump can be removed whilst the engine is in the car, oil-control rings can be fitted without a great deal of trouble, but if there is not space enough to remove the sump, the engine will have to come out first (see page 72).

If the oil-control rings are to be fitted with the engine *in situ*, disconnect the battery, drain the water, and remove and overhaul the cylinder head, referring to pages 53 to 57.

Next drain the engine oil, place the front of the car on firm supports, and remove the sump. On some cars, the front engine-mounting bolts must be removed and the engine raised to allow access to the front sump bolts and allow sufficient clearance for the sump to be 'dropped'. On Vauxhall Victors, the suspension assembly must be dropped slightly.

To facilitate the removal of the pistons and connecting rods, remove the carbon from the top of each cylinder bore. This will also act as preparation for checking the ring gaps on the non-wear ridge, and measuring the cylinder, to ascertain how much wear has occurred.

Mark the connecting rods and caps according to the position they occupy and remove the pistons and connecting rods from the cylinder block (see page 93).

Removal and reassembly of pistons from the bottom of the cylinders

On some side-valve engines, due to the fact that the bottom of the connecting rod will not pass through the cylinders the pistons have to be taken out and replaced from the bottom of the cylinders. Due to the cramped conditions of working, extra care must be taken to avoid ring breakage. Removal of the pistons and connecting rods from the cylinder is fairly easy, but the crankshaft must be rotated slightly to allow the piston assembly to be withdrawn from the crankcase.

When refitting the pistons and connecting rods to the cylinder, temporarily fit the connecting rod caps, fit a piston-ring clamp to the piston to compress the rings and when the piston is suitably aligned with the bottom of the cylinder, tap the cap gently with a hammer to encourage the piston to enter the bore then retrieve the piston-ring clamp, remove the connecting-rod cap and fit the connecting-rod big end to the crankpin, lubricate the bearing and fit the cap.

If it proves difficult to insert the pistons in the above manner, remove the piston assembly from the crankcase, take off the piston-ring clamp and remove all of the rings. Fit the piston into the cylinder without the rings and then push the piston up the bore until it protrudes from the top. Refit the rings and piston-ring clamp and re-insert the piston into the cylinder. Proceed with reconnecting the big-end caps in the usual manner.

Cylinder measurements

The normal way to measure cylinder wear is with an internal measuring micrometer. As this is unlikely to be available, recourse should made to the following method: scrape the carbon from the non-wear ridge at the top of the cylinder and enter a piston ring squarely onto the ridge and, using a set of feeler gauges, measure the gap. When this has been done, push the ring down the cylinder squarely, using a piston, until the ring is slightly below the ridge. Take another measurement with the feeler gauges.

To determine the overall wear, subtract the first measurement from the second, and divide by three (theoretically π or 3·142).

If there is more than 0·008 in. diameter bore wear, it is advisable to have the cylinders rebored and fit new pistons.

Piston ring gaps

The piston ring gap on water-cooled engines should, in the absence of manufacturer's instructions on the subject, be 0·002 to 0·003 in. per inch diameter. If special hardened-steel liners have been fitted, the gap should not be less than 0·003 in. per inch diameter.

When pistons are to be used again, new rings should always be fitted, and the side clearance in the ring groove checked (see fig. 5:1).

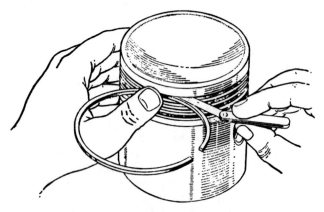

Fig. 5:1 Checking the piston ring to groove clearance

The side clearance of new piston rings in new pistons should not be more than 0·0015 in.

On used pistons the ring/groove side clearance should not be more than 0·003 in. In the event of excess clearance the piston should be renewed, or a piston ring-groove insert fitted.

The following table gives an indication of the ring gap measurement of various diameter cylinders.

	in.	in.	in.	in.	in.
Diameter of piston	2	2·5	3	3·5	4
Ring gap	0·003	0·005	0·007	0·009	0·012
Ring gap (diagonally cut)	0·002	0·003	0·005	0·007	0·009

Piston skirts

To ascertain whether the piston skirts require expanders (see fig 5:2), place the piston in the cylinder and take a measurement with a set of feeler gauges between the skirt and the piston wall. The ideal clearance for split skirt pistons in a new condition is:

	in.	in.	in.	in.
Cylinder bore	2½–3	3–3½	3½–4	4–4½
Skirt clearance	0·002	0·002	0·002	0·003

If the measurement does not come within 0·003 in. of the recommended clearance, it is advisable to have the piston skirts knurled (see Fig. 5:2b). If the measurement exceeds 0·004 in. piston skirt expanders should be fitted.

Renovating worn pistons

Providing pistons are not 'scuffed' or cracked, their life can be extended. Worn ring grooves may be corrected by fitting groove inserts. If the piston skirts, which play an important part in stabilizing the piston and eliminating piston-slap, are worn or collapsed, the skirt can be fitted with expanders.

Groove inserts

Ring-groove inserts are flat steel segments which contract inwards and are particularly valuable if the top compression ring grooves are worn, as they make it possible to restore the grooves to a new condition.

Fig. 5:2(a) shows the method of fitting.

Advantages of groove inserts are:

A piston with a worn groove can be repaired cheaply and effectively.

Ring flutter and hammering is eliminated and the groove being faced with steel results in a longer piston life.

It enables a standard width ring to be used in a worn and repaired groove.

A plain unstepped top compression ring can be fitted and will not contact the non-wear ridge at the top of the piston travel.

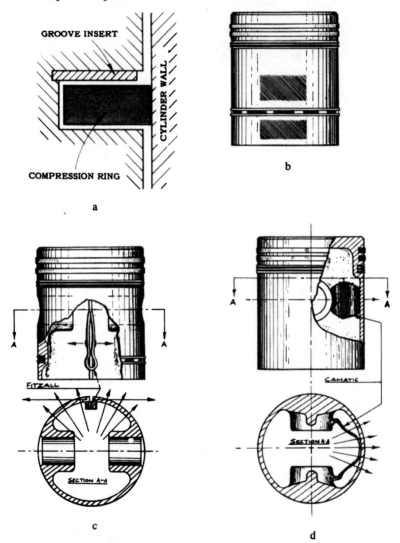

Fig. 5:2 (a) The Cords ring-groove insert, machining a register is preferable. Once fitted, this insert effectively prevents further wear on the piston groove. (b) Micro-knurling is carried out on the bearing surfaces of the piston and, in addition to expanding the skirt, also has the effect of slightly 'hardening' the surface so reducing wear. (c) The Fitzall expander is fitted in a hole drilled in the 'split' of a split skirt piston. (d) The Camatic expander is fitted with the curved ends braced against the gudgeon-pin bosses

Counteracting piston skirt wear

To prevent piston skirt collapse three methods are recommended (see fig. 5:2: (b)

micro-knurling which is ideal for counteracting slight wear, (c) fitting a Fitzall expander, or (d) Camatic expander).

Micro-knurling of the piston skirt

The knurling is performed by machining the piston with a special device so that a small area of the piston skirt is knurled (any Cord piston-ring agent will do the machining) but the method is not suitable for cast-iron pistons. On alloy pistons the knurling provides an oil-retaining surface that gives a degree of lubrication and damping.

Fitzall piston skirt expander

Suitable for any split skirt piston, the Fitzall expander is extremely easy to fit and is a U-shaped spring with special lugs that allow it to fit into a small hole which has to be drilled in the skirt split.

The Camatic piston skirt expander

Made of high quality spring steel, the Camatic expander is suitable for fitting to solid skirt pistons. When fitted, the radiused edges of the expander engage beneath the gudgeon pin bosses; the 'hump' of the expander then presses evenly against the piston skirt. The Fitzall, and the Camatic expander, expand the skirt from 0·003 to 0·005 in. On some pistons it is not possible to measure the amount of expansion when the piston is cold, but when heated the skirt expands and the expander permanently prevents it from contracting again.

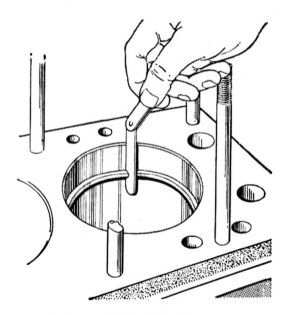

Fig. 5:3 Checking the piston ring gap. If new rings are fitted to old cylinders, the non-wear ridge at the top of the cylinder should be cleaned and the ring inserted on to the ridge and the ring gap checked to make sure that it cannot butt and break when fitting

Stepped rings or 'ridge dodgers'

Standard type compression rings should be used in the top ring grooves. Oil-control rings should not be fitted in this position. In the absence of stepped rings, often termed 'ridge dodgers', new top compression rings of the correct cylinder size should be obtained and should be chamfered slightly around the top outer edge.

Before fitting the rings to the pistons, place each ring in the top of a cylinder beneath the non-wear ridge and adjust the gaps. When the rings gaps have been checked (see fig. 5:3), place each ring on the non-wear ridge to check that the ends of the rings do not butt. If any rings butt, the ends of the rings must be filed until clearance is obtained.

If the rings are not checked on the ridge, they may break during piston replacement.

If the car manufacturers' standard-type top compression rings are chrome rings, unless they are of the 'coated' type such as the Hepolite Caragraph ring they should not be used, as chrome top compression rings take a considerable time to become fully bedded. (Caragraph is a special coating which assists chromium rings to bed in rapidly). If suitable rings are not available from the manufacturers' local agent, an engine specialist firm should be consulted.

If the compression ring groove wear is severe, it is advisable to have 'groove' inserts fitted as shown in fig. 5:2 (a) or to fit a set of 'overhaul' type pistons that have oil rings fitted to the skirt (see fig. 4:4). This is the type of piston that is supplied with engine reconditioning kits and it may pay to go to the extra expense of purchasing a kit.

If the piston oil-ring grooves are worn, it is unlikely that single-section rings will be satisfactory, but Cord rings may be the answer, for the oil ring grooves can also, if necessary, be recut to incorporate steel inserts (fig. 5:2(a)).

Before the oil-control rings are fitted, check the piston gudgeon pins for wear. If 'lift' is found, the worn part should be renewed (see also page 95).

Unless the wear on the pistons, or connecting rods, is severe enough to make renewal necessary, or unless the pistons have to be machined to widen the grooves, do not remove the pistons from the connecting rods. Take off the old piston rings, thoroughly clean the grooves and all the oil holes. Then clean the piston crowns and wash the pistons and connecting rods thoroughly and inspect them. Check the ring clearance as suggested for the top compression ring, and carefully fit the new rings to the pistons. Some oil-control rings require pressure rings beneath them and it is essential that these are not overlooked.

Note: Bevelled scraper rings should be fitted with the bevelled edge upwards; stepped compression rings with the step facing upwards.

When the rings have been fitted onto the pistons, space the ring gaps equidistant around each piston, lubricate them and tighten a piston ring compressor clamp around them.

Turn the crankshaft until No. 1 piston is at bottom dead centre and insert the connecting rod and piston (see page 111, 'Refitting the Pistons and Connecting Rods').

Refit the sump and cylinder head. Refill the sump with oil.

Cord rings

Cord rings can be fitted to almost any piston and one big advantage with Cords is that apart from the top compression ring, the likelihood of breakage is virtually non-existent. The rings are supplied ready for fitting in sets according to the make of car for which they are required. Full fitting instructions accompany each set and the rings are in coloured envelopes according to the position they occupy on the piston.

It should be noted that Cord oil rings should not be expanded over the piston lands but

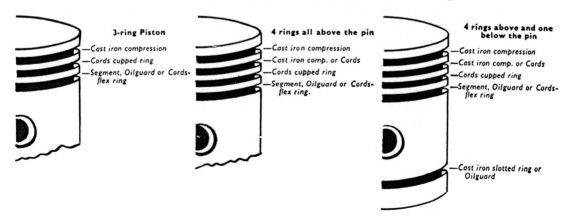

Fig. 5:4   Cords replacement ring sets. Typical ring arrangements

should be threaded into position. It is essential that each segment is fitted the correct way up. Typical installations of Cord rings are given in fig. 5:4.

If Cord rings or 'overhaul' type pistons are fitted it is important to remove the oil glaze from the cylinders.

Refitting sump gaskets

Make certain the crankcase face, and sump face, is completely free from old gasket material and sealing compound.

Hold the gaskets in position to see if they are the correct type and to see if the holes in the gasket align with the bolt holes in the crankcase.

Cork gaskets that have shrunk and have holes that do not align should be immersed in water. If they are removed after a few moments and dried they should then fit satisfactorily.

Before fitting the sump, it is essential to position the gaskets correctly. If a sealing cork fits around the front and/or rear main bearing cap, the ends of the crankcase gasket usually fit beneath the cork. On some cars difficulty may be encountered in fitting the sealing cork correctly, as it appears unduly long. In such an event, if working underneath a car, it is advisable to shorten the cork a little, rather than risk displacing it when the sump is positioned.

Oil or grease should not be used to hold cork gaskets in position; an adhesive should be used.

Before fitting the sump, check:

That all of the nuts within the crankcase are tightened
That the tabs on all lockwashers are bent over
That all oil pipes have been refitted and tightened
That the markings on the big-end caps are correct
That the engine rotates freely

Fitting the sump from beneath the car normally requires the help of an assistant. Usually, if one person raises the sump into position, the assistant can start a screw each side of the crankcase and screw them in a few threads to retain the sump in position. The remainder of the setscrews can then be refitted but none of them should be tightened until all of them have been started in their threads.

Before finally tightening the setscrews, examine the gaskets to ensure they have not been displaced. If all is well tighten the setscrews evenly.

Refill the sump with oil.

If the sump has been removed from the engine whilst it is still in the car, and the crankshaft ovality (see page 94) is found to be such that a reground crankshaft and bearings must be fitted, it will be necessary to remove the engine from the car.

Removing the engine

To remove the engine from a rear-wheel-drive car, one of two methods can be adopted; the first method is to remove the engine and gearbox as a complete unit (see fig. 5:5) and, provided suitable lifting facilities are available, it is seldom difficult. The alternative method, which may sometimes be adopted, is to remove the engine leaving the gearbox in the car. Where lifting equipment is at a premium and the engine has to be manhandled, the latter method often proves the least troublesome especially on cars such as the Triumph Herald, Austin A40 and Ford Zephyr.

Whenever major work is being carried out on the engine, always disconnect the battery. When this has been done, drain the oil and water and remove the bonnet.

On B.M.C. Minis and 1100s, the radiator should not be removed as this can be lifted out with the engine (see also fig. 5:6).

Fig. 5:5   Removing the engine from a Triumph Vitesse

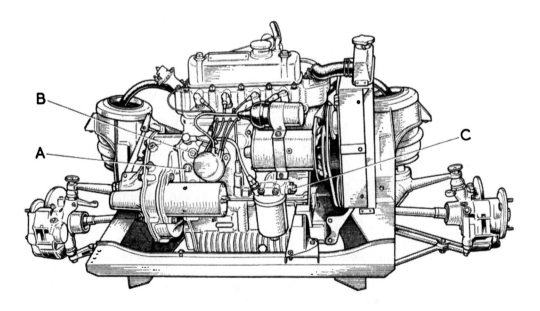

Fig. 5:6   Although the power unit of the B.M.C. Mini, 1100 and 1300 can be removed fairly easily from the engine compartment, the power unit and sub-frame can also be removed from the car as a complete unit. To do this however the hydrolastic suspension must be depressurized and the front pipes disconnected. **A.** Oilpressure relief valve. **B.** Low oilpressure switch. **C.** Oil filter differential switch

Disconnect the water hoses, take out the bolts which secure the radiator, and remove the radiator. On some vehicles it will be necessary to remove the fanblades and radiator tie-rods, etc. before the radiator can be lifted from the car.

Next, remove the air cleaners, carburetter and heat shield. Disconnect the exhaust pipe flange and remove the inlet and exhaust manifolds.

Disconnect and remove the generator, take the bolts from the front engine mountings and disconnect the thermal transmitter, oil pressure pipe, heater control cable and the heater hoses. Remove the starter motor and the distributor cap with the plug leads attached.

On some cars it may be easier to remove the engine, if the coil and the oil filter bowl (sump) is removed.

Ensure that when the engine is lifted out, it will not foul equipment fitted to the bulkhead and that no bracket has been left on the engine which could foul.

When taking the engine out of a Ford Zodiac, the large oil-filled air cleaner, fitted at the front near side of the engine, should be removed. The starter motor, too, should be taken off, and the fan blades and pulley. The clutch-return spring must be removed and a pair of circlip pliers will be required to remove the slave cylinder.

Attach lifting-eye brackets to the cylinder head nuts. If a B.M.C. 'B' series engine is being removed attach two B.M.C. engine-lifting brackets to the rocker cover, fitting them between the rocker cover sealing rubbers and the retaining nuts. Place a chain of a suitable length between the eye brackets and attach the lifting equipment.

If suitable eye brackets are not available, a rope may be used in sling fashion around the complete engine, but take care that the rope is strong and make certain that any knots tied are secure.

If the engine is to be removed without the gearbox, support the gearbox with a jack, take off the nuts and remove all bolts securing the clutch housing to the rear engine-mounting plate.

If there are not location dowels fitted between the engine backplate and gearbox, one of the bolts will probably have a machined shank, so as to provide correct alignment between the engine and gearbox. As the bolt must be refitted in the same position it is advisable that the bolt and the location it occupies are marked.

Tie the exhaust pipe to one side, otherwise it may prove troublesome. When all is ready, raise the engine slightly and pull it forward until the clutch is free from the gearbox first-motion-shaft. Continue lifting the engine slowly, tilting it at the front, if necessary, to allow the sump to clear the front crossmember. Once clear of the crossmember, pull the engine further forward, so that it clears the first motion shaft and can be lifted out.

If the engine and gearbox are being taken out as a complete unit do not, of course, remove the bolts securing the clutch housing to the engine plate, but remove the propellor shaft, first marking each flange with a light dab of paint to ensure that the shaft can be replaced in exactly the same position.

Disconnect the speedometer cable. Remove the clutch-lever tension spring and undo the clutch slave-cylinder bolts to release the cylinder from the clutch housing. Usually it is possible to tie the clutch cylinder to one side, to avoid disconnecting the hydraulic pipe. Subsequently, remove the push rod from the clutch withdrawal lever.

On many rear-wheel-drive cars, a detachable crossmember supports the rear of the engine and should be removed, and the weight of the engine taken by a jack or other support.

Disconnect the engine stay rod from the gearbox (if fitted) and the engine/body earth lead. On B.M.C. cars, the nuts securing the crossmember to the rear gearbox mounting rubbers must be removed before the rear crossmember can be taken off.

If the engine is being taken out with the gearbox attached it may be necessary to remove the rubber cover from the gearbox tunnel and take out the gear lever. On many B.M.C. cars, the lever must be lifted out vertically so as to retain the nylon cup on the end of the lever.

If the gearbox is fitted with a remote gear-control assembly remove it, but if the gear change is fitted to the steering column, disconnect the linkage from the gearbox.

On cars fitted with overdrive, the electrical wires to the overdrive unit must be disconnected.

Attach lifting eyes in a manner which will allow the engine to be tipped to a fairly acute angle. Before attempting to remove the engine, however, check that nothing has been overlooked, particularly in relation to wires and controls to auxiliary equipment.

Removing the engine: B.M.C. Minis and 1100s

To remove the engine from a Mini or an 1100 (see also fig. 5:6): raise the front of the car and place it on firm supports. Remove the exhaust pipe, the carburetter, and inlet and exhaust manifolds. Release both drive shafts by undoing the four nuts and removing the two U bolts which retain each rubber universal joint to its drive shaft. Push both the sliding joints onto their shafts to separate the couplings. Disconnect the engine-mounting rubbers from the subframe, release the engine steady rod and remove the two bolts which retain the clutch slave cylinder to the clutch housing. Disconnect the various electrical wires, marking them so they can be refitted without difficulty. Disconnect the engine/body earth strap and disconnect the gear lever. On 1100s and late Minis, disconnect the remote control assembly from the transmission case. Remove two cylinder-head nuts, one at the front of the engine and one at the rear, and attach two lifting eyes. Take the weight of the engine with the lifting equipment and lift the engine from the car.

CHAPTER 6

Partial Overhaul

Although parts for an intermediate overhaul can be obtained in kit form, they may also be purchased separately.

Sometimes an intermediate overhaul is possible with the engine *in situ*, but the extent of such an overhaul is usually severely limited and is governed to a great degree by the design of the engine.

If the engine is removed from the car, it is worthwhile completely stripping and rebuilding it as per chapter 7, the only difference being that reboring is avoided and the 'overhaul' pistons will be used. Unless the crankshaft is oval, this too may be used again and only the bearings renewed.

It is essential to obtain the correct-sized bearings and pistons. If the engine has previously been dismantled, a reground crankshaft may have been fitted and the cylinders rebored. In the latter event, the size will be found stamped on the piston crowns, but sometimes the bearings are not marked. In this event it is essential, when fitting new bearings, to check the bearing clearance.

To carry out an intermediate overhaul with the engine *in situ*, disconnect the battery, drain the oil and cooling system and remove the sump. Take off the cylinder head and overhaul it referring to pages 53 to 57. It is a good idea to have the battery on charge whilst the overhaul is carried out.

Withdraw the pistons and connecting rods (see page 93) and inspect the cylinders carefully. If the car was using oil it may have been caused by the pistons having partially seized to the cylinder wall at some earlier date or owing to a loose gudgeon pin fouling the cylinder wall. If deep scores are present, the cylinders will have to be rebored and new pistons fitted (see page 82).

Mark all of the main bearing caps, then remove the centre caps and inspect the bearings. If the bearings are partially 'run out' or cracked, and the rear, or the front, main bearing cap cannot be removed, the engine should be taken from the car and a complete set of main bearings fitted.

Fitting main bearings (engine *in situ*)

If the front and rear main bearing caps can be taken off, it is possible to fit a complete set of bearings whilst the engine is in the car. Make sure, however, to mark the caps so that they can be refitted in exactly the same position.

Remove all of the bearing caps and turn the crankshaft so that the oil hole in the shaft journal faces downwards. Insert a screw into the oilway, choosing a screw that has a head that will not foul the crankcase but will protrude sufficiently to butt against the main bearing shell. Rotate the crankshaft slightly, holding the screw in position until the head butts against the bottom of the bearing shell on the side that does not have a locating tab. Then continue to rotate the crankshaft and displace the bearing.

When the bearing has been removed, retain the screw and slide a new bearing into position. Make sure that the locating tab on the bearing fits into its recess in the crankcase. Next, place the bottom new shell bearing into its cap, lubricate the bearing and refit the cap to the crankcase. When tightening the bolts, check that the crankshaft does not bind. If it does bind remove the cap and investigate the reason. Fit the other upper and lower bearings in the same manner.

If there is any doubt regarding the crankshaft size and/or bearings after fitting the first upper shell bearing into the crankcase, place the bottom shell into its cap and lubricate the bearing surface. Next, assemble the cap onto the engine and screw up the retaining bolts (or nuts) very gradually, constantly checking that the crankshaft can be rotated. If rotation of the crankshaft becomes tight, the bearings are of an incorrect size. *Note:* if the bolts are fully tightened with an oversize bearing and the crankshaft is of standard size the bearing cap will probably crack.

Some indication of the bearing clearance must also be obtained, otherwise there is a possibility that standard bearings may be fitted to an undersize crankshaft. This would result in excessive clearance, no oil pressure and a ruined engine.

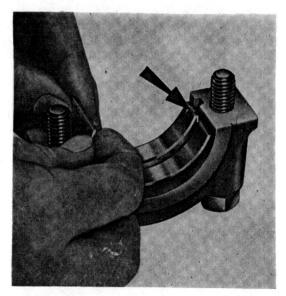

Fig. 6:1   Fitting a main bearing liner

Working on one of the centre mains, turn the crankshaft so that the oilway is *not* facing downwards. With the upper half of the bearing in place, insert the bottom bearing into its cap (see fig. 6:1), and place a ½-in.-long strip of soft lead about ⅛ in. wide and approximately 0·010 in. thickness, across the bearing surface.

Fit the cap onto the crankcase and tighten the retaining bolts. Move the crankshaft backwards and forwards through approximately 30°. While the initial movement of the crankshaft may be quite tight, after the first movement it should be possible to move it through 30° comparatively easily. If the shaft rotates freely, incorrect main bearing shells have been fitted, or the main bearing cap is not correctly seated.

Ascertain that the cap is fitted correctly, and then remove it from the crankcase and carefully retrieve and measure the lead strip. If the thickness has not decreased, the bearings are definitely wrongly sized. If the bearings are correct the lead strip should have been flattened to approximately the manufacturers' advised bearing clearance. The clearance may vary on different cars but a clearance between 0·005 and 0·002 in. should be satisfactory.

If the bearings have been purchased as a set, if the clearance is checked on one bearing this should normally be sufficient.

Make sure that none of the lead strip is left adhering to the bearing. Then, lubricate the bearing and refit the cap to the crankcase with its bearing in place, tightening firmly and bending the lockwashers in position. Fit the other bearings to their caps (see also page 109) and fit the caps to the crankcase.

If Plastigauge is available it will be better than using lead strip for checking the bearing oil clearance. Plastigauge is a plastic material, soft enough to be flattened by pressure when placed on a bearing and the cap is tightened. The amount of flattening is measured using a special gauge scaled to give the bearing clearance according to the width of the material. When Plastigauge is being used, it is advised that the crankshaft is not rotated or, if connecting rods are being checked, that they are not moved on the crankpin.

Checking big-end oil clearance

When the main bearing caps have been refitted and the lockwashers bent into position on the retaining nuts, or bolts, the big-end clearance may be checked in the same manner as the main bearings but the clearance should on no account exceed 0·003 in.

It is also possible to check bearing clearances with a cigarette paper. A cigarette paper is 0·001 in. thick. The method is to fit the bearing with a folded cigarette paper in place across the bearing surface. For instance, to check the big-end clearance which should be approximately 0·002 in. place a double thickness paper about $\frac{1}{2}$ in. square on the surface of the big-end bearing shell, in the connecting rod cap. Make sure that there is not any oil on the journal, or bearing, and tighten the bolts. If the connecting rod is tight on the journal there is less than 0·002 in. clearance. If the connecting rod is loose, try three thicknesses of paper. When the paper causes the connecting rod to lightly 'pinch' the crankpin the thickness of paper which has been placed in the bearing denotes the clearance.

It is essential that all traces of paper are removed when the clearance has been ascertained.

The next step is to fit the pistons to the connecting rods as per page 95 and insert them into the cylinders. 'Overhaul' pistons are designed so that the top ring will not foul the non-wear ridge.

Fig 6:2.   Ford Cortina timing marks and oil slinger

Fig. 6:3   Fitting the timing cover on the Triumph Herald

Fitting direct-metalled connecting rods

The Ford 8 hp and 1172cc engine used direct-metalled connecting-rod big ends until the introduction of the 105E engine on the Anglia in 1959.

Should direct-metalled exchange connecting rods be required for these engines, either the crankshaft size will have to be obtained by measurement, or the old rods taken to an engine specialist who should be able to determine the size from them, and supply correctly machined replacement remetalled rods from stock. The bearing clearance should be checked.

Fitting a timing chain (engine *in situ*)

To fit a timing chain with the engine *in situ*, remove the radiator, take off the fan belt, undo the crankshaft-pulley nut or bolt, and take off the pulley. Take out all of the timing-case screws and bolts. Turn the engine so that the piston on No. 1 cylinder is exactly at t d c with the timing marks on the timing wheels aligned (see fig. 6:2). Remove the oil thrower from

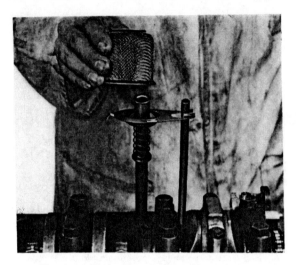

Fig. 6:4   Fitting the oil pump filter gauze on
the Cortina engine

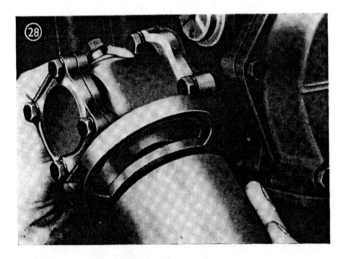

Fig. 6:5   Changing the oil filter is essential
after fitting new bearings

the crankshaft. Release the timing chain tensioner (see page 83) and take off the chain vibration damper if fitted. Remove the timing-wheel retaining bolts and nuts, check that the No. 1 piston is still at t d c and remove the chain wheels and timing chain together.

If rubber tension rings are fitted to the camshaft wheel, renew them. Fit the new chain and wheels and subsequently recheck the t d c position and the alignment of the timing marks.

If the timing tensioner slipper head or the vibration damper is worn, it should be renewed.

Tensioners of the type which fit to the timing case cover (see fig. 6:3) must be inspected to ensure that the springs are not broken before the cover is refitted. On this type it may be necessary to hold the tensioner in position with a stout piece of bent wire so that the tensioner does not jam on the timing chain when the case is refitted.

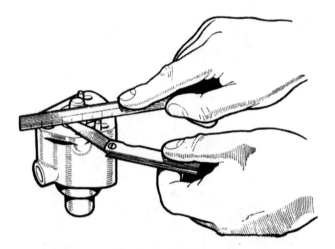

Fig. 6:6   End float on rotor type pumps should not exceed 0·005 in.

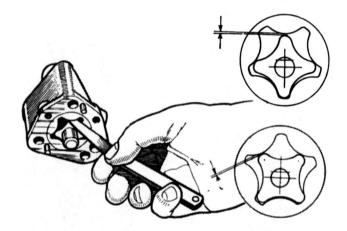

Fig. 6:7   Lobe clearance on rotor type pumps should not exceed 0·006 in.

Although reassembly is a reversal of the dismantling process, make sure to refit the oil thrower on the end of the crankshaft. It is also essential that a new oil seal is fitted to the timing case, and that the cover is refitted so that the seal is concentric with the crankshaft pulley. To do this, in the absence of a centralizing tool, the pulley should be inserted into the seal, and the cover and pulley should be fitted as one unit. It is important to tighten the timing-cover bolts evenly (see also page 116).

If a partial overhaul is being carried out and the main bearings, connecting rod and pistons have been refitted and tightened, fit the oil pump and any oil pipe and oil strainer necessary (see fig. 6:4), and fit the sump, making sure that the drain plug is tightened. Change the oil filter (see fig. 6:5) and refill the sump with oil.

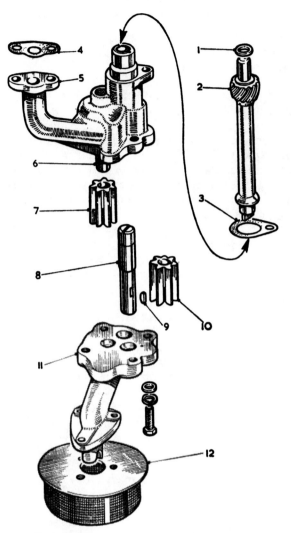

1. Thrust washer	7. Driven gear
2. Drive shaft	8. Shaft — driving gear
3. Joint washer	9. Key — driving gear
4. Joint washer	10. Driving gear
5. Pump body	11. Cover — pump
6. Shaft — driven gear	12. Strainer assembly

Fig. 6:8   Oil pump components

Typical 'gear' type oil pump (Austin); end float on this type of pump should not exceed 0·002 in. Radial clearance between the gears should not be more than 0·0025 in.

Fit the cylinder head, manifold and carburetter and any auxiliaries that have been removed and refill the cooling system. Connect the battery.

Make sure before starting the engine that the throttle is closed, otherwise the engine will 'race' before the oil has time to circulate and cause considerable damage to the newly fitted bearings and pistons.

REBORING THE CYLINDERS (Engine *in situ*)

Reboring the cylinders without taking the engine from the car is only possible if the sump and cylinder head can be taken off and there are not projections on the bulkhead of the car which will prevent the use of the boring equipment.

Before dismantling the engine it is advisable to make tentative arrangements for the car to be towed away and rebored, or arrange for the cylinders to be rebored whilst the car remains in the home garage.

When arrangements have been made, disconnect the battery, drain the cooling water, remove the cylinder head and overhaul it, referring to pages 53 to 57. Drain the engine oil, and remove the sump. Remove the oil pump and oil strainer, and then mark the connecting rod caps so that they can be refitted in the same position, and take out the pistons and connecting rods (see page 93) replacing the caps on the rods the correct way round as each connecting rod and piston is removed from the engine, otherwise the caps may accidentally be fitted to the wrong connecting rod.

Next, scrape the carbon from the 'non-wear ridge' at the top of the cylinder so that the ridge can be measured (see page 67). If the wear is less than 0·006 in., oil consumption problems due to oil penetrating the combustion chamber from the cylinders can be overcome without reboring (so long as the cylinders are not scored) by fitting an 'overhaul' type piston (see fig. 4:4).

Usually if the cylinders are being rebored, it is best to allow the person reboring the engine to supply the pistons for he will have the necessary equipment to measure the cylinders.

When a cylinder block is rebored, a considerable amount of iron swarf is created; therefore, before the cylinders are bored, wrap the crank pins with a band of masking tape, or other suitable material, to prevent the ingress of swarf into the crankshaft oilway.

When the cylinders have been rebored, thoroughly wash the cylinder block using petrol or a proprietary cleansing agent. A small stiff brush is handy and particular attention should be given to the crankcase webs above the main-bearing journal recesses.

All oilways in the engine should be blown through using an airline or suitable pump.

Before reassembling the engine, it is advisable to fit new big-end shells and to make sure that the oil pump clearances (see figs. 6:6, 7 and 8) are within acceptable limits. If possible new main bearings should also be fitted.

The engine should be reassembled, referring to chapter 7.

An engine with rebored cylinders and new pistons should be 'run in' carefully, avoiding high rev/min or pulling hard in a high gear. After 500 miles, the sump should be drained and refilled with fresh oil.

CHAPTER 7

Reconditioning the Engine

If the engine is to be reconditioned it is important, while dismantling it, to keep all nuts and bolts in a box or tray which should preferably be divided into various compartments. This saves a considerable amount of time and exasperation when rebuilding.

Dismantling the engine

If the engine has been removed from the car with the gearbox attached, remove the gearbox, the cylinder head (as detailed on page 50), the distributor, and all external accessories including the generator and water pump and set them aside for overhaul.

Next, tap up the lockwasher on the crankshaft pulley-retaining bolt, or nut (see fig. 7:1), and remove the bolt (or nut) using a socket spanner and holding the engine against rotation with a large screwdriver inserted in the flywheel-ring gear teeth.

Remove the pulley. Tyre levers may be used, but do not lever on the flange of the pulley and take care not to damage the timing cover.

On engines fitted with a Reynolds timing-chain tensioner, when the timing cover has been removed, take the plug from the bottom of the chain tensioner (see fig. 7:14), insert an $\frac{1}{8}$ in. Allen key and turn clockwise until the rubber slipper head is fully retracted and locked behind the limiting peg. The chain tensioner may now be removed complete by tapping the lockwasher back and undoing the two retaining bolts.

The next step is to unlock and remove the camshaft wheel nut, or bolts, subsequently levering the camshaft and crankshaft chain wheels forward. Ease each wheel a fraction at a time. On high-mileage engines they can be quite tight, but two small tyre levers and a little persistence usually do the trick.

After the camshaft-retaining plate and the engine-bearer plate have been removed, attention can be given to removing the clutch, flywheel and rear mounting plate. On most types of engines the clutch and flywheel removal is quite straightforward but the levers may be needed to assist in loosening the flywheel.

Removing the flywheel (B.M.C. Mini and 1100, see fig. 7:2)

Remove the inspection plate on the clutch housing and turn the engine so that the ignition timing mark is aligned with the pointer.

Remove the clutch/flywheel cover, knock back the tabwashers on the 3-point clutch thrust plate (see fig. 10:10), remove the three retaining nuts and take off the plate.

On 1100s and Minis that have a diaphragm clutch, the clutch thrust plate is held by a split ring. This can be 'hooked' out with a screwdriver to release the thrust plate.

Next, knock up the tabwasher on the flywheel retaining bolt, hold the flywheel against rotation as previously described and undo the bolt. Take the keyed washer from the crankshaft.

A flywheel puller will be required to remove the flywheel/clutch assembly. It is usually possible to hire one and preference should be given to a hydraulic puller, although a manually-operated screw type can be used if necessary. When using the puller, it is essential that the studs are screwed fully into the flywheel and that a spigot (usually part of the puller kit) is placed against the end of the flywheel so that the ram of the puller will butt against the spigot. Make sure to fit the puller squarely to the flywheel. Hold it so as to prevent rotation, and screw in the central ram and withdraw the assembly.

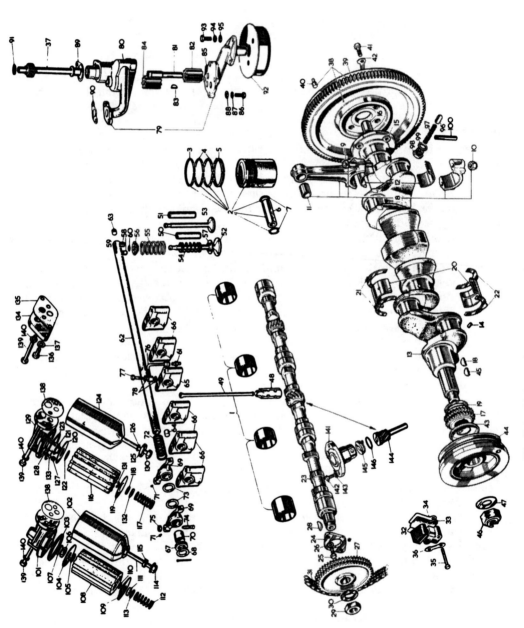

Fig. 7:1 (a) Internal components of B.M.C. type engine

KEY TO THE ENGINE INTERNAL COMPONENTS

No.	Description
1.	Camshaft liner.
2.	Piston assembly.
3.	Piston ring—compression.
4.	Piston ring—compression—tapered.
5.	Piston ring—oil scraper.
6.	Gudgeon pin.
7.	Gudgeon pin circlip.
8.	Connecting rod and cap.
9.	Cap bolt.
10.	Cap bolt nut.
11.	Small-end bush.
12.	Big-end bearing.
13.	Crankshaft assembly.
14.	Oil restrictor.
15.	Nut for flywheel bolt.
16.	Drive gear bush.
17.	Crankshaft gear.
18.	Key for gear.
19.	Gear packing washer.
20.	Crankshaft main bearing.
21.	Crankshaft thrust washer—upper.
22.	Crankshaft thrust washer—lower.
23.	Camshaft.
24.	Camshaft locating plate.
25.	Locating plate screw.
26.	Spring washer for screw.
27.	Camshaft gear.
28.	Key for gear.
29.	Nut for gear.
30.	Lock washer for nut.
31.	Timing chain.
32.	Timing chain tensioner.
33.	Chain tensioner body backplate.
34.	Chain tensioner joint.
35.	Chain tensioner bolt.
36.	Lockwasher for bolt.
37.	Oil pump driving spindle.
38.	Flywheel assembly.
39.	Starter ring.
40.	Clutch to flywheel dowel.
41.	Flywheel bolt.
42.	Lockwasher for bolt.
43.	Crankshaft oil thrower.
44.	Crankshaft pulley.
45.	Key for pulley.
46.	Starting nut.
47.	Lockwasher for nut.
48.	Tappet.
49.	Push rod.
50.	Inlet valve guide.
51.	Exhaust valve guide.
52.	Inlet valve.
53.	Exhaust valve.
54.	Valve spring—inner.
55.	Valve spring—outer.
56.	Valve spring cap.
57.	Valve spring collar—lower.
58.	Valve spring retainer.
59.	Circlip for retainer.
60.	Valve stem grommet.
61.	Rocker oil feed pipe union.
62.	Rocker shaft.
63.	Screwed plug for shaft.
64.	Plain plug for shaft.
65.	Rocker shaft bracket—tapped.
66.	Rocker shaft bracket—plain.
67.	Spring washer for rocker shaft.
68.	Plain washer for rocker shaft.
69.	Valve rocker assembly.
70.	Valve rocker bush.
71.	Rivet for valve rocker.
72.	Rocker spacing spring.
73.	Rocker spacing washer.
74.	Rocker adjusting screw.
75.	Locknut for adjusting screw.
76.	Rocker oil feed pipe.
77.	Oil feed pipe bolt.
78.	Washer for feed pipe bolt.
79.	Oil pump assembly.
80.	Oil pump body.
81.	Driving shaft.
82.	Driving gear.
83.	Key for gear.
84.	Driven gear.
85.	Oil pump cover.
86.	Oil pump cover screw.
87.	Spring washer for screw.
88.	Plain washer for screw.
89.	Oil pump joint.
90.	Oil pump filter flange joint.
91.	Thrust washer for oil pump driving spindle.
92.	Oil strainer assembly.
93.	Oil strainer screw.
94.	Spring washer for screw.
95.	Plain washer for screw.
96.	Oil pressure release valve.
97.	Release valve spring.
98.	Release valve plug.
99.	Washer for valve plug.
100.	Release valve drain pipe.
101.	Oil filter head assembly (Tecalemit).
102.	Oil filter sump.
103.	Filter sump to head seal.
104.	Filter element clamp plate.
105.	Clamp plate circlip.
106.	Felt washer for clamp plate.
107.	Dished washer for clamp plate.
108.	Filter element (Tecalemit).
109.	Filter element pressure plate.
110.	Pressure plate circlip.
111.	Felt washer for pressure plate.
112.	Pressure plate spring.
113.	Washer for pressure plate spring.
114.	Filter centre-bolt.
115.	Centre-bolt to filter sump seal.
116.	Filter element (Purolator).
117.	Pressure plate spring.
118.	Seal for pressure plate.
119.	Filter element pressure plate.
120.	Filter element clamp plate.
121.	Clamp plate gasket.
122.	Clamp plate snap ring.
123.	Filter sump to head seal.
124.	Oil filter sump.
125.	Centre-bolt collar.
126.	Centre-bolt seal ('O' section).
127.	Pressure relief valve seat.
128.	Pressure relief valve spring.
129.	Oil filter head (Purolator).
130.	Filter centre-bolt.
131.	Pressure plate circlip.
132.	Washer for pressure plate spring.
133.	Pressure relief valve ball.
134.	Oil filter adaptor plate.
135.	Adaptor plate joint.
136.	Adaptor plate bolt.
137.	Spring washer for bolt.
138.	Oil filter to adaptor plate joint.
139.	Oil filter to adaptor plate bolt.
140.	Spring washer for bolt.
141.	Distributor housing.
142.	Housing to crank case screw.
143.	Spring washer for screw.
144.	Distributor driving gear.
145.	Distributor to driving gear coupling.
146.	Coupling circlip.

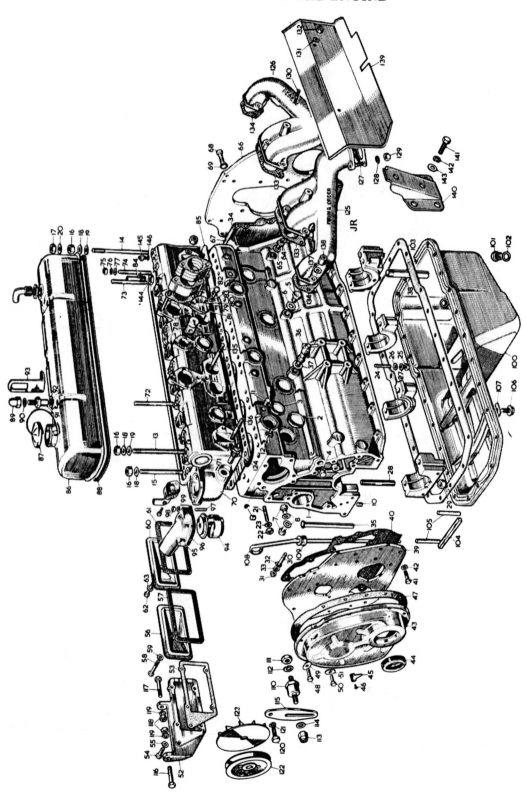

Fig. 7:1   (b) External components of B.M.C. early 'C' type engine

KEY TO THE ENGINE EXTERNAL COMPONENTS

No.	Description
1.	Cylinder block assembly
2.	Core hole plug—large
3.	Core hole plug—small
4.	Oil gallery plug
5.	Oil filter feed hole plug
6.	Feed hole plug washer
7.	Oil pump feed hole plug
8.	Feed hole plug washer
9.	Oil pump boss plug
10.	Tensioner feed hole plug
11.	Water gallery plug—large
12.	Water gallery plug—small
13.	Cylinder head stud—medium
14.	Cylinder head stud—long
15.	Cylinder head stud—short
16.	Nut for cylinder head stud
17.	Nut for cylinder head stud
18.	Washer for cylinder head stud (steel)
19.	Washer for cylinder head stud (bronze)
20.	Washer for cylinder head stud (steel)
21.	Water pump stud
22.	Nut for water pump stud
23.	Spring washer for stud
24.	Oil pump stud
25.	Nut for oil pump stud
26.	Spring washer for stud
27.	Plain washer for stud
28.	Main bearing cap stud
29.	Nut for main bearing cap stud
30.	Ignition coil stud
31.	Nut for ignition coil stud
32.	Plain washer for stud
33.	Spring washer for stud
34.	Rear plate to cylinder block dowel
35.	Oil level indicator tube
36.	Water drain tap
37.	Drain tap washer
38.	Rear main bearing drain tube
39.	Engine front mounting plate
40.	Mounting plate to crankcase joint
41.	Mounting plate to crankcase screw
42.	Spring washer for screw
43.	Cylinder front cover
44.	Front cover seal
45.	Timing pointer
46.	Timing pointer rivet
47.	Front cover joint
48.	Front cover screw—long
49.	Washer for screw
50.	Front cover screw—short
51.	Washer for screw
52.	Cylinder side cover—front
53.	Front side cover joint
54.	Front side cover screw
55.	Washer for screw
56.	Cylinder side cover—centre
57.	Centre side cover joint
58.	Centre side cover bolt
59.	Washer for bolt
60.	Cylinder side cover—rear
61.	Rear side cover joint
62.	Rear side cover bolt
63.	Washer for bolt
64.	Oil gauge pipe union
65.	Washer for union
66.	Engine rear mounting plate
67.	Mounting plate to crankcase joint
68.	Mounting plate to crankcase bolt
69.	Lock washer for bolt
70.	Cylinder head assembly
71.	Core hole plug
72.	Rocker bracket and cover stud
73.	Rocker bracket stud—long
74.	Rocker bracket stud—short
75.	Nut for rocker bracket stud
76.	Spring washer for stud
77.	Plain washer for stud
78.	Carburetter balance plug
79.	Balance plug cover
80.	Balance plug locating peg
81.	Balance plug cover joint
82.	Balance plug cover screw
83.	Spring washer for screw
84.	Brake screw vacuum take-off union
85.	Cylinder head joint
86.	Valve rocker cover
87.	Oil filler cap
88.	Valve rocker cover joint
89.	Rocker cover stud cap nut
90.	Washer for cap nut
91.	Rocker cover stud adaptor nut
92.	Washer for nut
93.	Engine sling bracket
94.	Thermostat
95.	Water outlet elbow
96.	Outlet elbow joint
97.	Outlet elbow stud
98.	Outlet elbow stud nut
99.	Spring washer for stud
100.	Engine oil sump
101.	Sump drain plug
102.	Drain plug washer
103.	Sump to crankcase joint
104.	Front and rear main bearing cap joint
105.	Front and rear main bearing cap plug
106.	Sump to crankcase bolt
107.	Washer for bolt
108.	Oil level indicator
109.	Washer for oil level indicator
110.	Dynamo adjusting link pillar
111.	Pillar to engine nut
112.	Spring washer for nut
113.	Dynamo adjusting link to pillar nut
114.	Plain washer for nut
115.	Dynamo adjusting nut
116.	Dynamo to side cover bolt—front
117.	Dynamo to side cover bolt—rear
118.	Dynamo to side cover bolt nut
119.	Spring washer for bolt
120.	Dynamo to adjusting link screw
121.	Spring washer for screw
122.	Dynamo pulley
123.	Dynamo fan
124.	Thermal transmitter
125.	Exhaust manifold—front
126.	Exhaust manifold—rear
127.	Manifold outlet flange stud
128.	Spring washer for stud
129.	Outlet flange stud nut
130.	Heat shield stud
131.	Plain washer for stud
132.	Heat shield stud nut
133.	Exhaust manifold gasket—inner
134.	Exhaust manifold gasket—outer
135.	Exhaust manifold stud—long
136.	Exhaust manifold stud—short
137.	Plain washer for stud
138.	Exhaust manifold stud nut
139.	Heat shield
140.	Engine front mounting bracket—L/H
141.	Mounting bracket to crankcase screw
142.	Spring washer for screw
143.	Plain washer for screw
144.	Brake servo vacuum take-off union locknut
145.	Distributor vacuum take-off union
146.	Distributor vacuum take-off union washer

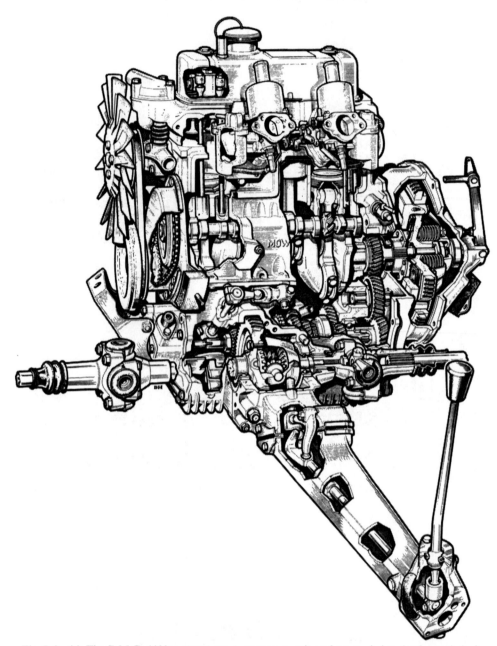

Fig. 7:2   (a) The B.M.C. 1100cc transverse type power unit and transmission (early type) (twin carburetter version). With B.M.C. transverse type power units, before the engine can be lifted from the transmission case, the clutch and flywheel must be removed (see fig. 10:10)

Removing the clutch housing (Minis/1100s)

Knock up the tabwashers and undo the nuts and bolts which retain the clutch housing to the engine and transmission housing. Remove the housing from the studs.

Thoroughly wash the housing in petrol, or a proprietary cleansing solution, and inspect the first-motion-shaft bearing race (fitted to the housing) and the roller bearing on the first motion shaft. Also, check the idler-gear bearing which is fitted to the housing, then remove the idler gear from the gear 'train' and inspect the bearing in the gear case.

Inspect the idler gear itself for broken or worn teeth. On early cars, the idler gear was pressed on to a shaft and the shaft should be inspected to ascertain whether the gear has moved on the shaft. If it has moved, the gear should be replaced by the later type that has a shaft integral with the gear.

Remove the primary gear by releasing the horseshoe type retainer and inspect the primary-gear bushes. If they appear worn, or have moved from their original position, they must be renewed.

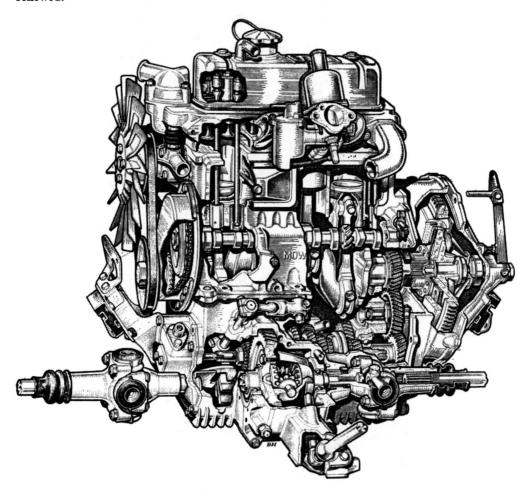

Fig. 7:2   (b) The Mini power unit and transmission assembly. 1966 Mini with diaphragm clutch and modified engine mounting rubbers. Later type Minis (Mk 2) are fitted with remote control gear lever

To remove the engine from the transmission case, take the bolts from around the flange of the crankcase, and lift the engine from the case.

Removing the distributor drive

On B.M.C. 'A' and 'B' type engines (see figs. 7:3 and 7:4), to release the distributor shank housing and drive shaft, remove the securing bolts and pull the housing from the cylinder block. Next, screw a 3 in. $\times \frac{5}{16}$ in. UNF bolt into the thread in the driveshaft head. The shaft may now be lifted out with the bolt, but on B.M.C. 'B' type engines with a 5-bearing crank-

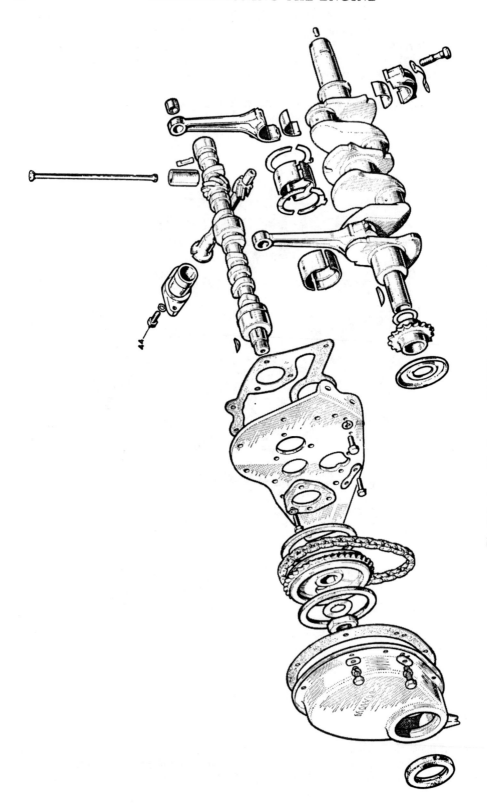

Fig. 7:3 Assembly diagram of B.M.C. early 'A' type engine

Chart 7:1

B.M.C. Transverse-Engine Flywheel Identification

Flywheel Identification Characteristics	HUB DETAILS	Part No.	
Flywheel diameter* 9 13/16" (249.24 mm)	Hub inner diameter 1 3/8" (34.93 mm) Hub depth 1 5/16" (33.34 mm)	12A 377	Mini Cooper (997 cc)
Flywheel diameter* 10 1/8" (257.18 mm)	Hub inner diameter 1 3/8" (34.93 mm) Hub depth 1 7/16" (36.51 mm)	22A 129	Austin Mini Range Morris Mini Range Wolsley Hornet (848 cc) Riley Elf (848 cc)
	Hub inner diameter 1½" (38.10 mm) Hub depth 1 5/16" (33.34 mm)	12G 96	Austin 1100 Morris 1100 MG 1100
Flywheel diameter 9 13/16" (249.24 mm)	Hub inner diameter 1½" (38.10 mm) Pressure spring recesses of unequal depth Hub depth 1 7/16" (36.51 mm)	12A 527	Mini Cooper (997 cc) Mini Cooper (998 cc)
	Hub inner diameter 1½" (38.10 mm) Pressure spring recesses of equal depth. Hub depth 1 7/16" (36.51 mm)	12A 642	Mini Cooper 'S' Models
Flywheel diameter 10 1/8" (257.18 mm)	Hub inner diameter 1 3/8" (34.93 mm) Hub depth 1 7/16" (36.51 mm)	22A 443	Austin Mini Range Wolsley Hornet (848 cc) Morris Mini Range Riley Elf
	Hub inner diameter 1½" (38.10 mm) Hub depth 1 7/16" (36.51 mm)	12A 669	Austin Mini MG 1100 Riley Elf (998 cc) Morris Mini Austin 1100 Mini Moke Morris 1100 Wolsley Hornet (998 cc) Van den Plas 1100
	Hub inner diameter 1½" (38.10 mm) Hub depth 1 7/16" (36.51 mm)	12G 250	Wolsley Hornet (998 cc) Riley Elf (998 cc) Austin 1100 Morris 1100 Van den Plas 1100 MG 1100
Flywheel diameter 10 1/8" (257.18 mm)	Hub inner diameter 1½" (38.10 mm) Hub depth 1 7/16" (36.51 mm) As with flywheel 12A 527, 12A 642, 22A 443, 12A 669 and 12G 250 an oil seal recess is not provided as the necessity for the seal has been avoided by using a primary gear assembly fitted with self lubricating bushes. † Clutch spring recesses are not provided as a diaphragm clutch is now used.	12G 424	Austin Mini Range Morris Mini Range Mini Moke Austin Mini Cooper (998 cc) Morris Mini Cooper (998 cc) Austin Mini Cooper 'S' Morris Mini Cooper 'S' Wolsley Hornet (998 cc) Austin 1100 Morris 1100 Van den Plas 1100 MG 1100 Riley Kestrel Wolsley 1100

* Recessed for oil seal.

† It is possible to convert old cars that are fitted with a flywheel oil seal to the later type so as to avoid oil leaks into the clutch. The work entails fitting a new primary gear assembly, plugging an oil way in the crankshaft and dispensing with the oil seal. A conversion kit is available.

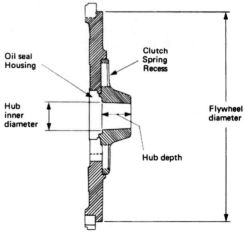

Oil seal Housing

Clutch Spring Recess

Hub inner diameter

Hub depth

Flywheel diameter

Fig. 7:4 Assembly diagram of B.M.C. 'B' type 3 main bearing engine (*Bottom*) Five main bearing crankshaft and oil seal

KEY TO THE ENGINE INTERNAL COMPONENTS

No.	Description	No.	Description	No.	Description
1.	Connecting rod and cap—nos. 1 and 3 cylinders	15.	Gear—tachometer driving	31.	Joint washer—tensioner
		16.	Key—tachometer gear	32.	Screw—tensioner to crankcase
2.	Connecting road and cap—nos. 2 and 4 cylinders	17.	Ring—spring—tachometer gear	33.	Washer—lock—screw
		18.	Plate—camshaft locating	34.	Bolt—flywheel to crankshaft
3.	Screw—cap	19.	Screw—locating plate to crankcase	35.	Washer—lock—bolt
4.	Washer—lock—screw	20.	Washer—lock—screw	36.	Nut—bolt
5.	Bearing—connecting rod—standard	21.	Gear—camshaft	37.	Flywheel
		22.	Key—camshaft gear	38.	Dowel—clutch to flywheel
6.	Screw—connecting rod clamping	23.	Nut—camshaft gear	39.	Ring gear—starting
7.	Washer—spring—screw	24.	Washer—lock-nut	40.	Tappet
8.	Crankshaft	25.	Gear—crankshaft	41.	Push-rod
9.	Bush—first-motion shaft	26.	Key—crankshaft gear and pulley	42.	Pulley—crankshaft
10.	Plug	27.	Washer—packing—crankshaft gear	43.	Nut—pulley
11.	Bearing—main—standard	28.	Thrower—oil—front—crankshaft	44.	Washer—lock—starting nut
12.	Washer—thrust—upper	29.	Chain—timing	45.	Crankshaft
13.	Washer—thrust—lower	30.	Tensioner—chain	46.	Oil seal
14.	Camshaft				

Items 6, 7, 8, 9 bracketed (18G/18GA)

Items 45, 46 bracketed (18GB)

Late cars have an electrical impulse type rev counter and the tachometer gear is not fitted

shaft, it may be necessary to have numbers 1 and 4 pistons halfway up their cylinders, otherwise the crankshaft may obstruct the drive shaft and prevent it from being withdrawn.

Withdrawing the cylinder studs

If the cylinder head is retained by cylinder-head studs and nuts, the studs may be withdrawn, either by using an extractor tool, or by locking two nuts together on the thread of each stud and using a suitably sized spanner on the bottom nut.

Removing the sump and oil pump, etc.

When removing the sump the engine should be turned upside down. Make sure to place a wooden board under the block to avoid scratch marks and other damage to the block face.

Remove the sump by undoing the retaining bolts.

After the sump has been removed, release the oil-pump retaining nuts, and withdraw the pump, the oil strainer and the drive shaft (if fitted).

The oil pump on the Ford Cortina (see fig. 7:5) and on a few other cars is an external fitment and the pump can be removed and refitted at virtually any time during the dismantling and reassembling procedure. Oil pumps which are fitted externally can also be removed and refitted whilst the engine is *in situ*.

Oil pumps should always be primed with oil before fitting.

If the complete engine is being stripped the camshaft may now be removed and the pistons and connecting rods taken out. When doing this, place the cylinder block on its side.

Fig. 7:5   On some cars the oil pump is an external fitting and can be removed from the crankcase without dropping the sump

Removing pistons and connecting rods

It is essential to keep any parts that are to be used again in such a way that they can be fitted to their original position. To avoid any likelihood of mistake with pistons and connecting rod assemblies, it is advisable to mark the cylinder numbers on one side of both the rod and the cap whilst they are in the engine, if not already marked (see fig. 7:26).

Turn the engine so that the connecting rods at the extreme ends of the engine are at bottom dead centre and remove the big-end caps. Push the pistons up the bore a little and remove the upper big-end shell bearings.

Pistons can usually be withdrawn from the top of the cylinder block, the caps and bearings being subsequently replaced on their respective connecting rods.

It is entirely dependent upon the amount of wear in the cylinders as to how easily the pistons and connecting rods come out. On high mileage units there will be a distinct ridge at the tops of cylinders, where the rings do not reach, in this case a fairly hefty push with a hammer shaft is the easiest method.

Removing the main bearings

When the pistons and connecting rods have been removed, the main-bearing caps should be marked to facilitate correct replacement and the tabwashers (if fitted) released, the nuts or the bolts retaining the caps undone and the caps removed.

The front and rear caps are sometimes tight but a copper mallet judiciously used will be a considerable help.

One of the centre main-bearing caps retains the bottom halves of the crankshaft thrust washers and it is important to note which 'faces' of the thrust washers go against the cap for, if wrongly replaced, the washers may disintegrate and cause the engine to be ruined after a few hundred miles.

Having removed the bearing cap, lift the crankshaft from the crankcase, retain the main bearing shells and the top thrust washers.

Before any further work is carried out, all parts must be thoroughly washed and laid out for inspection and measuring.

Cleaning and inspecting the parts

Although petrol may be used for cleaning, there are several excellent proprietary fluids available. Whichever is adopted, a stiff brush, plenty of fluid and a good syringe are essential.

Before washing the cylinder block, scrape all old gasket material from the sealing faces and, if the oil-relief valve is fitted to the crankcase, remove the cap, the spring and valve, and syringe all oilways in the block and crankshaft, being extra thorough if the engine has 'run' any bearings.

Whenever an engine is overhauled, unless one renews all of the parts, there are but few instances where a query does not arise regarding the future serviceability of some item, whether crankshaft, timing sprocket, or valve guide. Measurement of the components, and comparison with the measurements given in Appendix 2, may help in reaching a decision, but it cannot reveal flaws or fatigue. If any doubt exists regarding the serviceability of any part whatsoever, it pays to play safe and renew or, alternatively, consult a mechanic at one of the local garages.

Crankshaft wear

If the crankshaft shows signs of ovality or scoring on the main-bearing journals or crankpins, it is advisable to fit a reground crankshaft with the appropriate bearings. Reground assemblies are usually available in 0·020 in., 0·030 in. and 0·040 in. undersize. When used with the appropriate bearings of suitable size, the mains and big ends have a satisfactory diametrical clearance and the bearing must not be filed or scraped in any manner whatsoever.

If the crankshaft is in a fairly good condition, give it a thorough cleaning and polish it with a 1 in. wide strip of Corolite 320 grade abrasive, or a strip of exceedingly fine emery cloth, lubricated with oil.

Cylinder bore wear

Although the approximate amount of cylinder bore wear may be estimated by feeling the ridge at the top of the bore or by following the procedure suggested on page 67, it is preferable to measure the diameter of the cylinders with an internal micrometer or cylinder bore gauge. Take the measurement just below the ridge left by the rings and compare it with a measurement taken on the non-wear part. If the difference between the diameters is less than 0·006 in. it is unnecessary to rebore the cylinder block, provided of course, that the cylinders are not scored or have overheating discoloration patches. 'Overhaul' type pistons (see fig. 4:4) can be used instead of reboring the cylinders and are suitable for use in cylinders with up to 0·010 in. wear.

If an internal micrometer is not available, take the block to the agent for the make of car or a specialist firm that undertakes reboring. They will measure the block, rebore the cylinders and supply the appropriate pistons.

Normally pistons are available in oversizes of +0·010 in. +0·020 in. +0·030 in. and +0·040 in. The pistons are marked with the oversize dimensions. A '+040' usually requires a cylinder 0·040 in. above the standard cylinder size, the running clearance being allowed for in the manufacture of the pistons.

Usually, unless more 0·006 in. wear is found, the next size of piston will suffice, a cylinder block for instance, already +0·020 in. with 0·007 in. wear will often clean up satisfactorily if bored +0·030 oversize. The decision upon the size, however, should be left to the person carrying out the reboring for cylinders on some engines may not 'clean up' satisfactorily if rebored to the next oversize owing to the uneven distribution of the wear in the cylinder.

If the cylinder block is to be rebored, remove the pistons from the connecting rods (see fig. 7:6) and discard them. If reboring is not being carried out, however, examine the pistons carefully for scoring marks before deciding to use them again. New oil and compression rings may be fitted but in this case be sure to use stepped top-compression rings or to chamfer the top outer edge of non-stepped rings, otherwise the top ring may foul the 'non-wear' ridge with detrimental results. The same applies if new big-end shells are fitted for they restore full piston travel and, if the engine is assembled without stepped rings or chamfering the non-stepped type, a pronounced 'tap' may occur.

Pistons and connecting rods

If the engine is fitted with fully-floating gudgeon pins and bushed connecting rods (see fig. 7:7), before removing the pistons from the rods, check the gudgeon pins for 'lift' which will denote excessive wear. To dismantle the pistons from the rods, circlip pliers are necessary to enable the retaining circlips to be removed from the gudgeon pin bore.

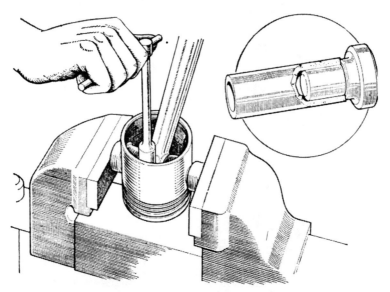

Fig. 7:6   Method of holding piston and connecting rod to release or tighten the gudgeon-pin clamp bolt

Many engines have connecting rods with clamp bolts in split little-ends, the bolts engaging with a groove in the gudgeon pin. To remove this type of piston, insert a small plug into each end of the gudgeon pin to allow the assembly to be held in a vice to undo the clamp bolt (see fig. 7:6). When the bolt has been removed, the gudgeon pin can be pushed out using a suitably sized drift.

When reassembling the pistons onto this type of connecting rod, always use new clamp bolts and make certain to line up the groove in the gudgeon pin correctly with the clamp bolt in the connecting rod otherwise damage to the clamp bolt threads will result. If the gudgeon pins show signs of wear they should be renewed. This involves a certain amount of selective assembly to get a satisfactory fit. Ideally, the pins when cold should be a thumb push-fit for 75% of their travel, and yet sufficiently tight to require lightly tapping with a hide-faced mallet for the remainder of the distance.

New pistons and gudgeon pins (piston pins) are usually all ready for fitting and only require a thorough wash in petrol. When fitting them, make sure to lubricate the pins with a little oil and place the pistons the correct way round on the connecting rods. Some pistons are marked 'front' to assist this. Split-skirt pistons may not be marked, but the rule is to fit the pistons to the rods so that the split skirt will face towards the non-thrust side of the engine.

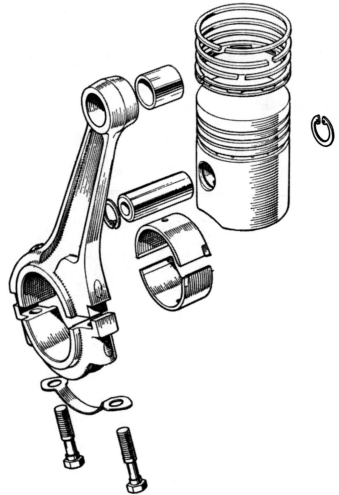

Fig. 7:7   Piston and connecting rods as fitted to some B.M.C. 'B' type engines

It is essential to fit the connecting rods and piston assemblies to the engine with the gudgeon-pin clamp bolt facing towards the non-thrust side of the cylinder. On transverse B.M.C. engines, the clamp bolt faces to the rear of the car. As an added safeguard, check that the marks previously made on the connecting rods are facing the correct way.

Be sure to renew the spring washer that goes under the head of each clamp bolt and, if any undue resistance is felt when screwing up the bolt, remove it, realign and try again. Tighten the bolts firmly and take care not to strip the threads in the rod or bolt. When tightening, hold the assembly in the vice again using plugs inserted in each end of the gudgeon pin so that undue stress on the connecting rod is avoided (see fig. 7:6).

On engines that have pistons with fully-floating gudgeon pins, the connecting rods have bronze bushes in the 'little ends' and these must be carefully examined for signs of wear even if it was impossible to detect any lift. New gudgeon pins and bushes, however, are not available for some engines and if signs of undue wear are apparent the complete rod and piston may have to be renewed. If new pistons are fitted to a slightly worn cylinder, obtain the largest grade possible. The grades are usually marked on the piston crown and possibly on the cylinder block face. (The markings are principally for use when assembling new parts together.)

When new pistons have been fitted, make certain that the circlips are seated correctly in their groove.

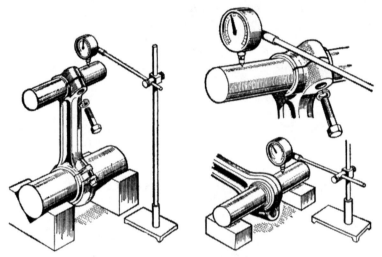

Fig. 7:8 Checking connecting rod alignment

Fully-floating gudgeon pins should be a hand push-fit into the rods and pistons at room temperature (20° C). Before the connecting rods are fitted to the pistons, however, they must be checked for twist and bow. As this involves the use of special connecting-rod alignment equipment, it is advisable to have this done by a garage or engine specialist. If, however, a surface plate, a suitably sized mandrel, V-blocks and a dial gauge are available, fig. 7:8 shows how to proceed. Should the connecting rods show any indication of twist or bow they must be realigned with a suitable tool.

It should be noted that very short connecting rods such as fitted to the Imp are not suitable for resetting and if any misalignment is found the connecting rod must be renewed. It should also be noted that the pistons and connecting rods on the Mini Cooper 'S' and some Vauxhall engines are only supplied as complete assemblies.

Piston rings

New piston rings require careful fitting and, even when the new rings are supplied already fitted to the pistons, they should be checked in the bore (see fig. 5:3). The piston-ring side clearance in the grooves is normally correct, but if the old pistons are being used it is essential to check that the clearance is not excessive (see page 68).

When taking rings from a piston, remove each ring from the top of the piston, using for preference a ring expander tool. Failing this, use a 0·012 in. feeler gauge, easing one end of the ring from the groove and inserting the steel feeler between the ring and the piston. Having reached this position, rotate the piston, applying a light upward pressure to the raised part of the ring until it rests on the land above the ring groove. Subsequently, it can be easily slipped off the top of the piston. Needless to say, great care must be taken during this operation particularly with oil-scraper rings.

If the pistons are to be used again, carefully clean the grooves and crowns, removing all carbon deposits.

When cleaning the grooves, be careful not to remove inadvertently any metal from the sides of the grooves otherwise the rings will have excessive side clearance with a consequent loss of gas tightness, which in extreme cases may lead to heavy oil consumption.

To check the piston-ring gaps, it is best to insert a piston in the bore and arrange so that it remains in one position. In this way, as each ring is inserted it will remain square with the bore and a true feeler-gauge reading can be obtained.

If any ring gap is found to be below the minimum clearance, ease the end of the ring with a small file, taking care to file square and evenly.

Rings that are even slightly above the maximum tolerance must be renewed.

When refitting the rings to the pistons, assemble them in the correct order: the top ring may be a parallel chrome ring, and the second and third are often taper rings which are marked with a 'T' to denote that this face goes upwards, so facilitating correct assembly. Before deciding to use any piston again, carefully examine the skirt, particularly on the thrust face, for any signs of slight seizure or scoring marks, discarding any piston that shows the slightest sign of deterioration.

If the engine has been rebored to the maximum and has more than 0·007 in. wear, it is advisable to fit 'overhaul' type pistons or to have cylinder liners fitted.

Oil pump, filter and relief valve

Although they seldom give trouble, the oil pump should be dismantled, the parts thoroughly cleaned and the gears, or rotors, examined and checked for wear. If the pump is a rotor type see figs. 6:7 and 6:8. If a gear-type pump, see fig. 6:9.

Service replacement oil pumps are available for most cars. The manufacturer's agent should be consulted.

It is important to change the oil filter every 5000 miles or when any major work has been completed on the engine.

Points to note when fitting a new oil filter element are:

To position the spring and pressure plate correctly, and the pressure plate seal (see fig. 7:9).

To renew the sealing ring, or rings (filter-sump to filter-head seal).

When refitting the filter-sump, to hold it in position on the sealing ring whilst screwing up the retaining bolt.

To tighten firmly; avoid excessive pressure otherwise the threads in the sump-head will be stripped.

Paper filter elements, as well as felt elements, are available. The paper type are cheaper, but the felt type better.

The oil relief valve and spring seldom give trouble but if, on examination, the valve does not appear to have been seating adequately, or wear is visible, the valve should be renewed. The free length of the spring should be checked and if necessary the spring renewed.

Flywheel and clutch

Flywheel starter ring gears usually have a long life, but should be renewed if the teeth appear worn, for otherwise unsatisfactory starter engagement may occur long before the life of the engine has expired. On high-mileage units it is well worth renewing merely as a precautionary measure for there is nothing more irritating than a car that gives starting trouble.

The best way of removing the old ring gear is with a cold chisel, taking care not to damage the flywheel. If a hole is drilled first, the ring will split more easily.

Before attempting to fit the new starter ring make sure the surface where the ring fits is free from burrs or chisel marks.

Fig. 7:9 Assembly view of typical oil filter with renewable element; the order of assembly is very important. Oil filters should be renewed every 5000/6000 miles. A new sealing ring must be fitted and care taken to ensure that the correct type of seal is used. Two types are usually supplied with each filter. The seals are of differing widths and can easily be identified by comparison with the old seal after it has been prised from its location with a sharp pointed instrument. After a new filter has been fitted always warm the engine and inspect for an oil leak; if leaking, slacken the bolt and carefully reseat the bowl.

The new ring gear must be heated according to the manufacturers' recommendation, usually in the temperature range of 300°–400°C, when the ring gear will slip easily onto the flywheel. Ideally, a thermostatically controlled furnace is required, but the job can be done without, provided that some heat resisting fire bricks and a welding torch or a blow lamp are available. Place the ring with the lead of the teeth (where the starter engages) facing upwards on the fire bricks. Play the torch around the gear slowly and evenly. In the absence of specific manufacturers' instructions the ring gear will have to be tried on the flywheel several times until it has expanded sufficiently to slip into position. To try the ring in position, pick it up, using long-handled pliers, and slip it over the flywheel with the starter motor gear teeth leads facing in the correct direction. On no account should the ring gear be heated until it is red, otherwise not only will the temper of the teeth be affected but the ring gear will not shrink sufficiently when cooled.

If the ring gear does not immediately slip evenly into place, a light tap around the periphery often helps position it correctly. Once the ring gear is in place, allow it to cool naturally; the 'shrink fit' is permanent and further treatment is not required.

If the ring gear does not fit over the flywheel the first time, it has not been heated enough and should be reheated and tried on the flywheel again, continuing the process until it can be seated in its correct position.

Diaphragm clutches

The diaphragm clutch on some of the later B.M.C., Rootes and other cars has a different method of holding the release plate from earlier models. The earliest type release plate is retained by a circlip which allows the plate to be removed from the cover. On later models, the release plate is an annular thrust ring affixed to the cover by three drive straps which are riveted in position, making the cover and release plate an integral unit. The oldest type is no longer available as a complete service unit. On the second type, the release-plate straps often break. Service parts are available but rather than go to the trouble of overhauling old types it is worthwhile fitting the very latest assembly (a third type, see fig. 7:10) on which the release plate is not held by straps.

On clutches, see also page 193, fig. 7:11 and charts 10:1 and 10:2.

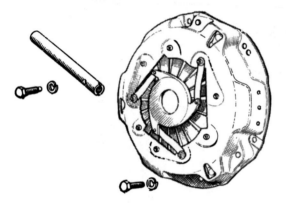

Fig. 7:10 (a) Early type diaphragm clutch assembly with the cover cut away to show the diaphragm spring arrangement. Note the release plate held by drive straps

Fig. 7:10 (b) Later type clutch assembly which dispensed with drive straps

The clutch driven-plate

Inspect the driven-plate for wear and freedom from oil or grease. The friction surface should be of a light, highly polished nature with the grain of the material clearly visible. Do not, however, confuse this appearance with the darker glazed surface which is caused by oil finding its way to the facing and subsequently being burnt off to leave a thin highly polished carbon residue.

If the presence of any oil or oil residue is observed on the clutch facings, the source of the oil leak must be found and the trouble rectified. The wear factor on the working faces of the driven-plate is about 0·001 in. per 1000 miles under normal operating conditions. A rough approximation of how long the plate will last can therefore be obtained by measuring the depth of material between the retaining rivets and the facing surface. Where a complete engine overhaul is being carried out, however, it is always worthwhile renewing the plate, for not only is wear of the friction material and splines involved, but also the torque reaction springs and seats (see chart 10:1). A rough check on the reaction springs is to try rotating them with thumb and forefinger, cases of slackness denoting a need for replacement.

The driven-plate has to accept such extreme stresses that renewal is worth serious consideration.

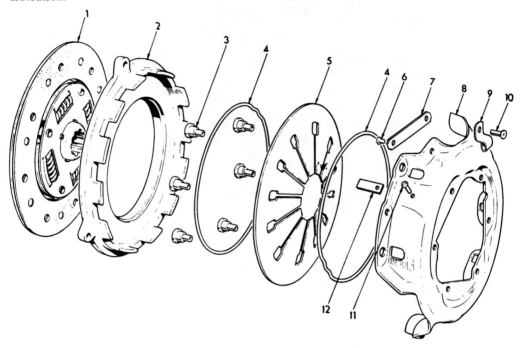

1. Driven plate	7. Drive strap
2. Pressure plate	8. Cover pressing
3. Rivet	9. Retaining clip
4. Fulcrum ring	10. Rivet
5. Diaphragm spring	11. Rivet
6. Rivet	12. Balance weight

Fig. 7:11 The diaphragm clutch fitted to Triumph cars. This 6½ in. diaphragm spring clutch unit was introduced on Spitfire Mk 2 and Herald 13/60 at the commencement of production and on the Herald 1200 from engine No. GA. 204020E and GB. 24121E; 12/50 GD. 4446E. In no circumstances should this type of pressure plate assembly be dismantled; should any fault develop, a complete replacement assembly must be fitted

If, when the car was being driven, the clutch was chattering or dragging, making smooth gear-changing difficult, and the friction faces of the driven-plate are good, spline wear of the driven plate may have been the cause and the splines should be carefully checked. In most cases of rapid spline wear some form of misalignment is responsible. Usually the fitting of a service clutch assembly and driven-plate will cure the trouble but, unless the engine is a transverse unit, in severe or persistent cases the 'run out' of the flywheel must be checked when it is bolted in position (see fig. 7:12). Should the 'run out' be more than 0·003 in. the cause must be found and the trouble rectified.

Thoroughly clean the clutch cover assembly and inspect the pressure plate for wear, ridging and surface cracks. Ideally the surface should be perfectly smooth and free from blue marks. Check the release plate for wear and the drive straps, if fitted, for damage and tightness.

Overhauling the clutch, Minis and 1100s

To remove the clutch plate from the flywheel/pressure plate assembly (see fig. 10:10) the

Fig. 7:12   Checking flywheel run-out

pressure springs must be compressed. Ideally, the tools depicted in fig. 7:13 should be used; but if they are not available use two 'G' clamps.

No special tools are required to dismantle a diaphragm-spring clutch as the springs are fully released before the bolts are out of their threads.

When the spring housing is released from the flywheel, the clutch drive plate can be removed.

Wash the springs, housing and flywheel in petrol and allow to dry. Inspect the parts thoroughly; if any deterioration in the driving pins or driving straps is observed, renew them.

Although reassembling the clutch is a reversal of the dismantling procedure, make sure to align the mark 'A' on the spring housing, and on the pressure plate, with the timing

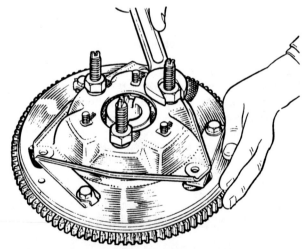

Fig. 7:13   Clutch pressure springs on an early Mini being compressed with the aid of service tool 18G 304M, with tool 18G 684 being used to keep the driven plate and flywheel centralized. Adjusting clutch, see page 194

marks on the flywheel, otherwise severe vibration may occur at high rev/min. When assembling the clutch to the flywheel it is essential to keep the driven-plate in alignment with the flywheel using a suitably sized mandrel. Usually a suitable mandrel is supplied as part of the flywheel puller kit.

Clutch driven-plate (Minis and 1100s)

Inspect the rivets around the central hub of the driven-plate. If they are loose, fit a new driven-plate.

Minis and 1100s often suffer from a condition varying from a fierce to a 'snatchy' clutch operation which may occur from a standing start or when changing gear if the car is being driven at low speed.

A modified clutch driven-plate was introduced with a splined hub of slightly less length and this should be fitted if the clutch is dismantled. The 'snatchy' condition can also be due to wear on the primary-shaft splines and if the splines are worn, the primary gear should be renewed. Dryness of the splines can also result in poor clutch operation and it is advisable when fitting a new clutch plate to place a light smear of hmp grease on the splines. Liberal use of grease must be avoided however, for it would collect clutch dust.

Clutch thrust bearing

If a graphite release bearing is fitted, although they have a surprisingly long life, if more than 30% worn it is worth fitting a new bearing. If a ball type thrust is used and feels rough when rotated it should be renewed.

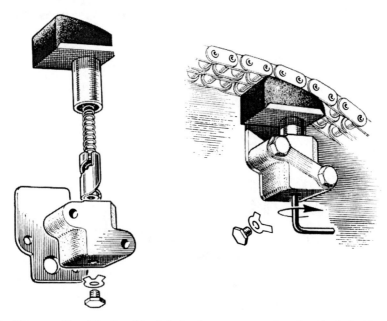

Fig. 7:14   The assembly of the Renolds chain tensioner components and method of setting the tensioner on refitment to the engine

Timing chain and tensioner

Renolds timing chain tensioners should be washed thoroughly and the cylinder in the adjuster body examined for ovality. If the ovality is greater than 0·003 in., when measured near the mouth of the cylinder, a complete new assembly must be fitted. If, however, the cylinder is within the limit it is sufficient to fit a new slipper head to the body.

To reassemble the tensioner components (see fig. 7:14), insert one end of the spring into the cylinder and place the plunger on the opposite end. Compress the spring until the plunger enters the cylinder and engages the helical slot with the peg in the plunger, and with an Allen key turn the plunger clockwise until the end of the plunger is below the peg and the spring is held compressed. Remove the Allen key.

Now insert the slipper assembly into the tensioner body, replace the backplate, fit the gasket and, with the two setscrews and lockwasher in position, fit the assembly into the front plate and bend the lockwasher over.

When the timing chain is in place, release the tensioner for operation by inserting the Allen key and turning it clockwise until the slipper head moves forward under pressure against the chain (see fig. 7:14). On no account turn the key anticlockwise or attempt to push the slipper into the chain by the use of force. Remove the Allen key and refit the bottom plug and lock with the tabwasher.

Camshaft wear

Close inspection of the camshaft may disclose a little wear on the apex of the lobes, but normally this does not cause appreciable loss of performance until excessive. Where top performance is required, however, renew the camshaft if the full lift cannot be obtained. The easiest way to check the lift is to replace the camshaft in its bearings in the cylinder block and insert the cam followers and measure the full amount of rise and fall of the cam follower when the camshaft is rotated. The measurement, however, can only be carried out accurately using a dial gauge.

Camshaft bearings seldom require attention and only in extreme cases are they likely to need renewal. The work is best entrusted to an agent for the make of car, who will have the necessary tools for pressing in the new bearings, as well as the cutters for line boring them in position.

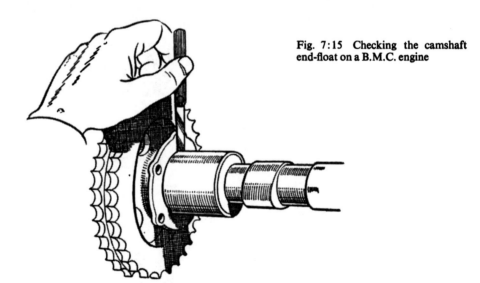

Fig. 7:15 Checking the camshaft end-float on a B.M.C. engine

Camshaft end-float and timing wheels

Before fitting the camshaft to the engine, if a new shaft is being used, check the camshaft end-float; on B.M.C. engines this must be done by fitting the camshaft retaining plate and the chain wheel onto the shaft (see fig. 7:15). Tighten the nut and measure, with feeler gauges, the end-float between the retaining plate and thrust face of the shaft. If the end-float is more than stipulated (see Appendix 2) renew the locating plate. On other types of engine, the end-float may be checked during normal assembly (see fig. 7:16).

The timing wheels should be inspected for wear and any damage that may have resulted when they were removed. The best way of checking the amount of wear is to compare the old wheels with new ones but if this proves difficult and the engine has done a high mileage renew them.

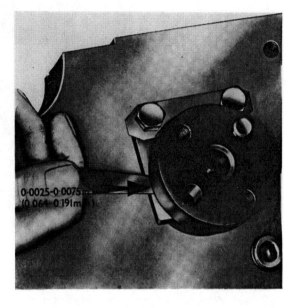

Fig. 7:16 Checking the camshaft end-float on a Cortina engine

Chain wheel alignment

If a new crankshaft, or camshaft locating plate, or camshaft, is being fitted, it will be necessary to check the alignment of the timing chain wheels. With the crankshaft and the front

Fig. 7:17 Location of crankshaft shims

plate fitted, the camshaft in position and the camshaft-retaining plate fitted, replace any alignment shims removed from the old crankshaft behind the timing sprocket (see fig. 7:17) and fit the chain wheels.

When the wheels are in position, check them with a straight edge (see fig. 7:18) to ensure they are in alignment. If they are out of alignment, shims should be added or removed as necessary.

If the timing sprocket location keys are removed during this procedure, make sure to refit them carefully, removing any ragged burrs.

Fig. 7:18 Checking sprocket alignment

Reassembling the engine

Before starting to reassemble the engine, clean and inspect all nuts and bolts, replacing those that show any sign of thread deterioration. Renew all spring washers, shakeproof washers and lockwashers.

When all of the engine components are clean, have been inspected, and the necessary new parts and gaskets acquired, reassembly may commence. Adequate bench space and cleanliness are essentials and, if the components can be placed in a position where they are ready to hand in the sequence in which they will be required, so much the better.

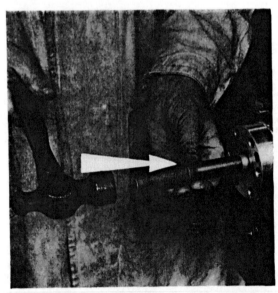

Fig. 7:19 Fitting a new clutch spigot bearing into the rear of the crankshaft

If a new crankshaft is being fitted, make sure that it has a bush to accept the gearbox first-motion-shaft spigot in the rear end (see fig. 7:19). If the crankshaft is of the type where the flywheel retaining bolts have to be fitted in the rear crankshaft flange, insert the bolts before fitting the crankshaft to the crankcase.

Fitting the crankshaft

Before fitting the crankshaft into its bearings in the crankcase make sure to check the oilways in both the crankshaft and the cylinder block. To fit the crankshaft it is necessary to turn the cylinder block upside down on the bench, making sure to stand it on a flat board to avoid damaging the block face.

The main-bearing caps on some engines (Vauxhall PB Velox and Cresta, Ford Zephyr and Zodiac, overhead camshaft B.M.C. engine once fitted to the MGA, etc), have oil-sealing strips which must be renewed, and the top section of the strip should be fitted before the crankshaft is positioned in the crankcase. It is essential that the sealing material is pressed well into position. Should the ends of this sealing strip protrude above the block/cap fitting face the excess material must be removed otherwise the bearing cap will not seat properly.

Insert the thin-wall shell bearings into their housings in the crankcase, taking care that the locating tab on each shell fits into the correct position and make certain that the oil drillings in the bearings align with those in the crankcase.

Lubricate the bearings with engine oil and carefully lower the crankshaft into position.

On some engines, the crankshaft thrust washers are an integral part of a main bearing but if the engine is of the type with separate thrust washers, slide the top semicircular thrust washers into position each side of the centre main, noting that the correct position for the oil grooves in the washers is away from the bearing, *i.e.* the face in which the oil groove is machined goes against the crankshaft (see fig. 7:20).

Fig. 7:20 When crankshaft thrust washers are fitted it is essential that they are placed so that the grooves face the crankshaft

When the thrust washers are fitted, check the measurement between the thrust washers and crankshaft with a feeler gauge (see fig. 7:21) and adjust the clearance as necessary (see also Appendix 2) by lightly rubbing the thrust face on a sheet of emery cloth placed on a surface plate or thick piece of plate glass.

Insert the remaining bearing shells into their caps. If thrust washers are used, fit them each side of the appropriate main-bearing cap and, ascertaining that they fit correctly, refit the cap, making sure that it is the correct way round according to the markings.

Fit the remaining caps but, before positioning the rear cap, coat the joint surface with a

Fig. 7:21 Checking the crankshaft end-float on a Cortina engine

sealing compound taking care not to apply any of the compound too near to the bearing or it may squeeze into the bearing as the cap is tightened. On many cars such as the Ford Zephyr and Zodiac and B.M.C. 'C' type engines, when the main bearing caps are refitted, new sealing strips which fit into vertical grooves each side of the main bearing cap must be fitted. On the B.M.C. 'C' type engine, the front main bearing cap is also fitted with sealing strips (see also fig. 7:22).

When the caps are positioned, fit the spring washers, or locking plates, to each bearing cap and insert the retaining bolts, or screw on the cap-retaining nuts and tighten firmly and evenly. If the engine is fitted with locking plates bend them into position. Tighten each cap separately and, after tightening each one, check that the crankshaft spins easily. If it does not, the cause must be investigated and, should the assembly prove in order with the correct parts fitted, a dimensional check should be made. It is extremely unusual to find causes other than incorrect assembly or incorrect parts; normally, new parts and service-reconditioned material can be fitted without any trouble whatsoever.

On the Ford Corsair and Cortina, the rear camshaft end-plate and gasket should now

Fig. 7:22a Fitting wedges to front sealing block (Triumph). See also next page

Fig. 7:22b Aligning front sealing block. On some cars it is also necessary to align the main bearing cap in the same manner before fully tightening the cap

be fitted. On Triumphs, the crankshaft rear oil retainer (see figs. 7:23 and 7:24) and the cam followers and camshaft are also best fitted at this stage. Insert the seven bolts with their spring washers and partially tighten. Centralize the scroll-type oil retainer by tapping it into

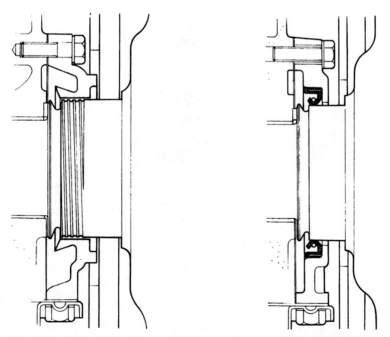

Fig. 7:23 Section view of the crankshaft rear oil seal arrangements on Triumph engines

Fig. 7:24   When fitting a lip type seal on a Triumph engine make sure to fit a gasket
beneath the retaining plate

Fig. 7:25   Centralizing scroll type seal on a Triumph engine

position with a hide-faced hammer (see fig. 7:25) until there is circumferential clearance of 0·003 in. around the scrolled portion of the crankshaft and the internal scrolling of the retainer.

Refitting the pistons and connecting rods

When the crankshaft has been fitted, the pistons and connecting rods can be assembled to the engine, provided of course that the ring gaps, etc. have received attention (see page 68).

When replacing pistons and connecting rods, ensure that they are fitted into the engine the correct way round, and that all marks match up (see fig. 7:26). If they do not, be certain to investigate the cause and rectify.

Points to watch are: that the gudgeon-pin (piston-pin) clamp bolts face the non-thrust side of the engine and, if the pistons are marked 'front', make sure the pistons are fitted to the rods so that the marking on the pistons do in fact face towards the front of the engine. If the pistons are of the split-skirt type, the split in the skirt must face the non-thrust side of the engine. As connecting rods and/or pistons on some engines are offset (see fig. 7:27) it is absolutely essential that they are refitted correctly. If the connecting rods and big-end caps were marked when they were removed from the engine, little difficulty is encountered.

Fig. 7:26 Make absolutely sure to refit the correct big-end cap to each connecting rob

If old parts are being used again they must be replaced in precisely their original position.

If piston and connecting-rod assemblies are being inserted from the top of the cylinder and the crankcase is on the bench, it is best to lay the crankcase on its side. Subsequently, remove the bearing cap from No. 1 big end, insert a shell into the connecting rod and the bearing cap, noting that the locating tabs are correctly engaged.

Space the piston ring gaps equidistant around the piston and lubricate generously with engine oil. Turn the crankshaft so that No. 1 crankpin is at bdc and check the cylinder bores for cleanliness. Fit a piston ring compressor tool to the piston, taking care not to trap and break any rings as they are compressed. Insert No. 1 piston and rod. Give a light push with a hammer shaft and guide the big ends into position on the crankpin.

If any difficulty is encountered, remove the piston and rod, retighten the ring compressor and try again rather than use excessive force. Make sure to lubricate the bearings and to fit the connecting-rod cap the correct way round.

Next, fit the big-end lockwashers and the bolts (or nuts) and make sure to tighten them firmly. Bend the lockwashers over, check that the crankshaft still turns freely and if all is well, fit the other rods and pistons.

When the connecting rods and bearings have been fitted, the rear engine-bearer plate and its gasket can go on but give the latter a coating of sealing compound on both sides.

Rear engine-bearer plate

If the rear engine-bearer plate is fitted with an oil seal, the sealing lip of which fits over the flange on the end of the crankshaft to prevent the ingress of oil to the clutch pit, whenever the bearer plate is disturbed a new seal should be fitted. Make sure, however, before fitting the bearer plate, to lubricate the lip of the oil seal to avoid any initial seal scorch and do not forget on B.M.C. 'B' and 'C' type engines (see fig. 7:1) to place the square-sectioned cork into its groove when bolting on the plate, otherwise the plate will have to be removed later to do this.

If the rear engine-bearer plate is fitted with lockwashers, these should be tapped over when the bolts have been tightened. The oil-seal retaining plate and lockwasher should also be fitted.

If the engine is fitted with a crankshaft rear oil seal with a retaining plate which is separate from the bearer plate, as on the Herald (see figs. 7:23 and 7:28), the seal and retainer must be fitted prior to fitment of the bearer plate.

On some engines the cam followers and camshaft will already have been fitted but if this has not been carried out, the camshaft bearing may be lubricated, and providing that the end-float has been checked (see fig. 7:15) the camshaft may be inserted.

Fit the front engine-bearer plate with its gasket, giving the latter a thin coating of sealing compound. Do not forget the square-sectioned cork (if fitted). Subsequently insert the cam followers and fit the camshaft-retaining plate using shakeproof washers under the head of each setscrew (or tabwashers, see fig. 7:16).

Setting the valve timing

Some engines are fitted with one or sometimes two tension rings (see fig. 7:3) around the perimeter of the camshaft wheel. Before the timing wheels are refitted, the rings should be renewed.

To fit the timing wheel and chain on B.M.C. 'A' and 'B' type engines, turn the engine so that No. 1 piston is at tdc and position the camshaft so that the keyway is at 1 o'clock when viewed from the front upright position. Now assemble the gears onto the timing chain with the two timing marks opposite each other as fig. 7:29. Keep the gears in this position and assemble them as a complete unit, engaging the crankshaft timing sprocket keyway with the key until it just starts. Rotate the camshaft slightly until the keyway and key are aligned and push the gears on a little at a time until they are both fully in position. Check that the crankshaft is still exactly on tdc, recheck the timing marks and subsequently fit the camshaft lockwasher. Fit the nut or bolts and tighten securely, bending the lockwasher over.

Some engines are fitted with bright links (or marked links) on the timing chain which when the gears and timing chain are assembled should line up with the marks on the timing wheels with No. 1 piston at tdc.

The cardinal rule when setting valve timing is to make absolutely certain that No. 1 piston is at tdc and that the timing marks align (see also fig. 7:30).

The valve timing may be checked by noting the opening and closing points of each valve according to the timing chart for the engine. Checking the timing this way can, of course, reveal whether the correct camshaft is fitted to the engine. If a full timing check is to be carried out, the next step in reassembly of the engine prior to fitting the timing case is to refit the oil pump, oil pan, and the cylinder head and valve operating components.

Unmarked timing sprockets

To set valve timing when the crankshaft timing sprocket and/or camshaft sprocket do not

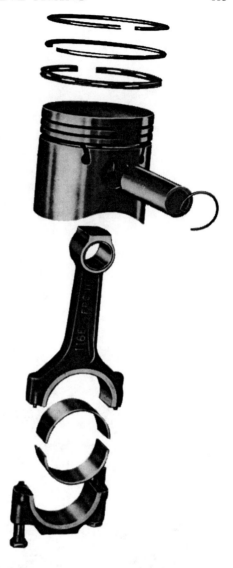

Fig. 7:27 When assembling pistons to connecting rods make sure the pistons face the correct way. The piston pin on this Cortina piston is offset in the piston 0·04 in. (1·0 mm) towards the thrust side of the engine. To facilitate correct assembly an arrow head is cast in the piston crown and must point forwards when the piston is fitted to the engine

have timing marks, the valve opening and closing sequence points and the timing valve-clearance must be ascertained from the manufacturer's workshop manual. It must be noted that the timing valve-clearance is distinct from the running valve-clearance.

To set the timing on single-camshaft push-rod engines, proceed as follows: Set No. 1 cylinder inlet valve rocker and No. 1 exhaust valve rocker to the correct valve-timing clearance. Rotate the crankshaft until No. 1 piston is exactly at the correct position for No. 1 inlet valve to open (see page 127). Turn the camshaft (the chain wheel may have to be fitted temporarily to do this) until No. 1 inlet valve is opening and just off its seat; be sure to turn the camshaft in the direction of normal rotation. Fit the chain wheels and chain tensioner. Now turn the engine ⅓ of a turn in the opposite direction to normal rotation and then turn it in the correct rotational direction and check the timing according to the chart. (see also fig. 7:41 and 'Checking the valve timing', page 125). If the timing is incorrect, the timing chain wheel must be removed and the timing reset accordingly. Some engines have timing sprockets that can be repositioned to obtain ¼ camshaft tooth variation (see also fig. 7:30).

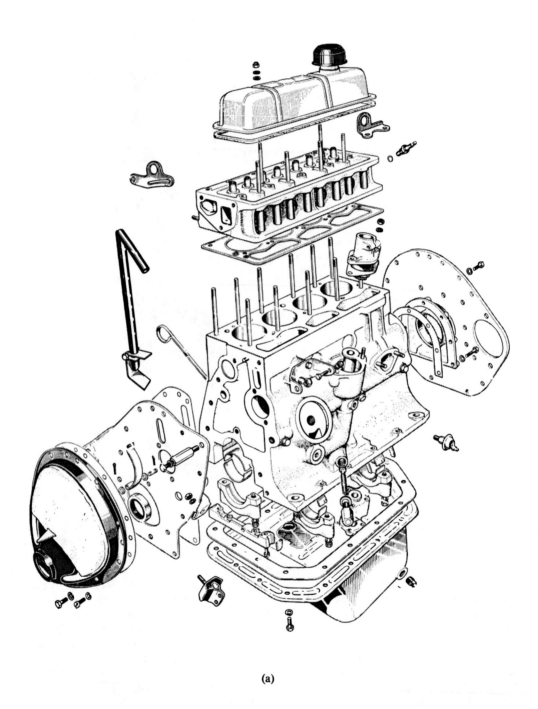

(a)

Fig. 7:28 Assembly details of the Herald and Spitfire engine. (a) Fixed parts, (b) moving parts

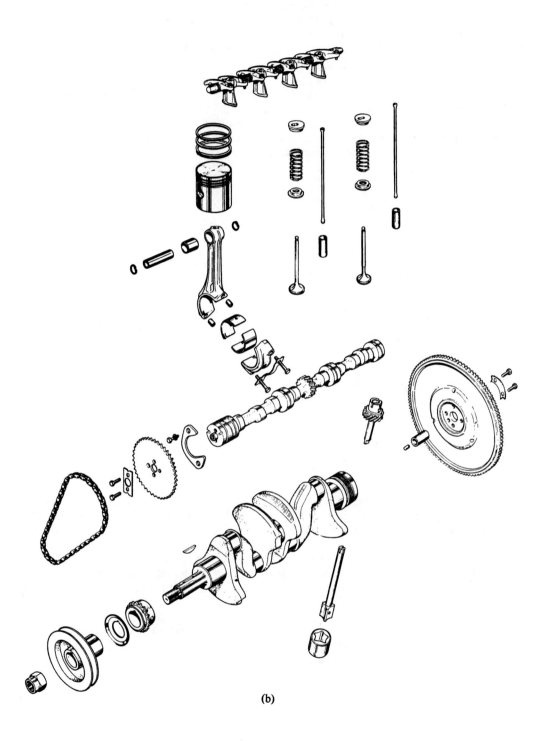

(b)

Fitting the timing case

On the Hillman Minx, prior to early 1962, the initial oil circulation tended to be slow and a timing-chain oil-bleed valve was introduced to help correct this. The necessary parts for converting early cars are available from most Rootes dealers and are well worth fitting; they consist of: an oil pipe jet assembly, oil pipe adapter, bleed valve spring, bleed valve ball.

If the engine is of the type that has a vibration damper and/or Renolds chain tensioner and/or damper, the tensioner should now be fitted.

When the tensioner is in position and locked up, fit the oil thrower to the crankshaft. On B.M.C. cars two types are used, depending on the type of timing cover. The early type oil thrower is fitted with the concave side away from the engine and must only be used with the early type front cover. The later type oil thrower is fitted with the face marked 'F' away from the engine and must only be used with the later type of cover.

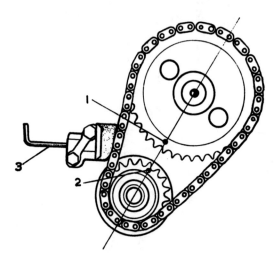

Fig. 7:29   B.M.C. 'B' type engine valve timing. (1) Camshaft gear mark, (2) Crankshaft gear timing mark. It is essential when resetting the timing that the marks are in alignment when No. 1 piston is at t d c. (3) Allen type key for releasing chain tensioner (see also fig. 7:14)

Before fitting the timing cover, fit a new oil-retainer felt or, on later models, fit a new oil seal. Coat the timing-cover gasket with sealing compound and place it in position. Apply a little grease to the annular groove of the timing-cover oil seal or felt, lubricate the hub of the crankshaft pulley and carefully push it into the seal, giving a slight twisting motion. With the key aligned, and the timing cover on the pulley, slide it onto the crankshaft.

On Triumph models it will be necessary to hold the chain tensioner (which is fitted to the timing case) in a retracted position with a piece of thick wire (see fig. 7:32) whilst the cover is fitted.

Line up the cover with the bolt holes, checking that the gasket has not moved, and subsequently insert the retaining setscrews. Tighten up evenly to avoid straining the cover against the flexibility of the oil seal, or the centralization of the seal to the shaft may be disturbed and an oil leak will ensue.

Having fitted the pulley and cover, insert the crankshaft-pulley bolt with its lockwasher and tighten the bolt securely, interposing a suitably sized block of wood between the crankshaft and crankcase to prevent the crankshaft from turning. When the bolt has been tightened, make sure to bend the lockwasher over.

Fitting the oil pump

Fit the oil-pump retaining studs.

Insert the pump drive shaft (if fitted) into the drive tongue, fit the gasket in position but check to make certain that the alignments of the oil holes in the gasket are correct, prime the pump with oil and place the assembly in position (see fig. 7:5).

Fig. 7:30 Valve timing marks on the Herald, Vitesse and Spitfire. Note: The camshaft timing wheel is provided with four bolt-holes which are equally spaced but offset from a tooth centre. A half tooth adjustment of the valve timing can be obtained by rotating the sprocket 90° from its original position. A quarter tooth adjustment may be obtained by turning the sprocket 'back to front' and three quarters of a tooth variation by rotating the sprocket 90° in this reversed position

When the pump is in position, exert a little pressure on it, turning the engine slowly to allow the teeth or drive tongue to mesh with the camshaft. Fit the washers and retaining nuts or bolts, and tighten. Refit the oil strainer.

Apart from transverse engines, at this juncture it is usually best to fit the flywheel, for whilst the sump is off it is much easier to prevent the engine from turning when tightening the retaining nuts.

After making certain that the mating faces of the crankshaft and the flywheel are clean, turn the engine until No. 1 piston is at t d c and fit the flywheel with the 1/4 marking (or 1/6 marking) on the periphery of the flywheel at the top. If a locking plate is fitted, place it in position, refit the nuts or bolts and tighten them, subsequently, if a locking plate is fitted, bend it in position against each nut.

If the clutch cover assembly has been carefully inspected, it may now be fitted, using an old first-motion shaft or a suitably sized mandrel to centralize the driven-plate.

When reassembling the clutch, place the driven-plate against the flywheel so that the wider portion of the central hub is to the rear of the engine. After locating the cover assembly

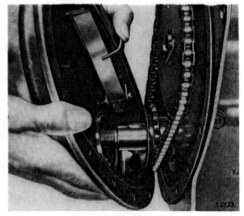

Fig. 7:31 The timing chain tensioner on the Cortina and, arrowed, the oil slinger

Fig. 7:32 Fitting the timing cover on the Herald

on the flywheel dowels, insert the centralizing tool, see fig. 7:33. Fit the bolts and tighten gradually by diametrical selection until they are completely tightened and then remove the centralizing tool.

After methodically checking to ascertain that all bolts are tight, and all lockwashers are bent over, the sump may be refitted. Lift the engine onto the cylinder block face placing a wooden board under the face to avoid scratching it.

If cylinder head studs have not been removed, support the engine with wooden blocks, otherwise the studs may get damaged.

Ascertain that the crankcase faces are clean, give the new sump gasket a coating of sealing compound, place the sump in position, insert the retaining setscrews and tighten evenly (see also page 72).

On B.M.C. Minis and 1100s, refit the engine to the transmission case.

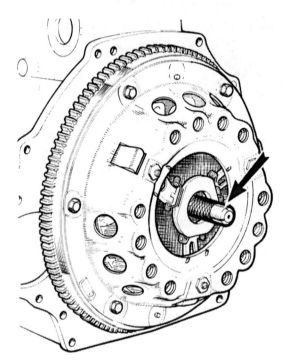

Fig. 7:33   Centralizing the clutch driven plate when refitting the clutch to the flywheel

B.M.C. Minis and 1100s

If a new clutch housing is being fitted, it should be temporarily assembled to the transmission case to adjust the idler alignment and thrust clearances prior to fitting the engine to the transmission case.

To refit the engine to the transmission case on B.M.C. Minis and 1100s place the gaskets in position, not forgetting the 'O' ring which fits into a recess cut around the oil port in the face of the transmission case.

Lower the engine onto the case carefully and evenly, insert the retaining set bolts and tighten evenly. Refit the clutch housing, flywheel and clutch cover.

Refitting the clutch housing (B.M.C. Minis and 1100s)

If a new clutch housing is to be fitted, it is necessary to fit a new first-motion-shaft bearing race to the housing and fit the roller bearing and its retaining clip to the first motion shaft.

A new idler-gear bearing will also be needed and it is essential that it is positioned with the face of the bearing slightly below the face of the housing. On no account must the bearing be

Chart 7:2

Assembly of B.M.C. Engine/Transmission, and Clutch Housing

Sharp edges will damage the oil
seal when clutch housing is fitted
unless a protective sleeve is used
or the instructions on page 120 are
followed.

Before fitting the clutch housing,
check the end float

Make sure a roller does not 'tilt'
as clutch housing is fitted.

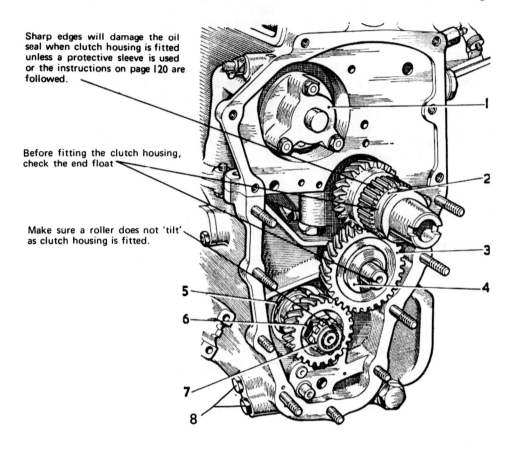

*The engine and transmission assembly of the B.M.C.
Mini, 1100 1300 c.c. engine with the flywheel housing
removed.*

1. Oil pump.	5. First motion shaft bearing.
2. Crankshaft primary gear.	6. First motion shaft driving gear.
3. Idler gear.	7. Roller bearing.
4. Idler gear thrust washer.	8. Detent spring retaining screws.

pressed into the housing as far as it will go, for this would blank off the oil hole with disastrous results.

The clutch housing should temporarily be fitted to the transmission case to adjust the idler-gear alignment before the engine unit is fitted to the case. Remove all old gasket material from the gasket faces of the transmission case and from the clutch housing. Place the new gasket in position on the clutch-housing retaining studs, remove the thrust washers from the shaft on each side of the idler gear and position the idler gear in its bearing in the transmission case. Make sure the location dowels are in position and fit the clutch housing and all of the retaining nuts and bolts and tighten them.

Should any resistance be felt as the housing is pushed into place, it may be due to a first-motion-shaft bearing roller becoming tilted in the bearing housing. In all cases of difficulty, remove the housing, check that the rollers are not out of place and try again; on no account must force be used.

The idler gear should be completely free to rotate, and it should be possible, as the thrust washers have not been fitted, to move it laterally in the case. Should there be any indication of binding, which cannot be overcome by slackening the retaining bolts and nuts and re-positioning the case on its studs, another housing should be obtained. If all is well, remove the clutch housing.

The housing must again be temporarily assembled to the engine to adjust the idler end-float. To do this, fit a thrust washer on the shaft each side of the gear ensuring that the chamfered bore of each washer is against the thrust face of the gear.

Assemble the clutch housing to the transmission case, refit and tighten the retaining bolts and nuts, and check the end-float on the idler gear with feeler gauges. The end-float should be 0·003 to 0·008 in. If it does not fall within this range, the housing must be removed and the thrust washers replaced with others until the correct end-float can be obtained. Thrust washers are available in sizes of 0·132 to 0·139 in.

When the correct end-float is obtained, remove the housing and gasket from the trans-mission case. Retain the gasket, idler gear and washers until required.

Fit the primary-gear oil seal to the clutch housing, taking care that it is pushed home squarely and goes fully 'home'. The oil seal is easily distorted but it may be tapped home with a hammer if a block of wood is interposed between the hammer and the seal.

When the engine has been fitted to the transmission case, the clutch housing can be finally assembled to the unit, provided the primary-gear end-float has been adjusted (see fig. 7:34).

Make sure the crankshaft primary-gear thrust washer is fitted with its chamfered bore against the crankshaft flange and fit the housing in the following manner:

Grease the sealing edge of the oil seal and carefully insert the primary-gear splines through the oil seal. Push the gear through the seal carefully so that the seal is not damaged, and then push the gear fully home until it is against the housing. With the idler gear positioned in the transmission case, line up the housing with the studs, place the bore of the primary gear over the end of the crankshaft and push the housing, complete with the primary gear, into posi-tion. As this is done it will be necessary to rotate the primary gear slightly so as to mesh the

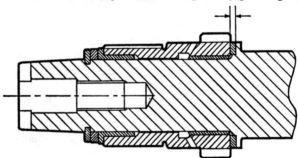

Fig. 7:34 The correct running clearance of the primary gear is 0·0035 to 0·0065 in. (0·089 to 0·165 mm). Measure the gap indicated and at the appropriate thrust washer as given below to obtain this clearance

When gap is	Use washer thickness
·1175 to ·119 in.	·112 to ·114 in.
(2.875 to 3·025 mm)	(2·848 to 2·898 mm)
·119 to 121 in.	·114 to ·116 in.
(3·025 to 3.076 mm)	(2·989 to 2·949 mm)
·121 to ·123 in.	·116 to ·118 in.
(3·076 to 3·127 mm)	(2·949 to 3·000 mm)
·123 to ·125 in.	·118 to ·120 in.
(3·127 to 3·18 mm)	(3·000 to 3·051 mm)

teeth with the idler gear. Force must be avoided. If the housing cannot be pushed fully home, remove it; make sure that a first-motion-shaft bearing roller has not become displaced and try again.

When the housing is in position, fit the lockwashers to their studs and refit all of the housing-retaining nuts and bolts and tighten them evenly. Bend the tabwashers against the flats of the retaining nuts and refit the thrust washers and retainer to the crankshaft.

The clutch/flywheel assembly, clutch cover and cylinder head may now be fitted.

Refitting the distributor drive

The method of fitting distributor drives varies according to the type of engine. Where specific instructions are not available, a method of trial and error should be adopted. The principal thing to make sure of is that when No. 1 piston is at t d c on its firing stroke the distributor rotor arm can be aligned with No. 1 segment in the distributor cap when the points are just breaking for the firing stroke, and the cap and distributor body is in the correct position. The simplest way is to pre-assemble the drive and distributor assembly to the crankcase and ensure that the above requirements are met (see also fig. 7:35).

Fig. 7:35 Fitting the distributor pedestal to a Herald engine. A gasket and packing washers must be used to obtain the correct drive gear end float (0·003 to 0·007 in.)

Refitting the distributor

With the engine on No. 1 compression stroke, rotate the crankshaft until the timing mark on the pulley or flywheel is opposite the appropriate position on the timing cover (see fig. 7:36), or clutch housing (see Appendix 4 for correct distributor settings).

It should be noted that on some engines t d c is indicated and the mark is not an ignition-timing mark.

Set the micro-adjuster on the distributor to the midway position, slacken the clamp-plate setscrew and insert the distributor until the driving dog, or gear, engages the slot of the drive spindle. Check that the rotor arm is opposite the correct electrode in the distributor cap for No. 1 cylinder, and position the distributor so that the contact points are just opening. Insert the two retaining bolts in the clamp plate, re-check the points, tighten the clamp bolt and then tighten the two clamp-plate retaining bolts.

While it is possible to get an approximation of the correct static ignition timing by ensuring that the points are just opening, it is advisable to carry out an additional check with a test lamp connected in parallel with the contact breaker points.

When the ignition timing has been set, fit the distributor cap and the spark plugs. Make sure, when fitting the HT leads, to get the firing sequence correct.

After fitting the valve cover and gasket, the distributor vacuum pipe and clip, the starter

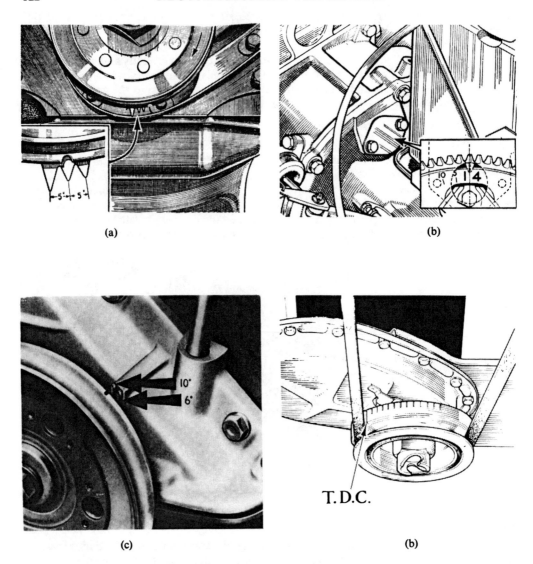

(a)

(b)

(c)

(b)

T.D.C.

Fig. 7:36 Timing marks (a) B.M.C. 'B' type engine. The notch in the pulley approaching the tdc position for pistons 1 and 4. The inset shows the timing set at 5° btdc. (b) B.M.C. Mini, 1100 and 1300. The timing marks on the flywheel, and the indicator, may be seen with the aid of a mirror after removing the inspection plate on the clutch cover. tdc position is indicated by the mark 1/4, and, in addition, 5° and 10° btdc marks are also provided. (c) Cortina ignition timing marks. (d) Timing marks on Rootes cars. Timing set at the 8° btdc position (engines with aluminium cylinder heads). The timing marks on other popular engines are very similar to those illustrated. On some Vauxhalls the ignition timing mark is a bright steel ball embedded in the flywheel. An inspection aperture is provided

and generator can be fitted. Make sure to adjust the fan belt tension properly and to tighten the bolts securely (see fig. 7:37).

If the gearbox has been overhauled, and the engine and gearbox are to be installed as a complete unit, the gearbox can now be fitted to the engine. Take care not to displace the clutch release plate as the box is lifted in position and after replacing the gearbox-retaining bolts and nuts, have a final inspection to ensure that nothing has been overlooked.

Fig. 7:37 Fan belt tension adjusting bolts, (Ford Cortina). On some cars that have long fan belts a slack belt can result in a distinctive squeal when the throttle is opened suddenly. A slack belt can also result in the battery receiving an inadequate charge, throwing suspicion on the condition of the battery, especially in winter weather conditions

Refitting the engine to the car

Although refitting the engine to the car is basically a reversal of the removal operation (see pages 72 to 75), a few added tips may come in useful. If, for instance, adequate lifting facilities are available, and the engine and gearbox are being refitted as a complete unit, the oil filter bowl (sump) can be fitted before the engine goes in. This may involve a little added manoeuvring when settling the unit into position but it is nevertheless well worthwhile.

When refitting the propeller shaft, ascertain that the flange faces are clean and free from burrs and if for any reason the front of the propeller shaft has been removed from the splines, reassemble it onto the mainshaft according to fig. 7:38. Failure to ensure that these conditions are met will result in excessive propeller shaft vibration.

Starting the engine

Make sure all tools have been removed from the engine compartment, that the battery is fully charged and that the carburetter float chamber is filled. Check that the ignition warning light comes on when the ignition switch is operated but before operating the starter:

Check the water level in the radiator.

If a heater is fitted, open the control valve to prevent an air lock forming in the water system.

Make sure the engine oil is up to the full mark on the dipstick.

Check that the LT leads are connected to the coil and distributor.

See that the generator leads have been refitted.

Ascertain that there are no bolts or nuts that have been left loose.

Check that the oil-gauge pipe or oil-light switch is refitted.

Ensure the battery leads have been refitted tightly and the connections lightly greased with petroleum jelly or some other non-corrosive paste.

Make sure that all control cables, e.g. 'choke' and starter have been reconnected and all linkages to the carburetter are properly secured and working correctly.

See that the spark plug leads have been refitted correctly.

Make sure that the gearbox oil level is correct.

Check the clutch adjustment and see that the clutch appears to function.

Make sure the car is not in gear and that the sump drain plug is tight and that there are no water or oil leaks.

Turn the engine over a few times without switching on the ignition and ensure that the fan blades do not foul the radiator. The engine may now be switched on and the starter motor operated. All being well, the engine should 'fire' on the second, or third time the starter is operated, and sometimes even the first time.

If the engine fails to 'fire' check that all electrical connections are tight and make sure that all wires have been connected to the correct terminals. If the distributor has not been disturbed during overhaul, *e.g.* for the fitting of points, condenser etc., there is no reason to suspect that it is the cause of non-starting.

As soon as the engine will run, adjust the throttle screw to give a fairly fast tick-over. Check the oil-pressure gauge or warning light. If oil pressure does not register on the gauge, or the oil light does not extinguish after 10 or 12 seconds, stop the engine and examine the oil feed pipes for an oil leak, a blockage or an air lock. If the trouble is not located, disconnect the oil-gauge pipe, or remove the oil-light switch, and check whether oil is ejected when the engine is started. If oil is not ejected, inject $\frac{1}{4}$ pint of engine oil into the gallery for priming purposes. If oil pressure still does not appear, remove the oil relief valve to see if oil is reaching this point. If it is not, the oil pump will have to be examined.

Fig. 7:38   If a propeller shaft is dismantled at the sliding joint, when the splined shaft is reassembled to the drive shaft, it is essential that the forked yokes on both shafts have their axes parallel to each other. Yoke (A) must be in alignment with yoke (B) and the flange yoke (C) must be in alignment with the flange yoke (D)

Whilst the engine is running, watch the engine temperature gauge; should the engine appear to be running too hot, switch off and allow the engine to cool before continuing the stationary running for approximately another 10 minutes. Whilst the engine is running, check for any leaks, *e.g.* hoses, cylinder head, and oil leaks from the sump and timing cover.

On ohv push rod engines, reset the valve clearance.

The first run

Before taking the car onto the road, allow the engine to reach normal working temperature and check that all instruments are operating satisfactorily.

The engine will not be at its best until quite a few miles have been covered for the components must be given a chance to 'bed in', but during this test-run the engine response to the throttle should greatly improve. If the response remains sluggish, check the timing and ignition before using the car again.

Ignition timing

It should be noted that the ignition timing mark on some Vauxhall engines is a steel ball embedded in the flywheel. When the ball is level with the pointer (in the aperture in the crankcase on a level with the petrol pump) the crankshaft is set 9° before t d c. This is the ignition setting if premium fuel is used.

Some cars do not have a distributor with a micro ignition adjuster. If the ignition setting has to be changed, slacken the nut on the distributor clamp at the base of the distributor and rotate the distributor body. A pencil mark on the base of the distributor and

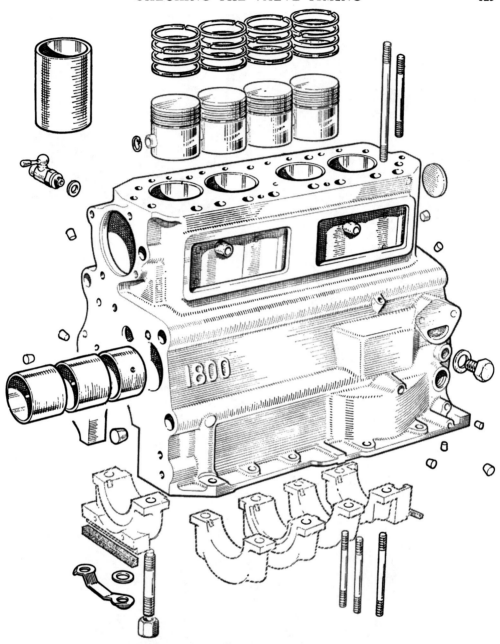

Fig. 7:39   The cylinder block of the B.M.C. 1800 transverse fitted engine. When overhauling the engine the small brass plugs should be checked for tightness

another in alignment on the clamp is a great help should the distributor be moved too much and the original timing lost (see also page 121).

Checking the valve timing

To check valve timing in the absence of the manufacturer's specific quick-check method (see page 128) and without removing the timing case to observe the timing marks on the chain wheels, the following method may be used: mount a circular protractor, or a circular card

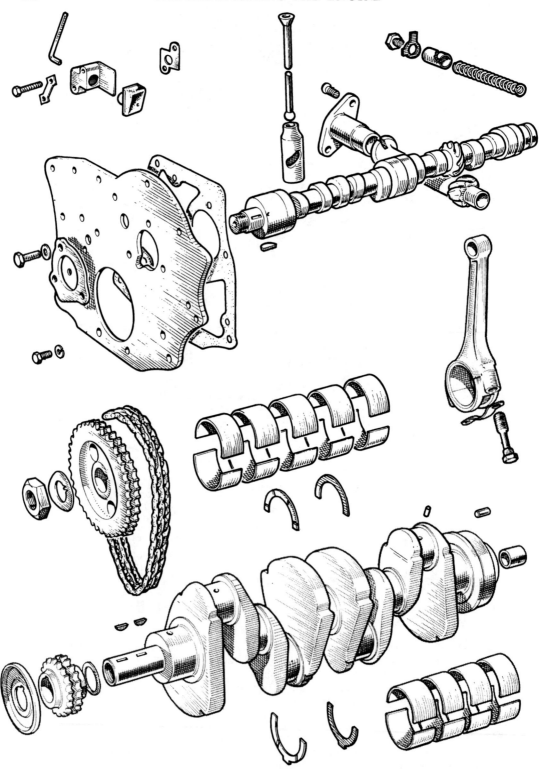

Fig. 7:40   Engine components of the B.M.C. 1800 transverse engine

marked in degrees, on the fan pulley (or flywheel if the engine is out of the car) so that the card will revolve when the crankshaft is turned but will not move in relation to it.

Turn the crankshaft until No. 1 piston is at t d c. If the flywheel or crankshaft pulley does not have the t d c point marked, No. 1 plug must be taken out to ascertain when the piston first reaches its maximum point of travel, and also the point at which the piston first begins to descend. These two points should be lightly marked on the card or protractor. Midway between these two points is t d c and this point should be legibly marked and a pointer arranged. A stout piece of wire will act as a pointer so long as it is fixed securely and cannot be moved accidentally.

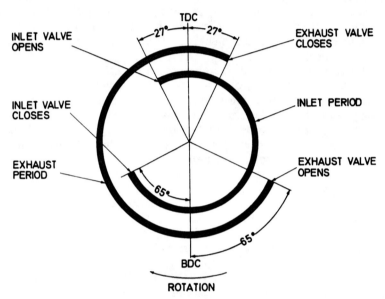

Fig. 7:41   Valve timing charts such as the above example for the Ford GT Cortina, depict the complete cycle of valve operation during two complete crankshaft revolutions. When checking the timing (see page 125) it is essential that the valve clearance is set to the 'timing' clearance. On the GT Cortina this is 0·016 in. (inlet) and 0·026 in. (exhaust). The running valve clearance is 0·012 in. (inlet) and 0·022 in. (exaust).

The standard Cortina 1300/1500cc valve timing is: inlet opens 17° b t d c and closes 15° a b d c. Exhaust valve opens 51° b b d c and closes 17° a t d c. The valve timing clearance is 0·015 in. (inlet) and 0·027 in. (exhaust). The running clearance is 0·010 (inlet) and 0·017 (exhaust)

The crankshaft should now be turned approximately 90° in the direction opposite to normal rotation and then be turned slowly in the direction of rotation and the valve opening and closing points marked on the card or protractor. The actual opening point of a valve can be determined by trying to rotate the valve cap backwards and forwards using the index finger and thumb. When the cap can be moved, the valve has lifted from its seating. Care must be taken that the engine is rotated very slowly otherwise a false reading will be obtained. The closing point of a valve can be ascertained by noting the precise point at which it is no longer possible to rotate the cap. It is advisable to have an assistant to turn the crankshaft and to have a practice run before noting the opening and closing degrees on the protractor.

The results should be checked against the timing chart for the engine in question. Should the timing prove incorrect make sure that the timing card or pointer has not been moved accidentally, that the valve clearances have been set to the valve-timing clearance and that the correct timing chart for the camshaft that is fitted to the engine is being used. Should the valve timing be incorrect, it must be reset but it should be noted that slipped valve timing on modern cars, especially those with single-camshaft push-rod engines, is extremely rare and virtually the only occurrence of valves found to be incorrectly timed is when an engine has been reassembled and fails to operate satisfactorily. Symptoms are spitting back through silencer and/or carburetter, severe overheating and failure to attain full power. Similar symptoms can result from slipped ignition timing and are by no means unusual.

Quick-check valve-timing methods

On single-camshaft engines, it is only necessary to check the opening point of one inlet valve but on twin-camshaft engines, the exhaust opening point must also be checked. If it is suspected that the incorrect camshaft is fitted to an engine, the complete sequence of valve operation must be checked. Unfortunately many manufacturers change camshafts during a production run and in case of doubt it is safest to write to the Technical Service Department of the company, giving engine and chassis number. The information is seldom found in the workshop manual.

Some manufacturers give special instructions in their workshop manual for checking valve timing quickly without the necessity of a chart or of removing the timing case. The basis of the method with single camshaft engines being the checking of the opening point of No. 1 inlet valve using a special valve clearance which has been determined so as to allow the inlet valve to open at t d c or some other well marked and specified point. It should be noted that the clearance for carrying out the check may be quite distinct from the valve-timing clearance used for checking the valve timing as per timing chart.

The Austin Westminster is an example; the procedure for a quick timing check is to set No. 1 inlet valve clearance to 0·030 in. (0·762 mm) with the engine cold and then to rotate the crankshaft until the valve is about to open. The t d c mark on the crankshaft fan-belt pulley should then be in alignment with the arrow on the timing case, i.e. No. 1 inlet valve should be about to open at t d c with No. 6 piston on its compression stroke. A valve timing diagram is also provided in the workshop manual for use with a 'timing' valve clearance of 0·021 in. (see fig. 7:42).

On the MGB, No. 1 inlet valve should be set to 0·055 in. (1·4 mm) and the engine turned until the valve is about to open. The notch on the flange of the crankshaft pulley should then be in alignment with the longest of the three pointers on the timing cover, i.e. the valve about to open at t d c and No. 4 piston on its compression stroke.

On the B.M.C. 1100, No. 1 inlet valve clearance must be set to 0·021 in. for timing and the crankshaft turned until the valve is about to open. At this point the 5° b t d c mark on the flywheel should be in alignment with the pointer in the clutch-cover inspection aperture.

To check the timing on the B.M.C. 1800 transverse mounted engine, fitted with the first-type camshaft which should be used with a running valve clearance of 0·018 in., set No. 1 inlet valve to 0·020 in. and turn the crankshaft until the valve is about to open. The indication groove on the crankshaft pulley should then be in alignment with the 5° pointer on the timing cover i.e. No. 1 inlet valve should be about to open at 5° b t d c with No. 4 piston at 5° b t d c on its compression stroke. On late 1800 transverse engine cars, the timing clearance is 0·021 in. with the inlet valve opening at t d c.

On most of the Rootes models, the crankshaft pulley is marked at 4° intervals and a valve clearance for timing purposes is given in the workshop manual with the appropriate valve-opening data according to the camshaft fitted to the particular model.

On Rover engines, the exhaust valve peak is used for checking valve timing. The E P mark on the flywheel, visible through the aperture on the right-hand side of the clutch housing (on the 60, 75, 90 and 105 engines) must be in alignment with the pointer when No. 1 exhaust valve is at its fully open position with the clearance set to 0·012 in.

Vauxhall Motors usually suggest the inlet maximum opening point for checking the valve timing. On the Victor and VX 4/90, the maximum opening point of No. 1 inlet valve is 109° after t d c and can be checked in the following manner: remove the clutch-housing bottom cover and mark the 109° point which is the 37th flywheel ring gear tooth from t d c (U/C marked on the flywheel) counting anticlockwise as from the front of the engine. Detach the timing aperture plug and then turn the engine slowly in a clockwise direction until No. 1 inlet valve is fully open (it is necessary to mount a dial gauge so that the plunger contacts the spring cap in order to ascertain the fully open point) when the 109° mark on the flywheel should be in alignment with the notch in the timing aperture.

On the PB Velox and Cresta, the inlet-valve maximum opening point for the high-compression engine is 113° after t d c (38th tooth) and 108° after t d c (between the 36th and 37th tooth) for the low-compression engine.

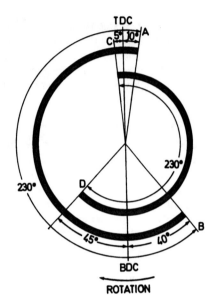

Fig. 7 : 42 The Austin A110, 6/110, and Princess 3-litre (Mk II) valve timing diagram. Exhaust closes at (A) and opens at (B). Inlet opens at (C) and closes at (D) with a valve rocker clearance of 0·021 in. (0·53 mm)

With any method of checking valve timing it is essential that the rocker arms are not pitted otherwise the valve clearance will be wrong, leading to a completely false timing check.

On some overhead-camshaft engines, the amount which the inlet and exhaust valve is open, at certain crankshaft degrees, is used for checking the valve timing and the appropriate workshop manual should be consulted.

Running-in the engine

When running-in the engine it is advisable gradually to increase the running-in speed but during the first 500 miles:

Do not excessively load the car with passengers.
Do not rev. excessively in low gear.
For the first 200 miles use only light throttle.
Check the oil level regularly, for oil consumption may be rather high for the first few hundred miles, but should gradually improve as the engine runs in.

Satisfactory running-in speeds are:

First 200	Maximum in top gear	35 mile/h
200/500	Maximum in top gear	40 mile/h
500/1000	Maximum in top gear	50 mile/h

On reaching 500 miles, retighten the cylinder head nuts, reset tappets and carry out any adjustments necessary to ensure smooth running.

CHAPTER 8

Steering, Suspension and Tyres

In the event of steering trouble, inspect the wheels, tyres, steering, front suspension and all associated parts. If this fails to reveal the fault, it is advisable to check the steering geometry. The castor and camber angles (see page 131) are best measured by the local distributors for the make of vehicle in question, for they will have the necessary equipment to do steering checks quickly and inexpensively. Setting the steering track is comparatively easy and should be carried out first.

Checking the track

A tyre worn on one side may be the first indication of incorrect track adjustment, but it is advisable to replace it with a more suitable tyre before adjustment is carried out, and to inflate all tyres to the correct pressures.

The ideal tool for measuring the track is an optical alignment gauge but if this is not available some method of measuring must be devised. On front-wheel-drive vehicles, the track should toe *out* unless radial-ply tyres are fitted when the track should be parallel, the measurements being taken between the tyre walls at wheel centre level. The track of rear-wheel-drive vehicles should toe *in* (see also Appendix 4).

The measurements should be taken as shown in fig. 8:1 and the car, in the case of rear-wheel-drive vehicles, should be pushed forward until the wheels have completed one half-

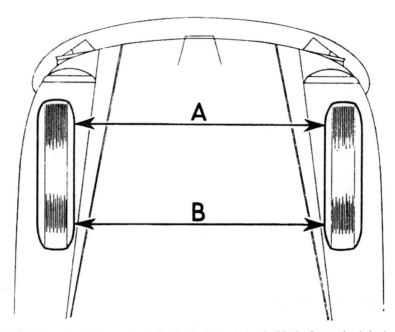

Fig. 8:1   The front wheel aligment must be checked at axle level with the front wheels in the straight ahead position. On rear wheel drive cars dimension B is greater than dimension A. The converse with front wheel drive cars (see Appendix 4)

turn and the measurements taken again. For front-wheel-drive vehicles, it is advisable to push the car backwards to take the second reading. The average of the two measurements must be calculated and adjustments made accordingly. With steering racks, or where there is an adjustable drag link on each side of the car, the adjustment may be carried out by slackening the locknut, or the clamps, and then turning each rod the same amount. On cars with a steering rack, make sure to release the clips which secure the rubber gaiters to the tie rods and, after the final adjustment, do not forget to retighten them as well as the track rod or tie rod locknuts (or clamps). On some steering assemblies, adjustment of the track is carried out by slackening the locknuts on a single track rod and turning the rod until the correct measurement is obtained.

Whatever method of adjusting the track, it will probably be necessary to check the alignment several times and make two or three adjustments until the track is absolutely correct. After each adjustment, give the steering wheel half a turn in each direction before rechecking the alignment.

On cars where tie rods have been adjusted to unequal lengths, this should be rectified by resetting them as near as possible to the same length. This may entail realigning the steering wheel by removing it from the inner column and replacing the wheel on a more suitable spline.

Whenever readjustment of the track becomes necessary, a thorough examination should be carried out to ascertain the reason why the adjustment is necessary.

When a vehicle has sustained accidental damage, examine it for bent and loose steering rods and arms.

On many cars, particularly B.M.C. Minis and 1100s, the bottom link pins become bent owing to impact with curbs, etc. Damaged tie rods, too, affect the track. Bent stub axles, worn king pins and bushes and worn track-rod ends are other common faults.

On the Vauxhall Victor the tie-rod ends, which connect the steering central tie rod to the steering lever and idle lever, are rubber bushed. On early models oil tended to get onto the bushes and ruin them and a shield-type modification was introduced. If undue steering-wheel play develops, the bushes should be inspected and if worn or soaked in oil they should be renewed and the track reset.

Steering wander

On vehicles with rubber-mounted sub-frames, inspect the mountings. Bent steering rods, seized steering idlers, king pins partially seized to the swivel bushes, or king pins with a great deal of wear, are other common faults. An incorrectly adjusted steering box, or a steering rack assembled too tightly, loose wheel bearings, or a low suspension on one side, are things to look for.

Castor angle

Insufficient self-centring action of the steering wheel is usually due to an incorrect castor angle. Usually the car will be found to pull to the side with the least castor angle. Examine the top and bottom suspension radius links for damage.

Wheel tramp and shimmy

This is due to the tendency of the wheels, when subjected to bumps, to turn right and left as they move up and down, and produces a feeling of jerkiness on the steering wheel. Excessive play in the steering rack, steering box or linkage, a badly fitted tyre, wheels out of balance, or worn-out shockabsorbers, are the principal items to look for.

Wheel-wobble

Wheel-wobble may be caused by almost any steering fault. Often the wobble occurs only within a certain speed range and the first step is to make certain that the wheel balance is correct. If this fails to cure the wobble, a complete investigation of the steering should be made. If the cause cannot be found check the castor and camber angles—a radius arm, steering arm, or tie rods may be slightly bent.

Some other common steering faults are: heavy steering due to lack of lubricant (especially steering idlers), steering wander owing to the front and rear suspension being out of alignment, a buckled wheel, uneven tyre pressures, a wrong-sized tyre, loose rear-axle bolts.

Minimizing tyre wear

To obtain the maximum mileage from tyres it is advisable to change the tyre position from the front of the car to the rear every 6000 miles. It should be noted, however, that this must not be done if the car is fitted with radial ply tyres on the rear only.

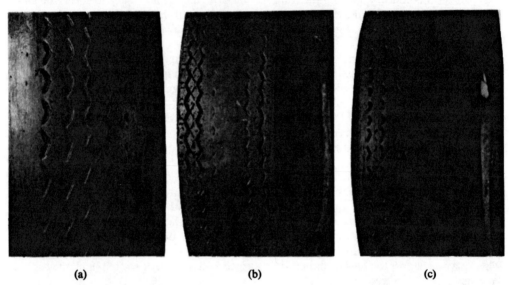

(a)   (b)   (c)

Fig. 8:2 (a) Tyre wear resulting from under-inflation. Under-inflation causes fast wear, excessive heating, and can bring about tyre failure through blow-out. There is an approximate average loss of 13% tread mileage for every 10% reduction in inflation pressure below recommended pressure figures. (b) Tyre wear resulting from over-inflation causes the fabric to be easily damaged, and seriously shortens tyre life by rapidly wearing the centre of the tread. (c) Wear due to to excessive front wheel camber or an incorrectly adjusted track. Excessive camber can be the result of impact damage to the suspension unit. Although most cars have wheels with suitable rims for radial-ply tyres, owners of older models should note that tubeless radials should not be fitted unless the wheels possess 'safety' rims to retain the tyres

Regular adjustment of tyre pressures and inspection of the tyres, to ensure that wear is taking place evenly, is important. Should a tyre wear more on one side of the tread than the other (see fig. 8:2), the cause must be investigated and the fault corrected. (See also page 317.)

If one tyre has constantly to be re-inflated more often than the others it may be due to a:

Distorted wheel (tubeless only).
'Gashed' tyre possibly due to impact with kerb (damage is not always externally visible).
Faulty valve core and lack of a valve cap (secondary seal).
'Lifted' patch on the tube.
Unsatisfactory puncture repair (synthetic tubes do not 'take' patches well).
'Chafed' tube.
Perished tube.
Very slow puncture.

If a slight bulge is observed on a tyre, the tyre should be renewed immediately.

Fig. 8:3   When lifting the bead over the rim, if a tube is fitted, care must be taken
to ensure the tube is not pinched

Removing and refitting tyres

When removing a tyre, always take out the valve core to fully deflate the tyre. Separate each tyre bead from the rim flange and push both beads into the base of the rim.

'Breaking the bead' can be frustrating without a proper tool and if it cannot be accomplished by tyre levers, 'run over' the tyre with the car, making sure to place the tyre so that the wheel (from which the tyre is being removed) does not damage the tyre on the car.

When the bead joint is 'broken', lubricate the beads with a soft soap and insert two tyre levers, 6 in. apart, close to the valve (see fig. 8:3) (if the tyre is fitted with a tube, use the 'hooked' end of the lever so that the lever will not pinch the tube), make sure that the beads on the opposite side are in the base of the rim (see fig. 8:4), and lever the cover edge over the rim of the wheel, working around the tyre until one side is completely free.

Fig. 8:4   Pushing the tyre bead into the well. Before the tyre can be lifted over
the rim it is necessary to push the tyre bead into the well. Note the position of the
valve

Tyres can be removed from the narrow flange side of the wheel only. This is particularly applicable to the B.M.C. Mini, 1100 and 1300.

If a tubed tyre is being changed, when the tube is removed, stand the wheel upright and lever the tyre from the wheel by inserting a lever between the bead and the flange. A few careful bangs with a rubber-headed mallet on the bead, and the tyre and wheel will be separated. If a mallet is not available, the tyre will have to be levered from the wheel.

When fitting a tubeless tyre, fit a new valve assembly as well, for the rubber body of the valve becomes perished and an old valve assembly may not remain reliably airtight for as long as the life of the new tyre.

Lubricate the tyre beads with soft soap, lay the wheel flat, place the tyre in position with the balancing spots aligned with the valve, kneel on one side of the tyre and lever the bead edge of the inner side of the tyre over the wheel. If a tube is being fitted, place it in the tyre and slightly inflate.

Complete the refitment of the tyre by levering the outer bead over the edge of the wheel working from the side opposite to the valve. Take care not to pinch the tube.

When a tubeless tyre has been fitted, it will usually be necessary to use a high-pressure air line to inflate it. If air escapes through the sealing beads before the tyre inflates, try inflating the tyre without a valve core inserted. If the tyre now seals satisfactorily, remove the air line, insert the valve and re-inflate to the correct pressure and fit the valve cap.

If the tyre refuses to inflate, a tourniquet must be applied around the tyre so as to squeeze the bead onto the wheel flange.

Tubeless punctures

Sometimes, due to poor sealing at the beads (probably due to a faulty wheel rim), a tube has to be fitted to a tubeless tyre and, provided the inside of the cover is completely smooth, it is quite in order. If, however, a puncture has previously been repaired using a plug which protrudes inside the tyre, the plug must be removed otherwise it will chafe the tube. Punctures in tubeless tyres should be repaired with rubber self-vulcanizing 'mushroom-headed' plugs which are usually inserted from the outer side of the tyre tread by a special pneumatic gun (unless the tyre is removed from the wheel). Non-vulcanized plugs which protrude inside the cover are best regarded as a temporary form of repair only and should never be used in the side walls. Never take risks with tyre repairs, consult a competent specialist.

Cross ply tyres

In a cross ply tyre, the cord casing is constructed with numbers of plies laid at an acute angle to the centre line of the tread, first from one side then from the other, so that in the resulting cover the plies criss-cross each other in both wall and tread area.

Radial ply tyres

In the radial ply tyre, the plies cross the centre line of the tread at a right angle, and thus the cords of the tyre walls are radiating from the tyre beads. Immediately below the tread area, a number of plies are laid in such a way that the tread is braced.

The advantage of radial ply tyres is that the tread maintains a stable contact with the road. Steering is light and more precise, slip angle is reduced resulting in exceptional road-holding characteristics and a high degree of stability.

Having less rolling resistance, radials also build up less heat, fuel consumption is reduced, and more mileage is obtained from the tyres due to the firm bracing of the tread on the road without distortion or tread 'shuffle'.

Radial ply tyres, however, should be fitted in sets of four, plus a spare, although in some circumstances it is permissible to fit cross ply on the front wheels. Tyres of different construction must in no circumstances be used on the same axle. The diagonal positional changing of tyres, to even out wear, must only be undertaken if the same type of tyre is fitted all round.

Adjusting the steering box

If there is no play in the steering linkage and the steering box is not loose (see fig. 8:5) but if more than 1 in. free play develops in the steering, as measured at the rim of the steering wheel when the wheels are in the straight-ahead position, the steering box may require adjusting. To do this, raise the car at the front, support it firmly and disconnect the steering box from the rest of the steering mechanism by disconnecting the drag link from the steering-box drop arm.

Adjustment, to remove the free play, must be carried out with the steering wheel in the straight ahead position (see also fig 8:6), taking care to ensure that a tight spot has not developed when the free play has been removed. It should be checked by rotating the steering wheel throughout its complete range.

When the adjustment has been carried out, refit the drag link and test the car on the road.

Lateral play on a steering wheel, see fig. 8:8.

Steering racks (see fig. 8:9).

On some steering racks, the felt bush (fitted in the housing at the opposite end to the pinion) becomes severely worn and causes a 'knock' which can be overcome by removing the steering rack, overhauling it, and fitting a modification consisting of a polyvon bush (see figs. 8:10 and 8:11).

Steering racks prior to 1955 did not usually incorporate a felt bush and were fitted with dampers (see fig. 8:19). On this type of rack, the housing becomes worn and a 'knock' from the rack in this event may entail a complete new rack housing.

Overhauling steering racks

Take the rack from the car, hold it securely in a vice, and release the tie-rod-end locknuts and remove the tie-rod ends. Place a container under the rubber gaiters, release the gaiter clips and allow any oil to drain. Some steering racks are grease packed.

If the steering rack has been removed from the car with a coupling flange attached (as would be the case with the B.M.C. 1800—see fig. 8:11) mark the pinion coupling, the pinion housing and the pinion, with the rack in the central position (to ensure correct reassembly) and take off the coupling flange.

Next, take out the bolts which retain the damper cover plate and packing shims. Remove the damper cover, extract the spring, or the disc-type spring washers, and the support yoke.

Remove the pinion end cover, joint washer and shims. Take out the outer thrust ring, ball cage, inner bearing ring, and withdraw the pinion. The top bearing must be removed when the rack has been withdrawn from the housing.

If the pinion oil-seal requires renewing, it may be removed by levering with a screwdriver.

On some cars (the early Triumph Herald for one), the pinion is retained by a nut which must be removed, together with the shims, bush and thrust washer before the pinion can be withdrawn. On late 1200 models and S type Heralds, the pinion is retained by a circlip (see fig. 8:12).

The pinion nut on the 948cc Herald is fitted with an oil-retaining 'O' ring which must be renewed when the pinion assembly is refitted.

To continue stripping the rack, unlock and dismantle the tie-rod/rack ball-joints. Each tie-rod ball-joint is locked to a slot in the end of the rack by means of indentations in a locknut. The best procedure is to remove the tie-rod from the pinion end of the rack, and then release the tie-rod at the opposite end when the rack has been removed from its housing.

Punch the indentations in the locknut clear of the slot in the ball housing. Slacken the locknut and unscrew the ball-housing to release the tie-rod ball-seat and the seat tension spring. Pull the rack from the housing, withdrawing it from the pinion end. Dismantle the other tie-rod ball-seat.

On early types of rack fitted with a felt bush, undo the bush-securing screw, prise up the

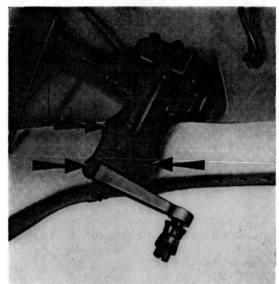

Fig. 8:5 If undue steering wheel play develops, always check the steering box/frame
bolts for tightness. On certain cars loosened
bolts are a notable source of trouble

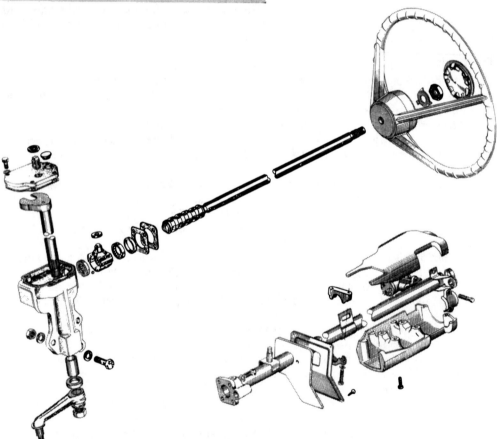

Fig. 8:6 The Cortina steering box provides for two adjustments (1) The rocker shaft by means of a
stud and locknut on the steering box top cover. (2) Steering shaft adjustment by means of shims
between the rear face of the box and steering column flange (see also fig. 8:7). Rocker shaft adjustment can be carried out with the steering box in the car

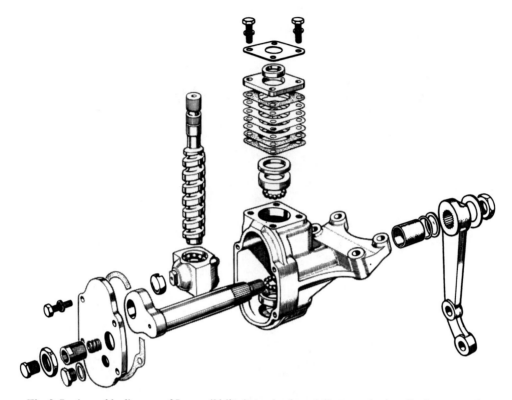

Fig. 8:7 Assembly diagram of Jaguar (2½ litre) steering box. Adjustment is virtually the same as for the Cortina steering box (see fig. 8:6)

Fig. 8:8 Lateral play on a steering wheel may be due to wear of the steering column top bush. On some makes of cars prior to 1965 the bushes were made of felt. If wear develops, the bushes should be replaced with a plastic type

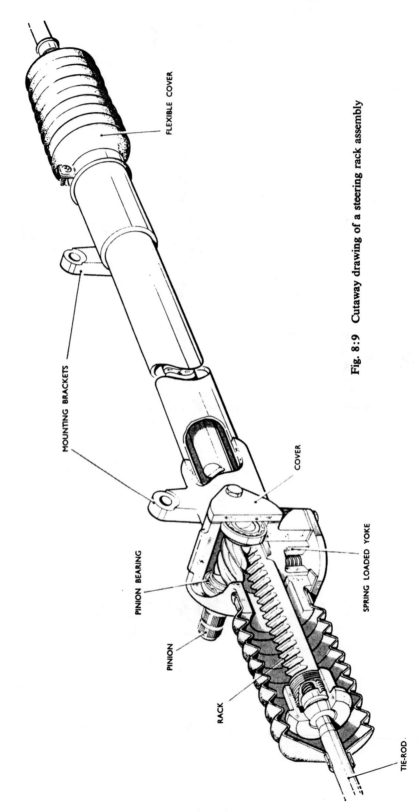

FLEXIBLE COVER

MOUNTING BRACKETS

COVER

PINION BEARING

SPRING LOADED YOKE

PINION

RACK

TIE-ROD.

Fig. 8:9   Cutaway drawing of a steering rack assembly

felt and extract it and remove the bush housing. If the rack is fitted with a polyvon bush (see fig. 8:10), the bush and housing usually comes out complete.

The parts must be thoroughly cleaned and inspected. If the teeth on the rack, or pinion, show signs of roughness, or hollows in the teeth more than 0·002 in. deep, the rack and the pinion are best renewed.

Carefully scrutinize the tie-rod ball-housing, ball-seat, the pinion bearings and races, and if they are worn, renew them.

If the oil-retaining rubber gaiters are perished or damaged, new ones must be fitted. In many cases, unless the vehicle has a high mileage to its credit, apart from readjustment and the fitting of a new polyvon bush, the rack will probably require little in the way of expenditure on parts but, of course, it is essential that parts that are worn are renewed.

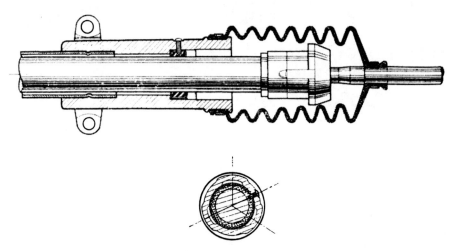

Fig. 8:10   A section through a steering rack and tie rod

Below. The correct position of the flats on the polyvon rack bush in relation to the securing screw and oilways

Reassembling the steering rack

If a felt bush has been removed from the rack housing, it must be replaced by a polyvon bush and housing, with a spacer.

Insert the spacer and then the bush and its housing so that the hole in the rack housing, which retained the felt bush, comes between the flats of the new bush (see fig. 8:10) and then drill a $\frac{7}{64}$ in. hole through the existing securing hole into the new polyvon bush housing and secure with the screw. Apply a sealing compound under the head of the screw before tightening.

Refit the pinion top bearing. Insert the rack into the housing and refit the pinion. Place the rack in a central position. Ensure that the marks on the pinion and housing are aligned and fit the lower pinion bearing.

Replace the pinion tail-bearing cover without shims and use a feeler gauge to measure the clearance between the cover and the housing (see fig. 8:13). Do not overtighten the cover screws or an incorrect measurement may be obtained. Remove the cover and refit it with shims to the value of the feeler gauge measurement minus 0·001–0·003 in. This produces the correct bearing pre-loading. Make sure to treat the joint faces with a sealing compound otherwise lubricant may leak from the joint.

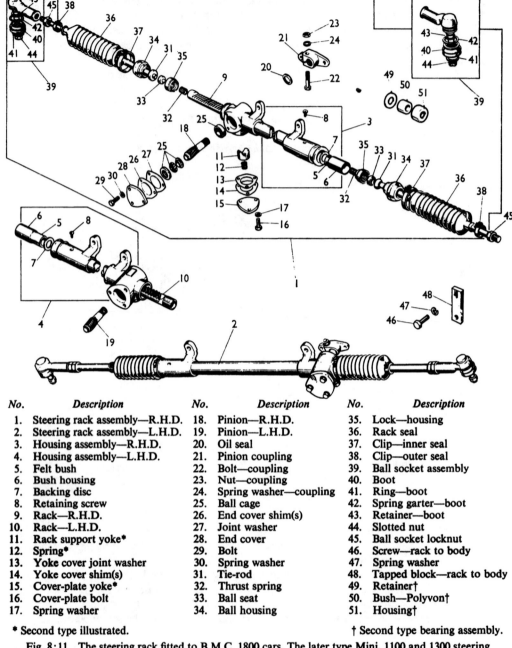

No.	Description	No.	Description	No.	Description
1.	Steering rack assembly—R.H.D.	18.	Pinion—R.H.D.	35.	Lock—housing
2.	Steering rack assembly—L.H.D.	19.	Pinion—L.H.D.	36.	Rack seal
3.	Housing assembly—R.H.D.	20.	Oil seal	37.	Clip—inner seal
4.	Housing assembly—L.H.D.	21.	Pinion coupling	38.	Clip—outer seal
5.	Felt bush	22.	Bolt—coupling	39.	Ball socket assembly
6.	Bush housing	23.	Nut—coupling	40.	Boot
7.	Backing disc	24.	Spring washer—coupling	41.	Ring—boot
8.	Retaining screw	25.	Ball cage	42.	Spring garter—boot
9.	Rack—R.H.D.	26.	End cover shim(s)	43.	Retainer—boot
10.	Rack—L.H.D.	27.	Joint washer	44.	Slotted nut
11.	Rack support yoke*	28.	End cover	45.	Ball socket locknut
12.	Spring*	29.	Bolt	46.	Screw—rack to body
13.	Yoke cover joint washer	30.	Spring washer	47.	Spring washer
14.	Yoke cover shim(s)	31.	Tie-rod	48.	Tapped block—rack to body
15.	Cover-plate yoke*	32.	Thrust spring	49.	Retainer†
16.	Cover-plate bolt	33.	Ball seat	50.	Bush—Polyvon†
17.	Spring washer	34.	Ball housing	51.	Housing†

* Second type illustrated.   † Second type bearing assembly.

Fig. 8:11 The steering rack fitted to B.M.C. 1800 cars. The later type Mini, 1100 and 1300 steering racks are virtually the same although of course of different dimensions

Triumph cars

On Triumph cars, when the thrust washer and the bush have been refitted, the pinion must be adjusted to have end-float. Fit the pinion nut with an excessive amount of shims and screw the nut down until the end-float is eliminated. With feeler gauges, measure the clearance between the pinion-nut hexagon flange and the housing. Remove the pinion nut and take out shims equal in thickness to the clearance that was measured plus an additional 0·006 in.

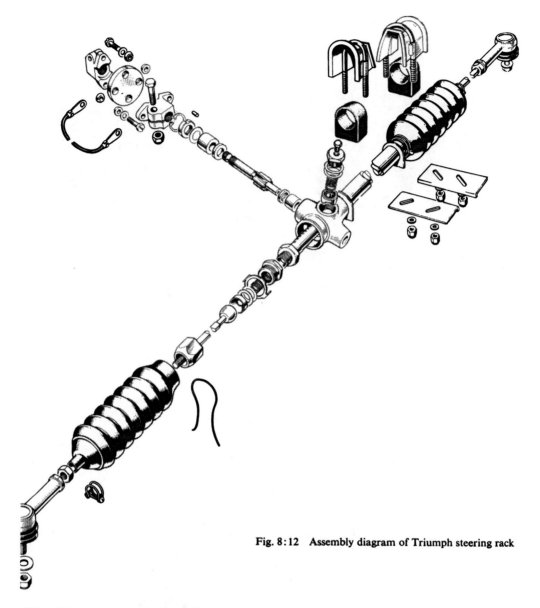

Fig. 8:12   Assembly diagram of Triumph steering rack

This will give the required end-float which should be the minimum required to allow free rotation of the pinion. Refit the pinion nut.

On 1200 and S model Triumphs, circlips are fitted to retain the pinion; end-float requirement was changed to 0·010 in. ('Measuring the end-float', see fig. 8:14.)

When reassembling the pinion, ensure that the distance tube and thrust ring are placed in the end of the pinion shaft. When the pinion is in position, follow with the thrust ring, distance tube, shims as required, and the retainer with the dowel correctly positioned and, finally, the circlip.

B.M.C. cars

To adjust the inner tie-rod ball joints on B.M.C. cars (see fig. 8:15 and fig. 8:16) screw the ball-housing lock ring (a new locking ring should be used) onto the rack-end to the limits of

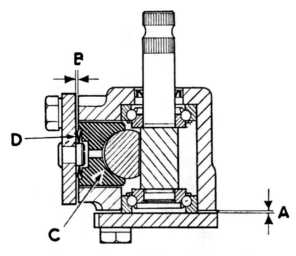

Fig. 8:13 A section through the steering pinion and rack damper. Adjusting pinion bearing preload: Take a feeler gauge measurement at A and fit the pinion end cover with shims to the value of the measurement minus 0·001 to 0·003 in. (0·025 to 0·076 mm) prior to fitting the damper yoke C. Subsequently measure the gap B and fit shims. (C) Damper yoke. (D) Dished spring washers

the thread. Refit the spring, seat, tie-rod and the ball housing and tighten firmly. Advance the locking ring to meet the ball housing, slacken the ball housing ⅛ of a turn and tighten the locking nut.

Test the pull required to move the tie rod by connecting a spring balance to the end of the rod. Three to five pounds should be sufficient to move the rod, if it is too tight this must be rectified.

The ball joint at the other end of the rack must now be assembled and tested in the same manner.

Next, pinch the lip, at the end of the locking nut, into the slot in the ball housing and then pinch the lip at the opposite end of the nut into the slot in the rack. (*For assembly of Triumph inner tie-rod ball joints see fig. 8:12 and for Vauxhall Viva fig. 8:17.*)

Steering racks fitted with a coil spring in the damper assembly (see figs. 8:11 and 8:17) should be adjusted by fitting the yoke and spring, and lightly screwing up the end cover. Centralize the rack and adjust the cover so that it is just possible to rotate the pinion by pushing the rack. Using a feeler gauge, take the measurement between the end cover and its seating. Remove the cover, apply a sealing compound and refit with shims to the value of the feeler gauge measurement.

If the rack has a damper fitted with disc-type springs (see fig. 8:16) refit the yoke, damper cover, and the disc-type springs, but leave out the shims. With the rack in a straight-

Fig. 8:14 Measuring the pinion end float on a Herald steering rack. Shims are available in 0·004 thickness

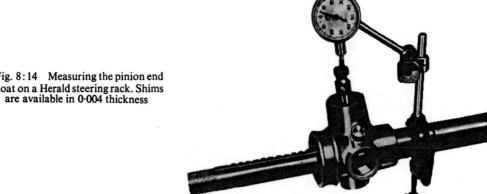

ahead position tighten the securing bolts, until it is only just possible to rotate the pinion shaft by pushing on the rack. Take a feeler gauge measurement between the damper cover plate and its seating, then remove the cover, treat the face with a sealing compound and refit, with shims to the value of the gauge measurement plus 0·003 in.

The torque load required to move the pinion after assembly must not exceed 25 lb/in., see also fig. 8:18.

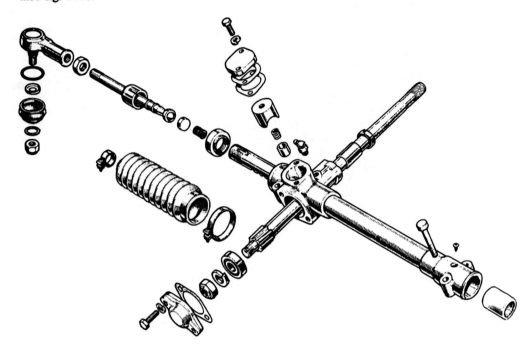

Fig. 8:15   The MGB steering

Fit a new pinion oil-seal and then refit the rubber gaiters to the housing and tie-rods, but on steering racks fitted to B.M.C. cars, before securing the clips at the pinion end, stand the assembly upright and pour into the gaiter approximately ⅓ pint of extreme pressure SAE 140 oil. Refit and tighten the clip. Triumph steering racks should not be charged with oil but should be packed with Retinax A grease.

If a flange was fitted to the pinion, refit it, according to the previous marks, before fitting the rack unit to the car.

On some early steering racks such as fitted to the MGA (see fig. 8:19) the ball tie-rods are adjustable by shims in the following manner: insert the ball end of the tie-rod into the female housing and assemble the ball seat, male seat housing and shims. Tighten the two housings together. The ball must be a reasonably tight fit without play. If too tight, insert another shim between the housings, varying the amount until the desired fit is obtained. (Shims are available in thicknesses of 0·003 and 0·005 in.)

When the ball joint is suitably adjusted, remove the assembly and fit a lockwasher. Then replace and tighten the ball housing into the rack, locking it in position in three places by bending the flange of the lockwasher.

To adjust the rack damper (the one by the pinion—see figs. 8:19 and 8:20) the plunger must be replaced in the cap. The cap should then be screwed in position without the spring, or shims, until it is just possible to rotate the pinion by drawing the rack through its housing.

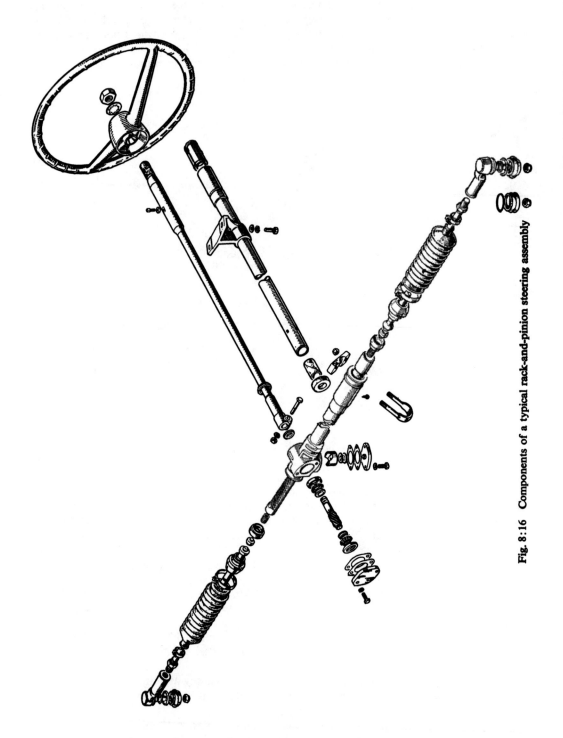

Fig. 8:16 Components of a typical rack-and-pinion steering assembly

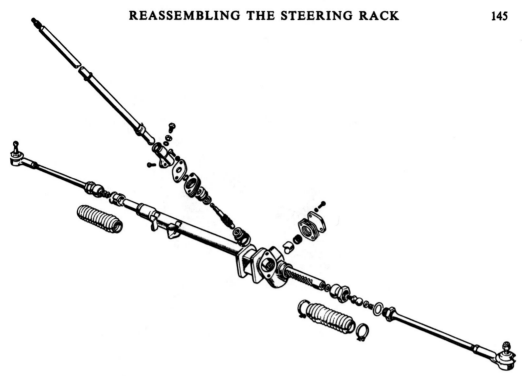

Fig. 8:17   General assembly of Vauxhall Viva rack and pinion

Use a feeler gauge to measure the clearance between the hexagon of the plunger cap and its seating. Add 0·004 in. to arrive at the correct thickness of shims needed and refit the cap with the shims in place, applying a sealing compound around the threads as a safeguard against lubricant leakage.

Some early racks were fitted with a secondary damper (at the opposite end to the pinion, see fig. 8:19). The dampers have been replaced on later racks by felt bushes and later by polyvon bushes or another type and are not usually adjustable, refitting merely being a matter of reassembling the plunger, spring washer and cap.

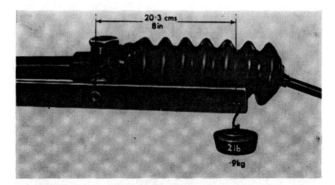

Fig. 8:18   When a Triumph steering rack is adjusted a load of 2 lb is required at a radius of 7–9 in. to rotate the pinion shaft. If necessary, re-adjust the unit by adding or subtracting shims from beneath the cap nut. The torque load required to move the pinion must not exceed 25 lb/in.

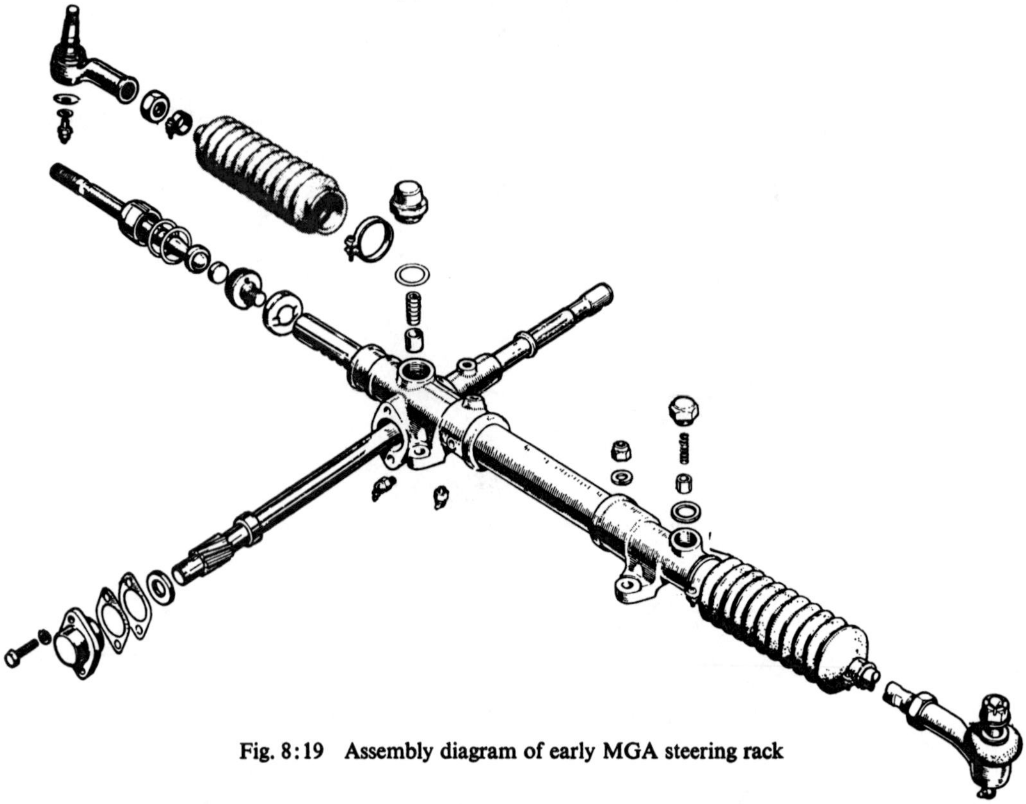

Fig. 8:19   Assembly diagram of early MGA steering rack

King pins and bushes

If new bushes have to be fitted to the stub axles, the bushes may require reaming or broaching to the correct size to accommodate the new king pin. When the stub axles have been removed from the car, it is best to take them to the local agent for the vehicle. The garage will have the necessary equipment to do the job; one cannot manage without it.

Many cars are fitted with 'screw' type king pin arrangements similar to the type depicted in fig. 8:21. Some types of steering have the bottom fulcrum pins and sleeves threaded as in the Austin Westminster (see fig. 8:24), while others have a plain bolt which fits through the lower trunnion and wishbone with a distance tube to form the trunnion bearing as in figs. 8:21 and 8:25.

To remove king pins, lower and top trunnions, take off the hubs, brake shoes, and brake

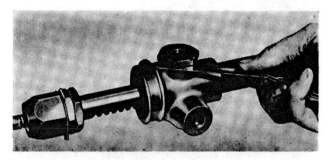

Fig. 8:20   Determining the shim thickness required under
the cap nut on a Herald steering rack

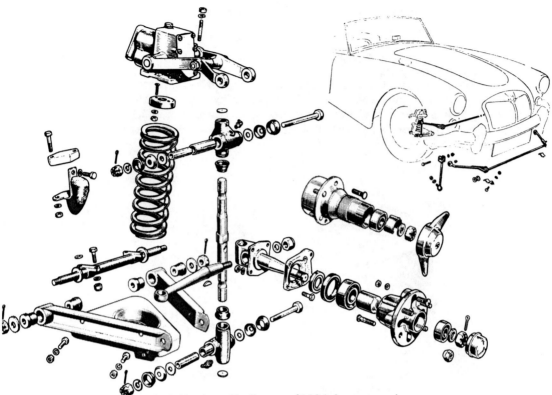

Fig. 8:21   Assembly diagram of MGA front suspension

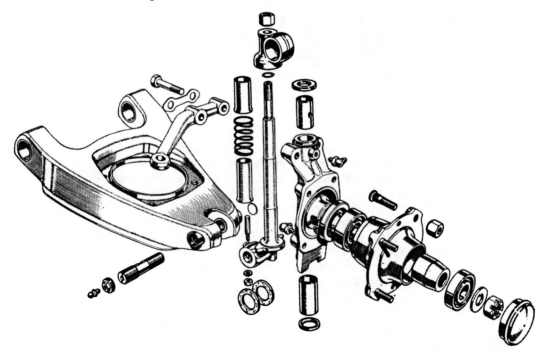

Fig. 8:22   Assembly diagram of front suspension components on the Austin-Healey Sprite, MG
Midget and A40

back-plate. If disc brakes are fitted, remove the calipers, hub disc unit and stoneguard. Disconnect the steering arm.

Make sure the car is firmly on stands, or blocks, placed under the main frame, and carefully place the jack under the suspension unit, so that it cannot slip. Raise the jack and take the full load off the spring.

On the Triumph Herald (see fig. 8:26) the front road spring and damper may be released by undoing the three nuts which secure the upper spring pan to the chassis sub-frame (see fig. 8:27), lowering the jack slightly and removing the bottom end attachment eye-retaining nut and bolt. The spring and shock absorber can then be withdrawn, see fig. 8:28.

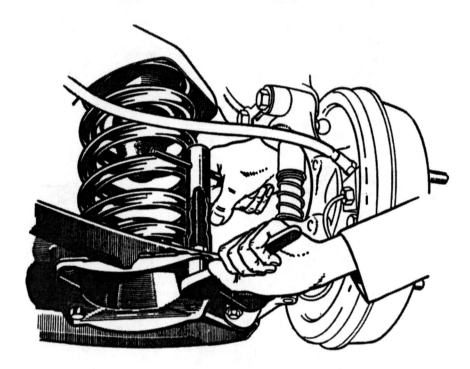

Fig. 8:23   Using a pair of slave bolts to remove or replace a front coil spring

On B.M.C. suspensions, as depicted in figs. 8:21, 24 and 25, disconnect the shockabsorber from the top link by slackening the nut and bolt in the centre of the shockabsorber arm and then slacken the clamp bolt where the arm is affixed to the shockabsorber. Next, from the top trunnion bearing bolt, remove the split pin, undo the nut and, easing the shockabsorber arm outwards (on no account remove the shockabsorber arm completely), tap out the top link bolt using a suitably sized drift.

On torsion bar suspension as fitted to the Riley 1·5, Wolseley 1500 and Morris Minor, remove the large nut which retains the trunnion pin to the shockabsorber, the pin can be slipped from the shockabsorber later in the dismantling procedure.

The next step is to release the king pin from the bottom wishbone; on some suspensions the front wishbone arm can be removed, which will allow the complete assembly to be taken off. On other types, such as the Morris Minor, MGB and Riley 1·5, the bottom pin can be tapped from the bottom bearing with a drift and hammer, so releasing the complete unit. Sometimes the bolts holding the wishbone arms to the spring pan may have to be slackened and on some models, the Morris Minor for instance, the tie-rod bolt must be taken out and the wishbone arm clamp bolt slackened.

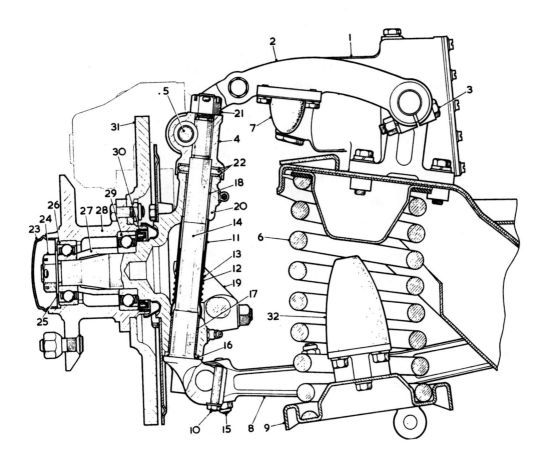

1. Hydraulic damper
2. Lever arms—damper
3. Clamp bolt and nut—lever arm
4. Upper trunnion link
5. Fulcrum pin—upper trunnion link
6. Coil spring
7. Rebound rubber
8. Lower arm
9. Spring plate—lower
10. Cotter pin and nut—lower fulcrum pin bush
11. Dust cover (upper)
12. Dust cover (lower)
13. Spring—dust cover
14. Swivel pin
15. Cotter pin and nut—lower fulcrum pin
16. Cork seal
17. Lower bush—swivel pin
18. Upper bush—swivel pin
19. Steering-arm
20. Swivel axle
21. Nut for swivel pin
22. Thrust washer assembly
23. Grease-retaining cap
24. Nut for swivel axle
25. Washer for swivel axle nut
26. Outer bearing—front hub
27. Distance piece—bearing
28. Front hub
29. Inner bearing—front hub
30. Oil seal
31. Brake disc.
32. Bump rubber

Fig. 8:24 Section through an Austin Westminster front suspension unit

On some models, Austin A50, Westminster, A90, Morris Oxford, etc. the bottom link pins are threaded and fit into threaded sleeves which are carried in the wishbone arms and held in position by half-moon cotter pins. The threaded fulcrum pin, however, is positioned in the bottom of the king pin by a cycle-type cotter which must be removed before the pin can be taken out. (The cotter should not be removed until the fulcrum sleeves have been removed.) To release the sleeves and allow them to be screwed out, the nuts on the half-moon cotters should be slackened a few threads and the cotter given a light blow with a hammer. (The pins cannot be removed until the sleeves have been removed from the wishbone.) When the sleeves are being screwed out, give the arm a few smart blows with a copper mallet, otherwise the sleeves may jam in the arms, with the result that they push the wishbone arms outwards, instead of the sleeves coming from the arms.

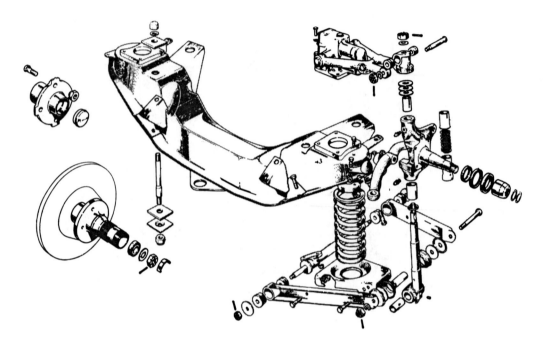

Fig. 8:25   Assembly diagram of MGB front suspension

With screw-type king pins, and threaded top trunnions, it is essential to observe how many turns the trunnion screws on to the king pin, for if the trunnions are replaced so that they do not operate on precisely the same part of the king pins the car may lean to one side.

On cars equipped with torsion-bar springing, it may be possible trim the height a little, but unless the links are fitted correctly, the trunnions will have to be disconnected and adjusted accordingly. (Normally, when fitting a new trunnion, it should be screwed down as far as possible and then back as near one complete turn as possible).

Austin A40 king-pin bottom-trunnion pins screw directly into the spring-retaining pan. They are often difficult to remove and it is essential to spray them with a rust solvent, a few days before the job is started.

Ford Cortina, B.M.C. 1800 and Vauxhall Viva suspensions are illustrated in figs. 8:29, 30 and 31.

Suspension ball joints

On B.M.C. Minis and 1100s, the hubs swivel by means of ball joints (see chart 8:1). Wear is compensated by removal of shims, or by renewal of the ball-retaining cups, pins and springs. B.M.C. supply new ball-pin cups and pins as a complete assembly. They are easy to fit and the job can be carried out on hydrolastic- and cone-type-suspension Minis without special equipment.

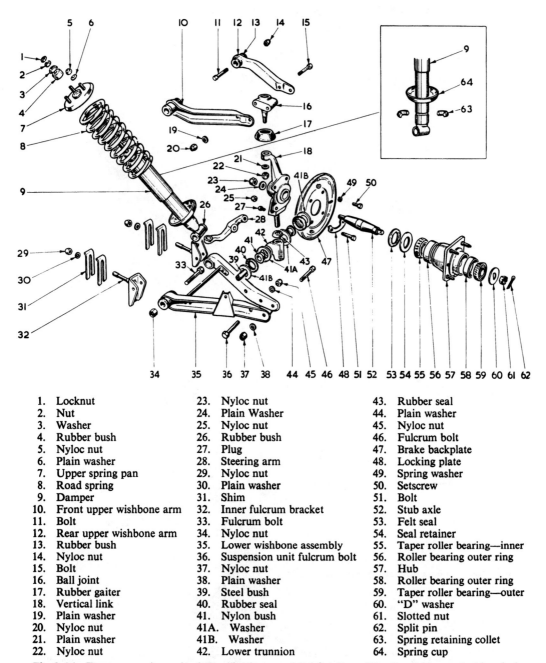

Fig. 8:26 Front suspension unit of Herald, Vitesse and Spitfire. Inset, Woodhead Monroe locking device

1. Locknut	23. Nyloc nut	43. Rubber seal
2. Nut	24. Plain Washer	44. Plain washer
3. Washer	25. Nyloc nut	45. Nyloc nut
4. Rubber bush	26. Rubber bush	46. Fulcrum bolt
5. Nyloc nut	27. Plug	47. Brake backplate
6. Plain washer	28. Steering arm	48. Locking plate
7. Upper spring pan	29. Nyloc nut	49. Spring washer
8. Road spring	30. Plain washer	50. Setscrew
9. Damper	31. Shim	51. Bolt
10. Front upper wishbone arm	32. Inner fulcrum bracket	52. Stub axle
11. Bolt	33. Fulcrum bolt	53. Felt seal
12. Rear upper wishbone arm	34. Nyloc nut	54. Seal retainer
13. Rubber bush	35. Lower wishbone assembly	55. Taper roller bearing—inner
14. Nyloc nut	36. Suspension unit fulcrum bolt	56. Roller bearing outer ring
15. Bolt	37. Nyloc nut	57. Hub
16. Ball joint	38. Plain washer	58. Roller bearing outer ring
17. Rubber gaiter	39. Steel bush	59. Taper roller bearing—outer
18. Vertical link	40. Rubber seal	60. "D" washer
19. Plain washer	41. Nylon bush	61. Slotted nut
20. Nyloc nut	41A. Washer	62. Split pin
21. Plain washer	41B. Washer	63. Spring retaining collet
22. Nyloc nut	42. Lower trunnion	64. Spring cup

To remove the ball-pin shank from the suspension arm, remove the retaining nut and if a ball extractor tool is not available hold a hammer to one side of the suspension-arm ball-joint eye and strike a few sharp blows on the opposite side of the arm to 'jar' the pin loose. With the pin loose in the eye, use a long lever to disconnect the pin. Each ball pin assembly should be overhauled individually and the pin reconnected to the suspension arm on completion of the overhaul of each joint.

Chart 8:1

Fault Diagnosis on B.M.C. 1100/1300 Front Suspension and Drive Shaft Assembly

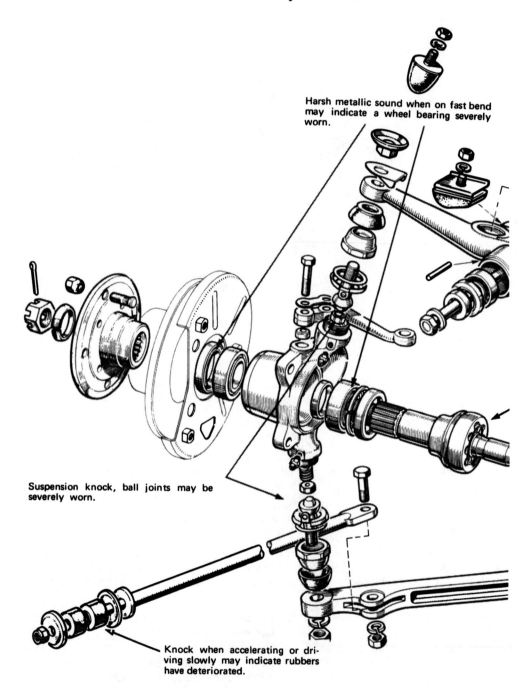

Harsh metallic sound when on fast bend may indicate a wheel bearing severely worn.

Suspension knock, ball joints may be severely worn.

Knock when accelerating or driving slowly may indicate rubbers have deteriorated.

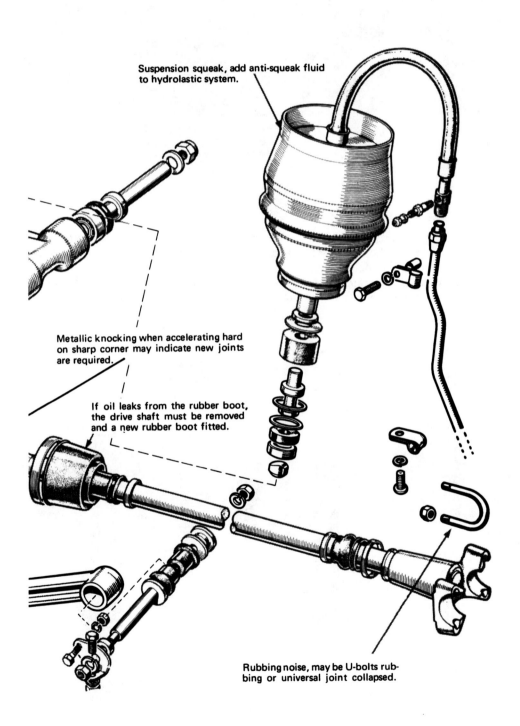

Suspension squeak, add anti-squeak fluid
to hydrolastic system.

Metallic knocking when accelerating hard
on sharp corner may indicate new joints
are required.

If oil leaks from the rubber boot,
the drive shaft must be removed
and a new rubber boot fitted.

Rubbing noise, may be U-bolts rubbing or universal joint collapsed.

Fig. 8:27 The spring and damper attachments to chassis sub-frame on a Triumph

Fig. 8:28 Removing the Herald spring and damper assembly from the front suspension. Before the damper can be removed from the spring assembly the spring must be compressed slightly

Hydraulic dampers (Shockabsorbers)

Suspension knocks that prove difficult to locate are sometimes due to loose shockabsorber retaining bolts; therefore, should a suspension knock occur, one of the first things is to check the shockabsorber retaining bolts for tightness.

Free play in a hydraulic damper may be caused through a shortage of fluid, aeration of fluid or general wear and when free play occurs in a front damper, it can result in severe wheel 'shimmy'.

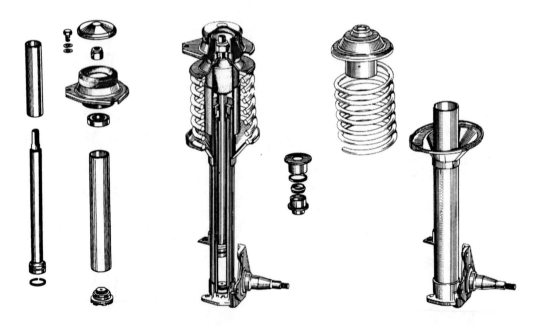

Fig. 8:29 The Ford Cortina front suspension assembly. Provided suspension units such as these have not received accidental damage, a repair kit can be used to overhaul them

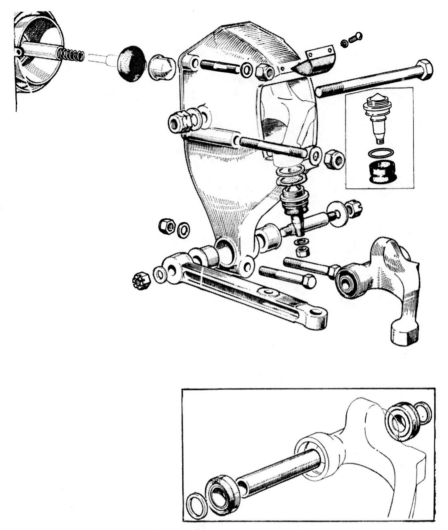

Fig. 8:30 The B.M.C. 1800 front suspension

To check a shockabsorber suspected of being faulty, it is best to remove it from the car and to place it in a vice in an upright position and move the arm slowly throughout its full stroke. On many cars, front shockabsorbers may be examined and 'bled' whilst still on the car if the actuating arm is disconnected. With rear dampers, this procedure is seldom possible and they usually have to be removed from the vehicle.

If the movement of the shockabsorber arm is found to be erratic or if free movement of the arm is felt, clean the grit from around the topping-up plug, remove the plug and fill the reservoir with shockabsorber fluid. Replace the plug.

Move the arm throughout its full stroke again. If the erratic, or free movement remains after half a dozen complete pumping actions, or if the damper shows any signs of a leak, or if the movement is exceptionally hard, the damper should be renewed.

When dampers have been checked it is not advisable to leave them upside down or to place them on their sides. They must be kept in an upright position and should be refitted as soon as possible.

Modern telescopic dampers are sealed. In the event of erratic, or undue movement, they must be renewed (see figs. 8:33 and 8:34).

Rear suspensions

Rear springs are often the most neglected part of a motor car, but the springs should be regularly inspected for leaf breakage. On the Vauxhall Victor, the 2nd and 3rd leaves of the semi-elliptic rear suspension springs are fitted at the outer ends with polythene buttons; when the buttons become worn squeaks tend to develop. The buttons can be replaced by

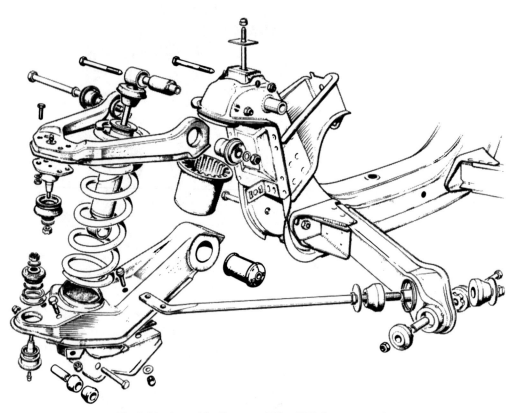

Fig. 8:31   Assembly diagram of Viva (HB) front suspension

jacking the car so that the load on the spring is removed and the leaves parted by inserting a screwdriver between them. It is also advisable to spray the leaves with a graphite-impregnated oil.

Rubber bushes on rear spring shackles often develop squeaks; if this occurs remove the shackle plate, lubricate the bushes with brake fluid, and refit the plates, see also chart 10:5.

Hydrolastic suspensions (see chart 8:2) have to be pressurized using special equipment, but if work has to be carried out on the rear suspension, the system can be depressurized by removing the brass cap from the valve holder and depressing the valve. The fluid will be ejected under pressure and some means should be adopted of saving as much as possible. It is possible to drive the car (at not more than 30 mile/h) with the system depressurized; the car can therefore be driven to the nearest suitable garage for evacuation and repressurizing the system.

On 1100s prior to 1966 the wear on rear radius arm bearings and shafts was found to be unduly severe and they were modified about this time by the fitment of an outer seal (part

Fig. 8:32 Typical suspension units which can be returned for reconditioning. An exchange service is usually carried out by most of the larger garages

no. BTA 750) and an inner seal (part no. BTA 751). To fit the seals to older cars and new bearings and pivot shaft, it is necessary to remove the radius arms from the car. To do this the rear sub frame can be taken from the car, or, alternatively, the radius arms can be removed leaving the sub frame in situ. Whichever method is decided upon the suspension must be depressurized on both sides.

The reason for depressurizing both sides, if only one radius arm is being removed, is because difficulty would otherwise be encountered in removing and refitting the bolts which retain the anti-roll bar to the inner side of the radius arm.

Prior to the removal of any rear suspension parts, it is advisable to treat all nuts and bolts with a corrosion dispersal solvent a few days before attempting to remove them.

To remove a radius arm, when the system has been depressurized and the car raised to a suitable working position with supports under the body, disconnect the brake hose, disconnect the handbrake cable from the lever on the brake backplate, remove the handbrake cable sector and slip the handbrake cable from the backplate bracket. Place both radius arms in the full bump position and remove the bolts securing the anti-roll bar and auxiliary spring bracket to the inner face of the radius arm using a $\frac{3}{4}$ in. AF socket spanner.

Remove the nuts securing the auxiliary spring to the sub frame.

Take off the nut and spring washer from each end of the radius arm pivot shaft and the setscrews from the radius arm support bracket and lift out the radius arm and displacer unit.

Fig. 8:33 Removing a rear telescopic type damper

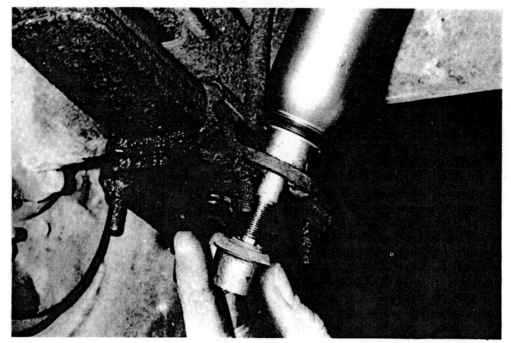

Fig. 8:34 When refitting a telescopic type damper it is essential that the mounting rubbers are correctly positioned and the retaining nuts fully tightened

Now dismantle the radius arm pivot as follows: drift the shaft, either way, from the bearings and collars; remove the bearing outer races from the arm, they cannot be drifted out from the opposite end and an expanding extractor is required (Service tool 18G704). Knock out the steel (or plastic) inner dust seals.

Reassembly is a reversal of the dismantling process but make sure to bleed the brakes and, when the hydrolastic system has been evacuated and repressurized, set the adjusting bolts on the centre auxiliary-spring bracket to allow a clearance of 0·125 in. (3·18 mm) between the inside face of the spring bracket and the bottom of the sub frame when the car is standing on level ground.

Make sure when renewing radius arms, pivot shafts and bearings to:

Push new plastic inner dust seals into position each end; drift the new outer races onto the arm; push an inner race on to the outer end of the pivot shaft followed by the fixed length collar (see Chart 8:2, pages 160–161), place the pivot shaft in the radius arm and then fit the inside inner race and a variable length spacer.

Now, temporarily place a large washer over each end of the shaft and fit the nuts; tighten the nuts and check the movement of the shaft. There must be no lateral movement whatever but it should be possible to rotate the shaft with the fingers (5 lb in. torque). Nine different lengths are available for the inner spacer and if possible a selection should be borrowed.

Once the desired pre-load is obtained, remove the nuts and temporary washers but before refitting the arms to the car, pack the bearings with molybdenum disulphide grease and fit new outer dust seals.

Removing a rear spring, see figs. 8:35 and 8:36.

Chart 8:2

Fault Diagnosis on B.M.C. 1100/1300 Rear Suspension

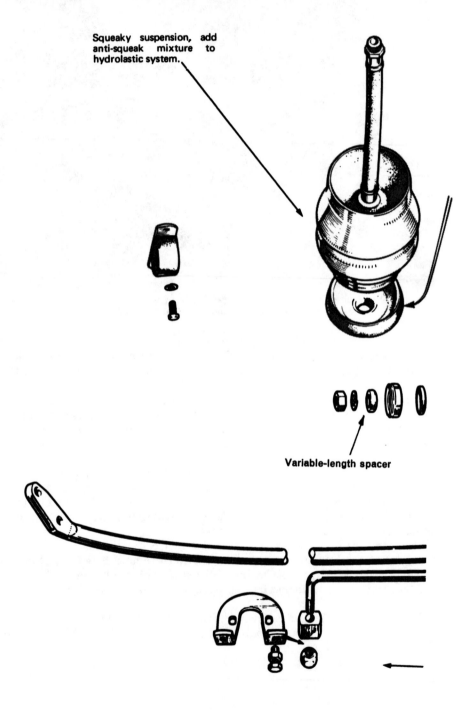

Squeaky suspension, add anti-squeak mixture to hydrolastic system.

Variable-length spacer

Note: Early cars were prone to sub frame mounting rubbers shearing. Later cars have modified rubbers and

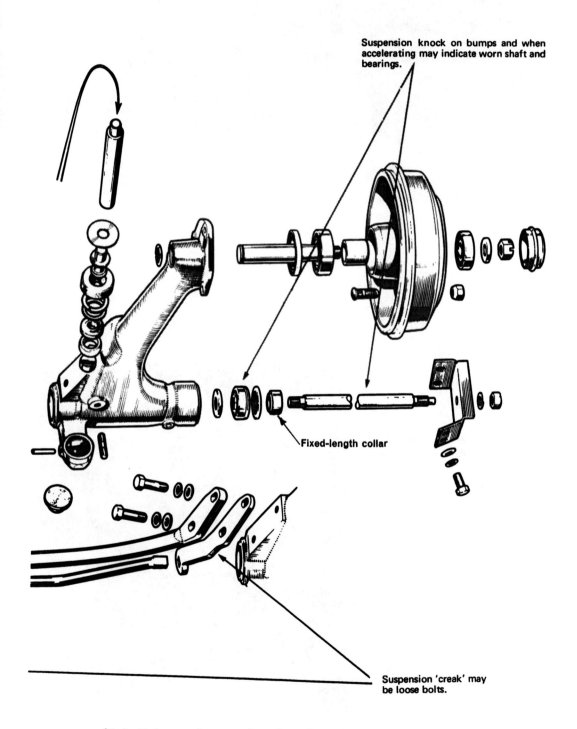

Suspension knock on bumps and when accelerating may indicate worn shaft and bearings.

Fixed-length collar

Suspension 'creak' may be loose bolts.

were fitted with inner and outer seals to the radius-arm bearings to protect them from dirt and water

Fig. 8:35 Jacking the vertical link on the Herald slightly to relieve pressure on the rear-spring eye bolt.

When working on any suspension unit, it is important to ensure that suspension spring tension is relieved before attempting to remove the bolt

Fig. 8:36 Removing a road spring on the Herald. When refitting a leaf spring, it is essential that the spring is repositioned correctly. Springs which have a centre bolt must be carefully positioned, so that the bolt head locates correctly, before the U-bolts are tightened

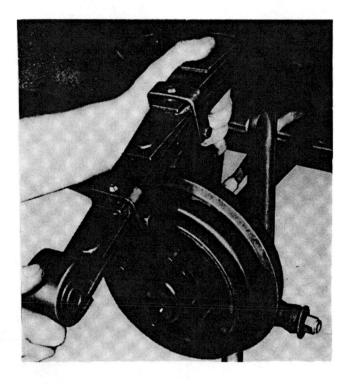

CHAPTER 9

Overhauling the Brake System

One of the difficulties that may be encountered when overhauling the braking system (see fig. 9:1), is removal of the rear brake drums, for if they are integral with the hub, a puller may be needed to remove them.

It should be possible to hire a puller from the local garage and this is much better than trying to manage with tyre levers, etc., for sometimes hubs can be difficult to remove. Before removing the rear brake drums, pull the handbrake on and off and ascertain whether the

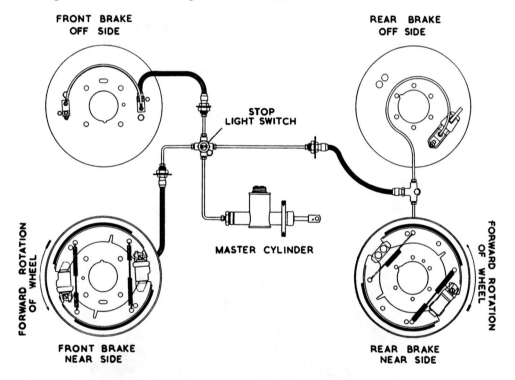

Fig. 9:1 Diagrammatic view of typical two leading shoe hydraulic brake equipment which consists of a master cylinder, of the integral barrel type containing a reserve supply of fluid, in which hydraulic pressure is generated; single-ended internal wheel cylinders which operate the two leading shoe front brakes; single-ended internal cylinders, incorporating handbrake operating levers, which operate the leading and trailing shoe rear brakes and the 'line' consisting of tubing, flexible hoses and unions interposed between the master cylinder and the wheel cylinders. In addition to the braking system, certain vehicles also incorporate a hydraulic clutch operating system consisting of a similar master cylinder connected to a slave cylinder which operates the release bearing mechanism of the clutch.

Pressure exerted on the brake pedal is conveyed to the brake shoes by a column of special fluid; the master cylinder has a single piston, as do the wheel cylinders, and all pistons are provided with rubber cups or seals to maintain pressure and prevent loss of fluid.

Hydraulic failure, due to brake pipe corrosion, is by no means unknown. Brake failure on one model of car has been found to be due to drips from the radiator tap falling onto the brake pipe which runs immediately below the tap. Examination of steel and flexible brake pipes should be carried out annually and is a feature of the M.O.T. test

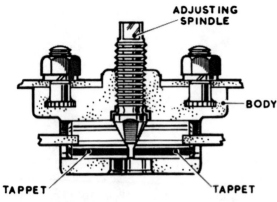

ADJUSTING
SPINDLE

BODY

TAPPET   TAPPET

Fig. 9:2 Lockheed spindle type adjuster and housing. If an adjuster is seized, the housing should be removed from the backplate and soaked in penetrating oil, after which constant backward and forward pressure with a spanner will gradually free it. A seized tappet may be freed by rotating it with a large screwdriver

brakes bind when the lever is released, for if they do, the cause must be located and rectified during the overhaul procedure. On B.M.C. 1100s and early Minis it is common for the cable sectors to seize on to their fulcrum pins on the suspension arms and they must be freed during the brake overhaul. A good solvent may do this, but unless the sectors are removed from the pivot pins and lubricated with a special grease obtainable from B.M.C. agents the trouble may occur again. On later cars the sectors are attached to the pivot pins and in the event of seizure the pins should be greased (see also fig. 9:37).

Removing the brake drums

When removing the brake drums, make sure the handbrake has been placed in the 'off' position and if the brake adjuster spindle (see figs. 9:2 and 9:3) is not seized to its pivot,

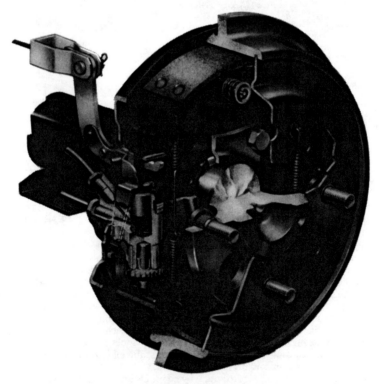

Fig. 9:3 Self adjusting (by handbrake application) rear brake assembly of the Cortina

Chart 9:1

Common Brake Faults and Their Remedy

EXCESSIVE PEDAL TRAVEL

Insufficient fluid in supply tank.	Top up.
Excessive wear on linings.	Adjust. If no further adjustment possible fit new linings.
Incorrect brake adjustment.	Check that the adjusters have been turned in the correct direction.
If discs on front, faulty adjusting mechanism.	Fit new pistons and/or seals.
Air in system.	Bleed brakes.
Cracked brake drum.	Fit new drum.
External leaks at pipe unions.	Locate and rectify.
Leakage past rubber seals in master cylinder, wheel cylinders or callipers.	Renew rubbers.
Fault in servo due to leaks.	Overhaul servo.
Worn clevis pin on pedal linkage.	Renew pin.
Incorrect pedal free play.	Adjust.

BRAKES GRABBING
(See also 'Unequal braking')

Incorrect type of linings.	Fit correct type.
Defective servo.	Overhaul.
Linings picking up.	Check for reason (possibly loose rivets).
Adjuster pin loose or shoe retainer loose.	Check location of brake shoes on adjuster pins and in location slots.
Dust in drums.	Blow out dust.
Retraction spring broken.	Renew.

UNEQUAL BRAKING

Air in system.	Bleed brakes.
Lining picking up.	Check that the lining rivets are secure.
Distorted drums or discs.	Renew.
Uneven tyre inflation.	Inflate to correct pressure.
Mixed types of tyres, or a smooth tyre.	Fit new tyres so that all are of the same type and tread pattern.
Brake calliper or backplate loose.	Tighten.
Worn steering.	Overhaul steering as necessary.
Brake shoes incorrectly fitted.	Refit correctly.
Rear spring/axle bolts loose.	Tighten.
Different type of lining fitted to one side.	Fit correct linings.
Loose wheel bearing.	Adjust or fit new bearing.
Grease or oil on linings.	Renew linings.
Sluggish wheel cylinder piston.	Overhaul wheel cylinders (see page 185).
Grease on discs.	Clean discs and fit new pads if necessary.
Wheel cylinder rubber leaking.	Fit new rubbers to all wheel cylinders. Overhaul complete system.
Clogged brake pipe.	Clear obstruction. Flush system.
Flattened brake pipe.	Renew.

SPRINGY PEDAL

Linings not bedded in.	Remove drums and take off high spots on linings. Further mileage necessary.
Brake drum cracked.	Renew.
Master cylinder loose.	Tighten cylinder.
Bulkhead moves.	Strengthen bulkhead.

SPONGY PEDAL

Insufficient fluid in supply tank.	Top up, examine for fluid leak.
Master cylinder main cup worn.	Renew master cylinder rubbers.
Leak past master cylinder secondary cup.	Renew master cylinder rubbers.
Defective hose.	Renew.
Air in system.	Bleed system.
Wheel cylinder rubbers leaking.	Renew all hydraulic rubbers.
Faults in servo.	See 'Vacuum servo faults'.

HARD PEDAL

Partially seized piston in master cylinder or wheel cylinder.	Overhaul hydraulic system.
Oil, or brake fluid, on linings.	Renew linings. Check hub oil seals and wheel cylinders, renew if leaking.
Binding pedal.	Seized calliper or wheel cylinder pistons or pedal fulcrum pin.
Inadequate servo action.	Check vacuum and servo mechanism.
Incorrect type of lining.	Fit correct type.

VACUUM SERVO FAULTS

Pedal travel excessive.	External leak between master cylinder and servo or between servo and wheel cylinder. Leakage past rubber cup on servo valve-operating piston, or between servo push rod and push rod seal.
Grabbing brakes.	Excessive friction between servo-vacuum piston and vacuum shell, or between servo valve operating piston and body.
Hard pedal.	Low vacuum from engine. Collapsed, restricted or disconnected vacuum line from manifold. Air valve and/or vacuum valve not seating properly. Air leak past leather cup or seal on vacuum piston. Faulty gasket under diaphragm, or between slave cylinder and vacuum shell. Blockage of the slots through which air enters the rear of the vacuum shell.

slacken the adjuster, otherwise the shoes may 'grab' as the drum is being removed. If the adjuster spindle is seized to its threads, soak the spindle with a corrosion solvent. If the adjuster still cannot be loosened the drum will have to be removed without slackening it and the adjuster overhauled later (see page 172).

Inspecting the brake linings

If the brake linings are saturated with gear oil, the hub seal will have to be renewed. Inspect the wheel cylinders for damage to the dust caps and traces of hydraulic brake fluid. If fluid is present, the wheel cylinders are leaking and will have to be renewed or overhauled using a kit of seals (see page 185).

Brake shoes should be renewed if the rivet heads are likely to foul the brake drums or if the linings are more than 75% worn. If the linings are of the bonded non-rivet type the thinnest part of the lining must not be less than $\frac{1}{8}$ in. Although it is possible to reline brake shoes (riveted type—see fig. 9:3), owing to the low cost of replacement shoes, it is not worthwhile; moreover, brake shoes from a reputable manufacturer have the linings ground to such a fine tolerance that a 'satisfactory brake pedal' can be obtained much more readily than with shoes relined at home.

It is quite normal to find one brake shoe of each pair worn less than its partner, but they should not be swapped in an effort to even out the wear, for this results in unsatisfactory braking.

Removing the brake shoes

Before removing the brake shoes, mark each shoe, for they are easy to refit wrongly. Take careful note, too, where the various retraction springs and steady springs are attached (see figs. 9:4 and 9:5). Some systems have springs in different colours, some red and others green, but in all cases it is important that they should be refitted in the correct position.

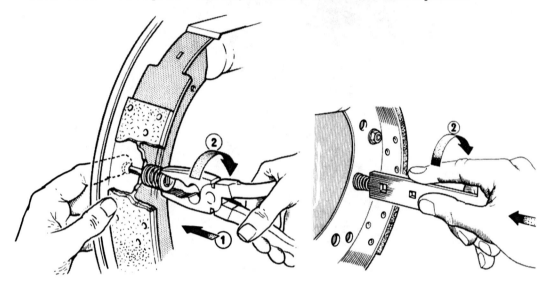

Fig. 9:4 Method of removing the steady springs. (1) Push. (2) Turn. Make sure to hold the pin firmly

Fig. 9:5 Some types of shoe steady spring can be removed using a tool which is supplied with Lockheed replacement brake shoes

Sometimes each shoe can be slipped from the wheel cylinder or pivot to reduce partially the loading of the retraction springs; the shoes can then be removed as a pair. On systems fitted with 'Micram' adjusters (see fig. 9:6), before removing the shoes first remove the adjuster by pulling against the retraction springs and slipping the adjuster from the cylinder.

When the shoes have been removed, place a thick elastic band around the wheel cylinder pistons or wind a piece of wire around the cylinder, so as to ensure that the pistons cannot inadvertently be ejected if the brake pedal is accidentally pushed.

Before fitting the brake shoes, thoroughly clean the backplates and check each shoe adjuster to ascertain that it is not seized in its housing.

If it is not intended to renew the wheel cylinders, they should be checked to ensure the pistons are not seized to the cylinder by removing the dust caps and pushing on the pistons with the finger. On some systems, before this can be carried out, the outer piston and handbrake shoe-activating lever have to be removed (see fig. 9:6), by disconnecting the lever from the cable and removing the lever fulcrum pin. If the lever is seized to the fulcrum pin or the piston to the cylinder, proceed as suggested on page 186.

Fitting new brake linings

Unless the car is being used for competitive events such as rallying or racing, which call for severe regular braking (when harder linings may be desirable), it is advisable to fit the type of brake lining recommended by the manufacturer. It is imperative, however, that all brake linings on the same axle are of the same make, grade and condition so as to ensure even braking. Old brake shoes should never be assembled in conjunction with new ones but it is permissible to fit new shoes to both of the rear brake assemblies, and retain the old shoes on both of the front assemblies, and *vice-versa*.

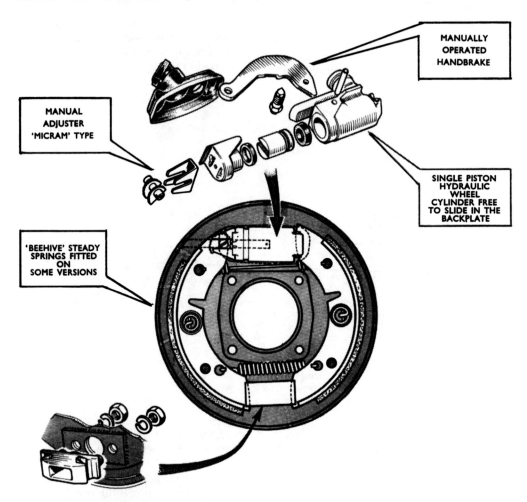

MANUALLY OPERATED HANDBRAKE

MANUAL ADJUSTER 'MICRAM' TYPE

SINGLE PISTON HYDRAULIC WHEEL CYLINDER FREE TO SLIDE IN THE BACKPLATE

'BEEHIVE' STEADY SPRINGS FITTED ON SOME VERSIONS

Fig. 9:6   A Lockeed brake fitted with 'Micram' shoe adjusters showing the outer and inner, wheel cylinder, piston and handbrake lever, assembly. Leading/trailing shoe in both directions of wheel rotation

Before fitting the shoes, turn the brake adjuster(s) to the fully 'off' position and smear a light trace of white grease on the sliding surfaces. If the wheel cylinder is of the 'floating' type (see page 185) check that it moves satisfactorily within its location. With most brakes it is possible to insert the brake-shoe retraction springs into their location holes in the brake-shoe webs and, exerting a little pressure to keep the springs in tension, to slip the whole assembly into position.

In some cases it may be necessary to use a screwdriver to lever the shoes into their final location.

On some assemblies one end of the retraction spring fits on to the backplate. To fit this type, position the spring on the backplate and then in the correct hole in the brake shoe.

Place one end of the shoe on the adjuster, lift the shoe at the other end and position it on the wheel cylinder.

On some types of brakes, one of the springs may have to be fitted when the shoes are in position. A good pair of pliers and a firm grasp does the trick. Make sure that all of the springs are refitted correctly and be sure to centralize the shoes on the backplate, so that the drum will go on unhampered.

When the drum has been refitted, rotate the hub to ensure that the drum or hub does not foul anything. If fouling occurs a spring or shoe is wrongly fitted, and must be relocated and the trouble rectified.

Fig. 9:7   The front hub and brake assembly on the GT Ford Cortina

If it is difficult to fit the brake drum over the shoes, check the assembly to ensure that the wheel cylinder, if of the fully-floating type, is fitting correctly into its location slot and that the handbrake link rod, if fitted, is in the correct position. If trouble is still encountered, check that the shoes have not slipped out of position and, with rear assemblies, ensure that the handbrake mechanism is allowing the brake shoes to retract sufficiently. If in doubt, slacken the handbrake cable adjustment. If the drum will still not go on, incorrect brake shoes or shoes wrongly positioned are likely to be the cause.

Fig. 9:8   On some cars the wheels have to be removed before access to the drum aperture can be obtained. On wheels which are drilled for access to the drum, should the wheel be changed, make sure to align the access holes when refitting the wheel

With the drums refitted, if shoe steady posts are fitted, the alignment of the new shoes with the drum must be checked and the steady posts adjusted, if necessary (see fig. 9:31) before adjusting the brake-shoe/drum clearance and the handbrake (see page 176). If the hydraulic system has been disconnected, the brakes must be bled. Brakes that have been relined must be tested carefully before taking the car onto the road; it is important to 'bed in' the linings by using the brakes as much as possible for the first 100 miles. They should then be readjusted.

Adjusting the brakes
Excessive brake pedal travel usually indicates the need for brake adjustment. Disc brakes

Fig. 9:9   Showing the adjustment through a backplate on a front brake

(see fig. 9:7), however, are self-adjusting and vehicles fitted with discs on the front and drum brakes on the rear require only adjustment of the rear brakes.

Jack up the vehicle and support it with stands, or blocks, so that the wheels remain clear of the ground. Make sure the handbrake is 'off' and turn each adjuster (see figs 9:8, 9 and 10) until the wheel is locked, then slacken the adjuster a notch so that the wheel rotates freely. If new linings have been fitted and a central adjuster is fitted (see fig. 9:2), the adjuster housing must be centralized. Slacken the stud nuts holding the housing to the backplate and turn the adjuster so that the brake shoes are tight against the drum. Retighten the housing retaining nuts and slacken the adjuster spindle to give clearance between the shoes and drum.

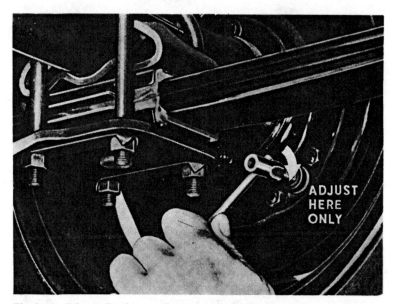

Fig. 9:10 When adjusting rear brakes, a quick check on the rear spring/axle mountings is not amiss. If the brake adjuster is seized use a good corrosion solvent

Several types of brake adjusters are in use, but they usually involve either adjustment through the brake backplate or through the brake drum, as in the case of the Morris Minor, or by an adjuster spindle protruding through the rear backplate, as in the case of the Mini.

With braking systems which have single adjusters on the backplates (see fig. 9:11), the adjusters may consist of a square shaft protruding from the backplate or be of the serrated-wheel type, access to which entails removal of a grommet (see fig. 9:12), or in the case of twin leading shoe systems where each shoe has to be adjusted independently, the adjusters may be 'Micram'. Other types may be seen as largish bolt heads on the backplate, which operate 'snail' type cams that butt against a pin on each brake shoe.

Adjustment through the drum may be carried out by using a screwdriver to turn the 'Micram' adjuster (see fig. 9:8), or by turning a serrated wheel on a threaded spindle (see fig. 9:14) carried between the wheel cylinder piston and the end of the brake shoe. The latter type of adjuster is also actuated by inserting a screwdriver through a hole in the brake drum (or on some cars through the backplate) but the tool is used as a lever to turn the serrated adjuster wheel in the required direction.

Some Girling and Lockheed brakes are designed so that the brake shoes rest in the slots on the end of plungers carried in a housing. The plungers are actuated by screwing in an adjusting spindle which protrudes through the backplate and at the opposite end bears a conical head which abuts against the wedge-headed adjuster plungers (see fig. 9:2). The conical head is shaped in a series of flats, which allows the pull-off springs to lock the mechanism in position.

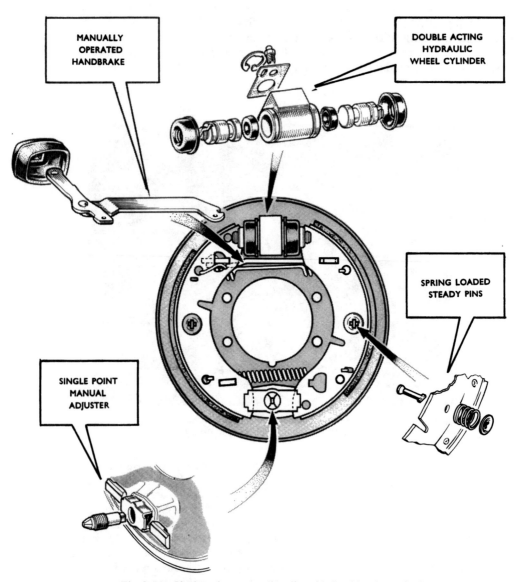

Fig. 9:11   Single point, manually adjustable Lockheed rear brake
Adjuster tappets and adjuster screws should be washed in ethyl alcohol (industrial methylated spirit) and coated with polybutylcuprysil (PBC), Shell corrosion resistant grease SB 2628 or an equivalent. PBC is an anti-seize lubricant and protective manufactured by K. S. Paul Products Ltd, Nobel Road, London N.18

Some cars, most notably the Austin A30, A35, early A40, $1\frac{1}{2}$ litre and $2\frac{1}{2}$ litre Rileys produced between 1945 and 1951, have a hydro-mechanical brake system. The rear brakes are operated by pull rods. One of the weak points of the system on the Mk I A40 is the cable which attaches to the mechanical compensator, for, in spite of the adjustment possible, the cable stretches to the extent that the slack in the cable cannot be taken up and the cable has to be renewed occasionally.

Girling Duo-servo automatically adjusted brakes (see fig. 9:13)

Some Girling Duo-servo brakes are automatically adjusted when the vehicle is moving in reverse and the brakes are applied. When new rear brake shoes have been fitted manual adjustment of the new shoes will be necessary, access to the toothed adjuster wheel being gained by removing a grommet from a hole in the backplate.

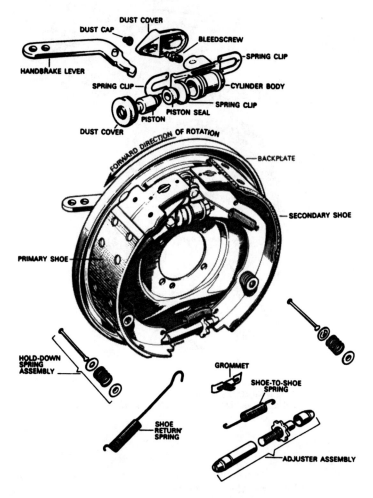

Fig. 9:12   The Girling Duo-Servo brake
The Duo-Servo brake is similar in appearance and operation to the Duo-Servo
Automatic Adjustment brake, the main difference being the deletion of the automatic adjustment mechanism from the secondary shoe. Shoe adjustment therefore has to be done manually

To adjust the shoes, rotate the adjuster until the linings are tight on the drum, and then slacken off until the wheel runs free.

After adjusting the brakes, if the handbrake cable or linkage has too much free movement, lock the shoes in the drums by turning the adjusters clockwise, correct the handbrake linkage adjustment and finally click back the brake adjusters until the wheels run free.

Girling HL3A automatic adjustment brake (see fig. 9:14)

On the automatic version of the Girling HL3 brake it is not possible to lock the shoes in the drum. To adjust the brakes, operate the handbrake, listening to the click of the adjuster ratchet; when the clicking ceases, the brakes are fully adjusted.

When adjusting the handbrake linkage it is particularly important to ensure that the handbrake cables are not over-adjusted and the brakes partly held on.

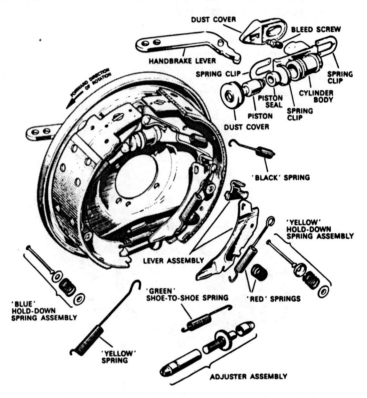

Fig. 9:13 Girling Duo-Servo Automatic Adjustment brake
The construction and arrangement of the Duo-Servo Automatic Adjustment brake. Take particular note of the position of the retraction springs

HLSSA Girling brakes

The lining/drum clearance on the automatic version of the HLSS brake (see figs 9:15 and 9:16) is adjusted when the foot pedal is applied but it is possible to adjust the brakes manually and this must be done when new linings have been fitted by turning the hexagon-headed adaptor-retaining bolts on the backplate. Although the bolts may be exceedingly tight (owing to the presence of corrosion) they must never be lubricated. To free them off a corrosion solvent should be used and they should be rotated back and forth. The torque setting should be between 24 and 30 lb ft.

Disc brakes

Disc brakes are self-adjusting and beyond an inspection at normal servicing periods they are usually trouble-free. If the discs appear scored, provided that the scoring is concentric and not excessive it will not be detrimental but heavy or uneven scoring will impair brake efficiency and a new disc (or discs) must be fitted or the offender reground. Reground discs, however, should only be resorted to if new discs are unobtainable. If regrinding takes place, it is important that it is carried out accurately, removing a total of not more than 0·040 in. thickness of metal. The amount may be removed by taking off 0·020 in. from each side of the disc or by removing 0·040 in. thickness from one side only. After the discs have been ground the faces must run true to within 0·002 in. and be parallel to within 0·001 in.

Whether reground or new discs are fitted it is important to ensure before fitting them that the faces are absolutely free from burrs. When the discs are in position they should be checked for 'run out' during rotation by a dial indicator (see fig. 9:19). The 'run out' on the periphery of the braking surface must not exceed 0·004 in. In the event of this being exceeded, the disc should be removed and repositioned until a more satisfactory measurement is obtained.

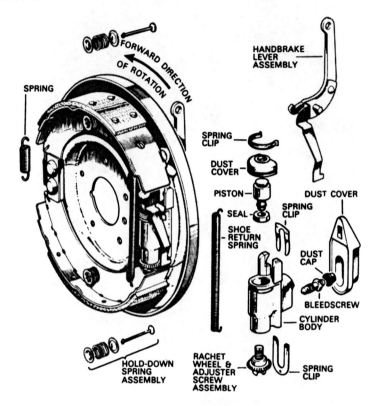

Fig. 9:14 The Girling HL3A rear drum brake incorporates automatic adjustment of the shoes. When the linings are worn to a predetermined thickness, the brakes no longer adjust and excessive pedal movement indicates the need to fit replacement brake shoes or new linings

Fitting new brake pads

It is important to carry out a regular inspection of the brake pads at every servicing period for although, as a rough guide, the brake-pad life is 10,000 to 13,000 miles it is essential to renew the pads before wear reduces the thickness of the friction material to below $\frac{1}{16}$ in.; indeed brake pads should preferably be changed just before the friction material is reduced to a thickness of $\frac{1}{8}$ in.

On most cars, a brake-pad change entails extracting the pad-retaining pins, the spring retainer, see fig. 9:18, and removing the friction pads. As each friction pad is removed, the new pad should be fitted.

Before replacing each new pad, clean the exposed end of each piston with methylated spirits (petrol must not be used) and ensure that the calliper recess, which accepts the pad, is free from rust and dirt.

Before a pad can be replaced, it is necessary to carefully push the calliper piston further back into its bore, using a small clamp or a lever. During this operation the brake-fluid level will rise in the master-cylinder tank and, if the reservoir is fairly full, before starting to fit new pads a little fluid should be siphoned off to prevent it overflowing through the breather hole in the reservoir cap and damaging the paintwork.

Fig. 9:15 Girling HLSSA brake
The Girling HLSS brake is similar to this automatically adjusted version. Adjustment of the non-automatic type is by snail cam. It is made in several different versions, the principal differences being the type of shoe retaining springs and steady posts

When replacing brake pads, always ensure that they are free to move easily in the calliper recess and do not bind. If they prove tight, remove the high spots on the pad pressure plate with a file. When all the brake pads have been inserted and the retaining springs and pins fitted, pump the brake pedal several times to readjust the calliper pistons. Subsequently top up the master-cylinder reservoir with disc-brake fluid.

On some cars (the Austin Westminster is one) it is necessary to remove each complete calliper assembly to re-pad the brakes. If callipers have to be removed it is essential to make absolutely certain that the correct bolts are used (see fig. 9:20) when the callipers are refitted and that they are fully tightened and new lockwashers are fitted.

Adjusting the handbrake
Before adjusting the handbrake, inspect the linkage mechanism and cables to ensure that

each part of the actuating mechanism is free to operate. Brake cables often cause trouble due to lack of lubrication or owing to cable strands breaking. If the inner cable is seized to the outer cable, remove the cable from the car and soak it in paraffin for a few hours. When the inner cable runs freely in the outer, give the complete cable a thorough clean and lubricate it with a graphite-impregnated grease. If a wire strand in the cable has broken, renew the complete cable.

To adjust the handbrake, position the lever in the 'off' position and adjust the rear brakes so that the linings are tight against the brake drums. Tighten the cable, or cables, so that the slack is taken up and then adjust the lining/drum clearance so the wheels rotate freely. Apply the handbrake and count the ratchet notches. The handbrake should generally be fully on after four or five clicks but this varies according to the design of the ratchet assembly. The principle is that the wheels are locked when only a third of the total handbrake travel is taken up.

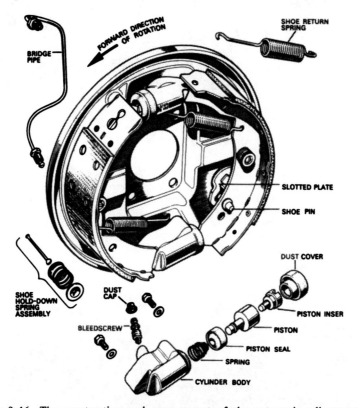

Fig. 9:16 The construction and arrangement of the automatic adjustment version of the Girling HLSS brake

On cars fitted with a separate adjustable cable to each rear brake, release the handbrake and then apply it three notches so that the road wheels are almost locked. Try to turn the wheels, applying the same amount of pressure to each wheel. If one of the wheels is too tight and cannot be turned release the handbrake and slacken the cable adjuster on that side, one complete turn. If one of the wheels fails to be held by the handbrake, the cable adjuster to that side brake assembly requires tightening. After two or three adjustments the same amount of pressure should be needed to turn each rear wheel. At this stage, place the handbrake lever in the 'off' position and check that both wheels are free to rotate. If the wheels 'bind' slacken both cable adjusters an equal amount. Recheck the equalization and make any necessary final adjustment.

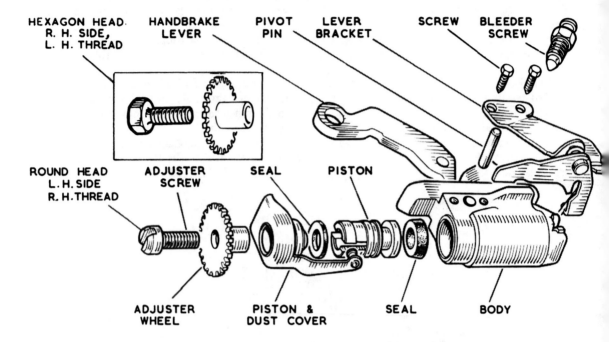

HEXAGON HEAD.
R. H. SIDE,
L. H. THREAD

HANDBRAKE
LEVER

PIVOT
PIN

LEVER
BRACKET

SCREW

BLEEDER
SCREW

ROUND HEAD
L. H. SIDE
R. H. THREAD

ADJUSTER
SCREW

SEAL

PISTON

ADJUSTER
WHEEL

PISTON &
DUST COVER

SEAL

BODY

Fig. 9:17 Before refitting the brake drum on Rootes Group vehicles fitted with automatically adjusted Lockheed brakes, if the vehicle is equipped with the above type of wheel cylinder, back off all adjustment by turning the adjuster screw in the appropriate direction, then unscrew 2½ turns. The wheel cylinders are handed to suit l-h side and r-h side.

When the brake drum has been refitted and the brakes bled, the foot pedal must be operated at least 50 times to set the automatic adjuster to the correct operating position

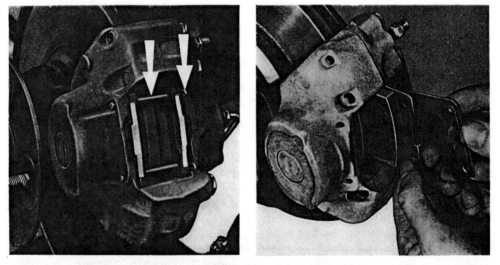

Fig. 9:18 Fitting brake pads with their backing shims to the Cortina. It is essential to check the colour coding on the brake pads, thereby ascertaining that the friction material is correct. The colour coding of standard Cortina pads is red and yellow, but on the GT model it is green

Fig. 9:19 Checking the 'run-out' of brake discs. The 'run-out' must not exceed 0·004 in.

Fig. 9:20 Disc brake calliper attachment bolts are of a special material. If the caliper has to be removed it is essential to ensure that the correct bolts are refitted

Removing the master cylinder

Master cylinder and wheel cylinder kits cost a few shillings. The fitting of them is so simple that the manufacturer's suggestion of renewing the rubbers every 30,000 miles should be adhered to. If a second-hand car has been purchased with more than 30,000 miles on the clock, this should be one of the first jobs.

Before removing the cylinder, drain the brake fluid. Attach a rubber tube to a brake-unit bleed screw, place the end of the tube into a draining receptacle and slacken the screw. Get an assistant to depress the brake pedal and then tighten the bleed screw. Allow the pedal to return unassisted and repeat the operation until the fluid is drained.

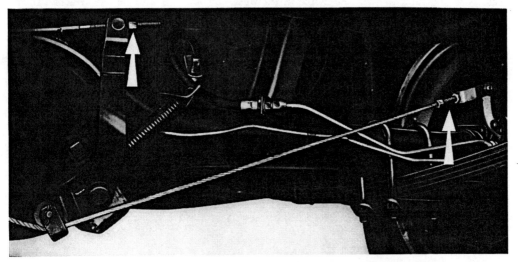

Fig. 9:21 The handbrake linkage adjustment point on the Ford Cortina.
Left: Primary cable adjuster. *Right*: Transverse cable adjuster

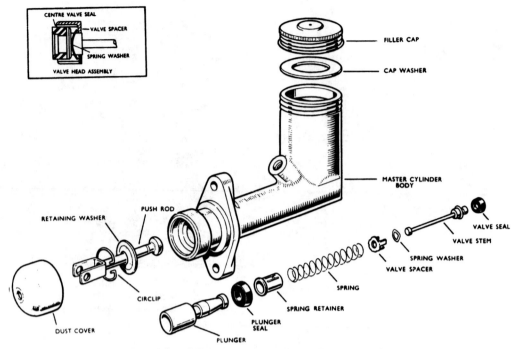

Fig. 9:22 The Girling C V (centre valve) master cylinder which is made in two
styles (see also fig. 9:28). If the plunger is of the early type and is grooved for an
end seal, this groove must be left vacant on reassembly. The latest type of plunger
seal (included in Girling repair kits) does away with the need for an end seal

On the Morris Minor, the master cylinder is carried beneath the floor boards and a small
plate in the floor has to be removed. If the screws are rusted, the correct sized tool and a
can of rust solvent will help to ensure success.

If the master cylinder is of the type shown in fig. 9:22, detach the push rod from the pedal
linkage by removing the clevis pin. On master cylinders where the fluid reservoir is of the
cast type and integral with the cylinder, the push rod may have to be left attached to the

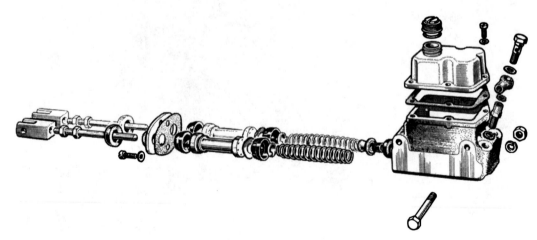

Fig. 9:23 The brake and clutch master cylinders are sometimes housed in the
same unit. Note that no valve is used in the clutch m/cylinder

pedal linkage and the rubber dust seal detached from the cylinder and left on the push rod.

Some types of master cylinder have only one pipe leading from them, others have two or three. Disconnect the pipes by unscrewing the union nuts. As each pipe-nut is undone, pull the pipe from the cylinder taking care not to strain any of the pipes unduly otherwise difficulty may be encountered when refitting them (see page 188).

Unscrew the master-cylinder fixing bolts and remove the unit from the vehicle. On some cars, the bolts fit into caged nuts, but on other cars the nuts must be held with a spanner.

Types of cylinder which fit into the bulkhead may have spacing shims between the cylinder and bulkhead. Take care not to lose them, for if they are not replaced the brakes will bind due to the master cylinder piston not returning fully.

On Morris Minors, difficulty in removing the cylinder-retaining bolts is encountered for the suspension torsion bar is in the way. There is sufficient springiness in the bar, however, to allow it to be levered downwards enough for the bolts to be removed but the services of an assistant will be required.

Removal of brake and clutch cylinders which are housed as one unit (see fig. 9:23) is basically the same as other types but it is a sound idea to take the opportunity of also overhauling the clutch master cylinder.

Overhauling the master cylinder

Slip the rubber boot from its position and slide it along the push rod. Make certain the cylinder is completely drained and depress the piston slightly and remove the circlip from the end of the barrel. Withdraw the push rod and piston and take out the piston washer, the main cup, the spring, the retainer, the check valve and rubber valve washer (if fitted). Carefully lay them on a clean piece of paper in the sequence in which they are removed from the cylinder.

Next, clean the parts with methylated spirit aи d ensure that the entire cylinder and reservoir is completely free from fluff, sediment and grit. Ensure too, that the by-pass port in the cylinder is not blocked by probing the hole with a fine wire. The port is only 0·028 in. in diameter and if blocked, pressure may build up in the system. On types of master cylinder with a detachable lid, the lid should be removed, the reservoir cleaned and the lid refitted using a new gasket.

When fitting the new rubbers (see figs 9:24, 25 and 26) lubricate them with brake fluid. Remove the secondary cup from the piston and fit the new cup onto the piston with the small end towards the drilled head. Make sure that the rubber engages with the groove in the piston, working the fingers around the cup to ensure that it is seated correctly. On units that have a valve washer fitted (see fig. 9:25), insert it into the bore and set it squarely against the end face of the cylinder. If the check valve incorporates a rubber valve cup (see fig. 9:25), fit the cup to the valve ensuring that it seats into the housing correctly.

Locate the spring retainer in the small end of the return spring and fit the check valve unit at the opposite end of the spring. Hold the cylinder so that the pressure outlet is uppermost, insert the spring (with check valve leading) into the barrel. Hold the spring in position and turn the cylinder so that the mouth is uppermost, insert the main cup with the lip leading and insert the piston washer (see fig. 9:26) so that the convex side is towards the piston. Insert the piston into the barrel and push the piston along the bore, follow with the push rod and stop washer, and secure it in position with the circlip. If the push rod stop washer is separate from the push rod, do not overlook it and check that the circlip locates properly into its groove. If the rubber dust cover is not perished or perforated it may be refitted. If it is faulty, renew it.

With the cylinder assembly refitted to the vehicle, check the clearance between the push rod and piston with the piston fully against its stop. If there is not clearance, hydraulic pressure may build up causing the brakes to bind. The pedal adjustment is set when the vehicle is manufactured, and though slight adjustment may be necessary, any excess of free movement, or alternatively, complete lack of free movement should be investigated. The

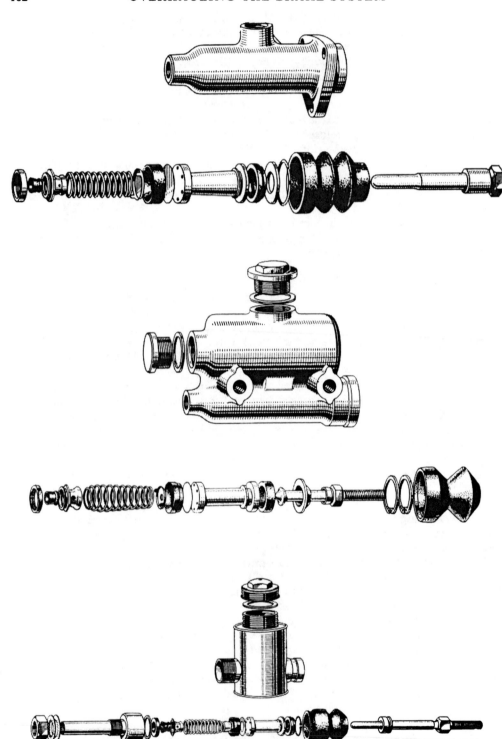

Fig. 9:24   Various types of Lockheed master cylinders. When reassembling cylinders observe absolute
cleanliness and lubricate the seals with brake fluid

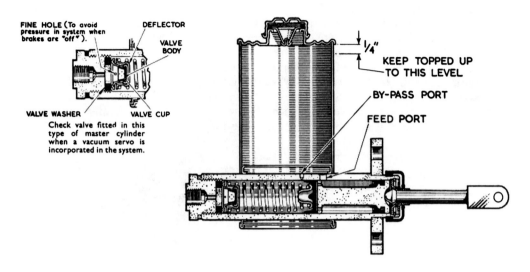

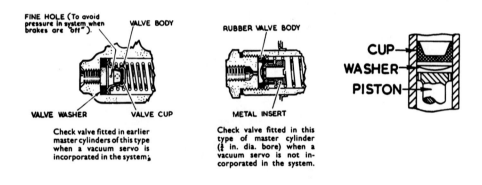

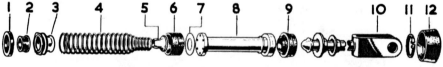

1. Rubber valve-washer
2. Valve cup
3. Check valve body
4. Spring
5. Spring retainer
6. Main cup
7. Piston washer
8. Piston
9. Secondary cup
10. Push rod assembly
11. Circlip
12. Boot

Fig. 9:25   The correct assembly of the Lockheed tank type master cylinder.

free play should be approximately $\frac{1}{32}$ in. and is best felt by moving the pedal up and down slightly with the hand (see fig. 9:27). If the push rod has an adjustable yoke and locknut, adjustment may be carried out by screwing the push rod in or out, the amount being determined by trial and error.

Where the master cylinder fits to the bulkhead as with Girling CV types, see figs. 9:28 and 9:29, shims are sometimes used to obtain the correct clearance but seldom require changing. If undue clearance is found, the pedal linkage is probably worn. Clevis pins which connect the brake pedal to the hydraulic-piston push rod yoke are common offenders.

If the brake cylinder is fed from a supplementary tank, the tank should be removed and flushed before being reconnected.

Before connecting the outlet pipe or pipes, test the cylinder. Fill the fluid reservoir and place some rag adjacent to the master-cylinder fluid outlet to prevent any ejected fluid from damaging the car paintwork, and pump the brake pedal several times.

If fluid is not ejected, remove the cylinder and strip it again for examination. If all is well, reconnect the pipe and bleed the system (see page 188).

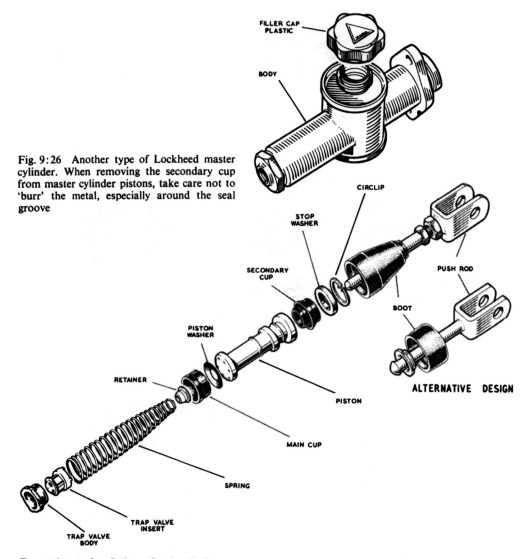

Fig. 9:26 Another type of Lockheed master cylinder. When removing the secondary cup from master cylinder pistons, take care not to 'burr' the metal, especially around the seal groove

Removing and refitting wheel cylinders

To remove a wheel cylinder, drain the fluid from the system, remove the brake shoes and undo the hydraulic pipe, or pipes, by unscrewing the retaining tube-nuts or banjo-type bolt. Where banjo-type bolts are used, or a flexible hydraulic pipe connects direct to the cylinders, note the copper washer, or washers, for replacement, in their respective positions.

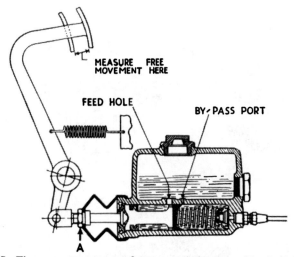

Fig. 9:27 There are many types of master cylinder, but with all of them it is essential that there is always a little free pedal movement. On the above type, A. is the adjustment point

On twin leading-shoe front brakes, each cylinder is bolted firmly to the backplate. With single cylinder brake assemblies, the cylinder 'floats' or slides, within a location slot, on the backplate. It is essential when refitting such a cylinder to make certain that the sliding movement is satisfactory and that the area is free from dirt and corrosion otherwise the reactive movement of the cylinder, when the brakes are applied, will be hampered and the brakes less effective.

Sliding hydraulic cylinders are usually retained in position by a spring clip, see fig. 9:30. When the clip is removed, the cylinder is released, but it may be necessary to take out the bleed screw before the cylinder can be completely removed from the backplate. Mechanically-operated brake-shoe expanders are normally held in position by two nuts, but care must be exercised when refitting them to ensure that they can 'float' satisfactorily.

Overhauling wheel cylinders

Although wheel cylinders differ in design according to the type of brake assembly with which they are used, whatever type is fitted, the rubbers should be renewed every three years, 40,000 miles or at every third change of rear brake shoes, whichever occurs first.

New rubbers can sometimes be fitted to cylinders without removing the cylinder assembly from the backplate (see fig. 9:11). If the cylinders are worn enough to have a ridge, the complete wheel cylinder must be renewed.

Fig. 9:28 Another version of the Girling CV (centre valve) master cylinder. (See also figs. 9:22 and 9:29)

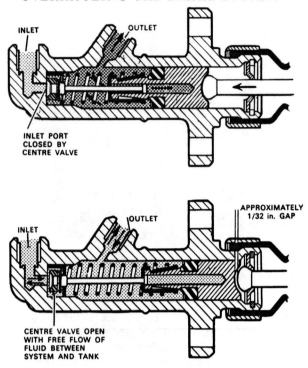

Fig. 9:29   Diagrammatic view of Girling centre valve master cylinder

On Lockheed cylinders fitted with a handbrake lever, see fig. 9:6, remove the outer piston and tap out the lever-retaining pin from its housing and remove the lever. Disconnect the brake pipe and apply a low air pressure to the fluid inlet so as to expel the inner piston, rubbers and spring. If it is difficult to remove the piston, fit a grease nipple to the cylinder (where the hydraulic pipe normally fits) and with the bleeder valve tightened use a grease gun to fill the cylinder with grease. The hydraulic pressure will remove the most stubborn piston. With wheel cylinders fitted with two pistons the grease gun method cannot be used, but a suitably sized drift carefully inserted against the offending piston and a few sharp hammer blows does the trick. Make sure to remove all burrs from the piston and cylinder with fine emery cloth.

When the cylinder has been stripped, the cylinder and components should all be examined. If the cylinder bore has a pronounced ridge, indicating wear, it is advisable to renew the cylinder. In most cases the removal of corrosion with fine emery paper is all that is necessary.

Before reassembling the wheel cylinders, wash all of the parts in methylated spirit and place them on a sheet of paper to dry.

Reassembling the wheel cylinder is simple, and is merely a matter of reversing the dismantling sequence. Make sure to position the rubbers correctly and avoid any grit or foreign matter contaminating them and lubricate them with hydraulic brake fluid.

When assembling rear cylinders, with a lever mechanism, make sure to insert the slotted pistons correctly.

Always fit new rubber dust seals and, when refitting the cylinder to the backplate, make sure to tighten the hydraulic pipe carefully for the threads strip very easily.

If the cylinder is of the 'sliding' type, ensure that the assembly moves on the backplate easily.

If the bleed screw has been removed, refit it, reassemble the brake shoes (see also fig. 9:31), adjust the brakes, and bleed the system (see page 188).

Fig. 9:30 *Above*: Removing plate from a real wheel cylinder. *Right*: Removing the cylinder

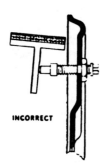

INCORRECT

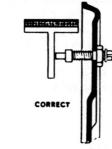

CORRECT

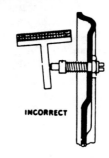

INCORRECT

Fig. 9:31   Steady posts

The function of the steady post is to keep the brake shoe square with the drum. The web of the shoe bears against the steady post on the backplate thereby counteracting the offset pull of the shoe pull-off spring. When relined brake shoes have been fitted, if adjustable steady posts are fitted on the backplate, the locknut on each shoe steady post should be released and each post 'backed off'. The brake shoe adjuster should be expanded until the linings are locked against the drum. The steady posts should then be screwed in to butt fully against the web of its brake shoe. The locknuts must then be tightened, the brake expanders released, and the brake adjusted

Hydraulic pipes and nuts

A flattened hydraulic brake pipe, a seized wheel cylinder or a choked brake hose may cause unequal braking and during a brake overhaul it is advisable to check that all of the pipes are completely clear of blockages by blowing through with a high pressure air line, or foot pump.

While the removal of flexible hoses is straightforward it is important to use the correct-sized spanner. An adjustable spanner, or even a slightly loose-fitting spanner only results in the corners of the hose nut being taken off, for the metal is very soft.

When fitting a new hose, do not forget the copper washer which forms a leak-proof joint with the wheel cylinder, and make certain the hose is of the correct length, and that it will not foul the suspension, especially when the latter is in its fully retracted position. Check too, that the hose cannot be trapped when the steering is on full lock and that the hose is not kinked or twisted.

All brake hoses should be inspected every 12,000 miles for signs of leakage and deterioration. The manufacturers recommend that the hoses are renewed every three years, or 40,000 miles, whichever occurs first. Metal brake pipes should be inspected every 12,000 miles and renewed if any sign of corrosion is found.

Pipe-nuts, which attach steel brake pipes to wheel cylinders and master cylinders, swell at the inner end when the nuts are tightened. When the nut is undone, if excessive swelling has taken place, it will be difficult to pull the nut from the cylinder. When the nut is free from the thread, however, a careful tug whilst turning the nut with a spanner will do the trick. Refitting brake pipes and nuts that have swelled can be troublesome for the nuts tend to go 'cross-threaded'. To avoid this happening, hold the swollen part of the tube-nut with a pair of pliers and whilst keeping a little pressure on the nut turn it about two complete revolutions. The idea is to allow the serrations in the plier jaw to cut a little metal from the swollen part of the nut. Take care to wipe the end of the nut to remove any specks of metal, align the pipe and nut with the cylinder and screw the nut in the first few threads with the fingers.

Do not use a spanner to start the nut. This invariably results in the nut going 'cross-threaded'.

Refitting the brake drums

Before brake drums are refitted, the friction face must be completely free from oil or grease and from any serious 'scoring'.

If scored brake drums are refitted, the brakes take a long time to bed in and an acceptable brake will not be obtained until several hundred brake applications have elapsed. Light scoring is permissible, but if numerous grooves more than 0·002 in. deep are present, have the drums skimmed in a lathe, or renew them. On drums that are rather thin, if the grooves are more than 0·007 in. deep, a new drum should be fitted, particularly on front brakes for a thin-walled drum is likely to distort under pressure when hot and may well cause trouble.

As well as a smooth friction surface, for first-class braking it is essential that drums are truly circular, as any ovality causes the brakes to grab. If an oval drum is suspected, due to one brake grabbing (flats on the tyres are a good indication) change the brake drum to the opposite side. If the grabbing effect is changed to the same side, renew the offending brake drum.

Fierce or grabbing brakes may be due to an accumulation of brake dust in the drum. This should be one of the first things to check and is much more common than an oval brake drum.

Bleeding hydraulic brake systems

To ensure complete safety of brake hydraulics, manufacturers advise that the complete hydraulic system is overhauled every three years and that it is completely drained and refilled with fluid every 18 months. This involves bleeding the system (see fig. 9:32).

If the system becomes contaminated by the use of incorrect brake fluid, the entry of water, or the fluid becoming gummy, or dirty, the entire system must be flushed and cleaned.

To flush the system, remove the fluid as suggested on page 179 and top up the fluid reservoir with industrial methylated spirit and pass at least a quart of spirit through each bleed screw.

To bleed the system, a 12 in. length of rubber tube to fit over the bleed nipple is required, a clean glass jar to recover the fluid, and at least two pints of brake fluid.

If a vacuum servo is fitted, do not have the engine running during the bleeding operation and, if the car has been used recently, pump the brake pedal to destroy any vacuum left in the system.

Fig. 9:32 If disc brake callipers are fitted with two bleed screws, fluid must be taken from each. When bleeding brakes, the floor carpet should be pulled back so that the full stroke of the pedal is obtained. If the brake fluid is clean it may be used again provided that it has not been in the system for more than eighteen months and it is allowed to stand until completely free from air bubbles

If a bleed screw is fitted to the master cylinder, this is the place to commence bleeding. Attach the tube to the nipple and slacken the nipple half a turn. Immerse the end of the tube in a small quantity of brake fluid and get an assistant to depress the brake pedal slowly, allowing it to return unassisted. Repeat the pumping action, allowing a slight pause between each stroke. Watch the glass jar and, when all air bubbles cease, hold the pedal down firmly and tighten the bleed screw.

During all bleeding operations, the fluid level in the reservoir must not be allowed to fall below $\frac{1}{3}$ of the total reservoir capacity, otherwise air may be drawn into the system and the complete bleeding operation will have to be repeated.

If the master cylinder does not possess a bleed screw, start bleeding from the front steering-wheel side and follow with the non-steering-wheel front side, and then repeat the same sequence at the rear. Some cars (*e.g.* Cortina), due to the layout of the system, do not have a bleed screw on the steering side at the rear.

If disc brakes are fitted to the front of the car, the same order of procedure applies. Where callipers have a bleeder on the mounting half of the calliper and another bleeder on the rim—or outer half, bleeding should be carried out from the inner bleeder first, and then the outer bleeder, of each calliper.

When bleeding Girling CV master cylinders, which can be recognized by the aluminium body instead of cast iron body (see figs 9:22 and 9:28), push the pedal through a full stroke followed by three short rapid strokes, allow the pedal to return to its stop and repeat the sequence until the air bubbles cease.

When the system has been bled, if the pedal feels spongy, the system should be bled again, indeed, it is by no means unusual to have to 'bleed' twice before the system is completely free from air.

Provided the fluid removed from the system is clean, it may be used again, but must be allowed to stand until completely free from air bubbles.

On completion of the bleeding operation, top up the reservoir to ¼ in. below the filler neck and replace the cap. Check the system by applying pressure to the brake pedal for

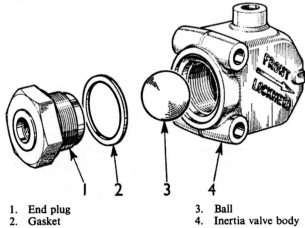

1. End plug   3. Ball
2. Gasket   4. Inertia valve body

Fig. 9:33   This inertia valve unit is mounted so that it is 28° from the horizontal. The valve cannot be adjusted. Failure of operation of this type of valve may be indicated by the rear wheels locking.

two or three minutes and examine the entire system for leaks. If the pedal gradually sinks, a fluid leak is denoted which must be found and rectified. If the pedal feels spongy, repeat the bleeding operation.

Hydraulic pressure-limiting valves

Cars with a high transference of weight from the rear wheels to the front wheels during braking are sometimes fitted with proportioning valves to reduce the locking at the rear

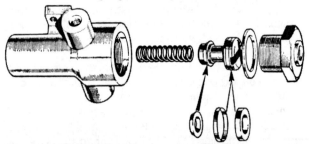

Fig. 9:34   The Mini pressure regulating valve. When new hydraulic seals are fitted to the master cylinder etc., this component should not be overlooked. It is also important that a *copper* sealing washer is fitted; aluminium types on no account should be used and must be replaced at once if found

wheels. Some systems are also self-compensating for braking on gradients. The Girling proportioning system in fig. 9:33 is fitted to the Austin Westminster. When bleeding the system, the proportioning valve can cause a little puzzlement if the car is jacked high at the rear for it is impossible to obtain a good flow of fluid from the rear brake bleed screws as the valve restricts the flow.

Several types of limiting valves have been tried on front-wheel-drive cars but the most usual is the Mini-type pressure-regulating valve (see fig. 9:34), the seals of which should be renewed during periodic overhauls. Simple to remove, overhaul and refit, it is essential to make sure that a copper seating washer is fitted (the aluminium type should be replaced by a copper washer) and to ensure that the plug is fully tightened.

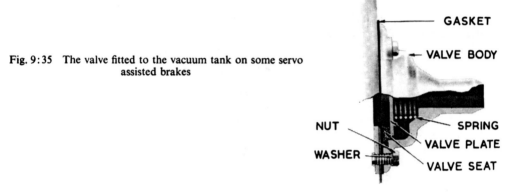

Fig. 9:35   The valve fitted to the vacuum tank on some servo assisted brakes

Vacuum tanks, servos and check valves

Servo assemblies are remarkably trouble-free, but if the hydraulic system becomes contaminated with mineral oil, a service exchange servo is advisable. Faults in the servo can result in an unusually hard brake pedal. A quick check to see whether the servo is working can be made by allowing the engine to idle and pressing the brake pedal firmly. If the engine

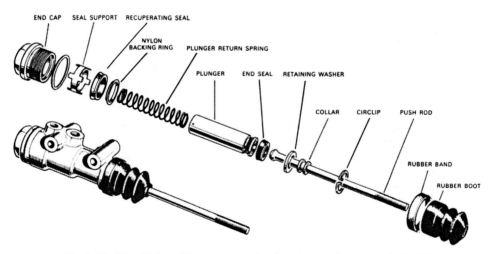

Fig. 9:36   The Girling CB or compression barrel type of master cylinder. To dismantle, the end cap must be removed

note does not change, the vacuum line to the manifold will probably be found to have chafed through or the vacuum check valve (see fig. 9:35) is not functioning. If the leak is difficult to trace, oil squirted over the suspected area will be visibly sucked in where the leak is occurring.

Cars fitted with vacuum reservoir tanks often incorporate a check valve in the vacuum tank. The reservoir is linked by means of a pipe so that it is between the inlet manifold and the servo, to smooth out fluctuations of manifold depression and provide a reserve of vacuum when the engine is not running. Sometimes a gummy deposit clogs the valve resulting in a brake pedal which has to be pushed much harder than usual to obtain the same degree of braking. In this event the valve assembly, which is basically a valve plate held on to a rubber seat by a spring, must be removed by undoing the retaining nuts and the valve cleaned and refitted. Sometimes the valve is separate from the tank, while in some cases it is fitted to the servo.

To check the valve operation, place a hand on the brake pedal and start the engine. Movement of the pedal denotes the suction valve is not closing correctly.

With the engine running, press the brake pedal lightly and, keeping the pedal in one position, switch off the engine. A definite reactionary kick on the pedal should be felt.

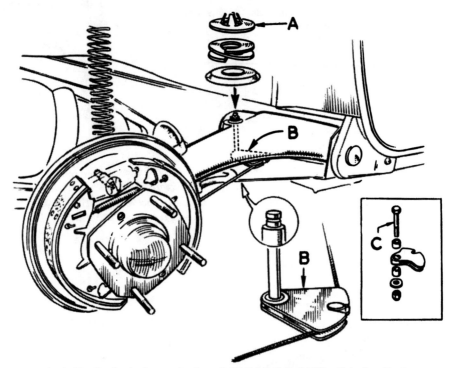

Fig. 9:37   During brake overhauls on B.M.C. Minis and 1100s, if the handbrake cable sectors are seized on their pivots (see B Hydrolastic Mini and C dry cone suspension Mini) the sector pins should be withdrawn from the radius arms, thoroughly cleaned and the pins greased with polybutylcuprysil (PBC lubricant). On hydrolastic suspension Minis a new spire clip A will be required. On 1100s and 1300s the cable sector is similar to 'C' but swivels on a stud screwed into the underside of the radius arm

CHAPTER 10

Curing Clutch, Gearbox, Overdrive and Universal-Joint Failure

Apart from B.M.C. 1100s and Minis, which are dealt with separately on page 101, access to the clutch entails removal of the gearbox, or engine from the car, and sometimes both. However, once access to the clutch assembly has been obtained, a clutch overhaul can be carried out easily and rapidly by fitting service replacement parts.

The diagnosis of clutch troubles, by reference to the following pages, is straightforward but, in all cases of poor clutch operation, before attempting any diagnosis it is worth adjusting the clutch-arm free movement.

If a clutch overhaul has to be carried out, take the opportunity to do other jobs at the same time. On older cars it may be worth replacing the flywheel ring gear and possibly fitting new brushes to the starter motor and cleaning the starter pinion drive. If the clutch slave-cylinder shows any sign of a leak, or if the tie-rod rubbers require renewing this, too, should be done.

Adjusting the clutch

Mechanically operated clutches are provided with an adjustment on the linkage mechanism (see fig. 10:1). Access is usually from underneath the car. Adjustment is usually carried out to allow 1 in. free movement at the pedal pad.

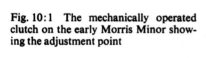

Fig. 10:1 The mechanically operated clutch on the early Morris Minor showing the adjustment point

The adjustment provided on hydraulically operated clutches varies according to the make of car. On some, an adjustable master-cylinder push rod is provided (see fig. 10:2) on others the correct clearance between the clutch thrust bearing and the release plate is obtained by adjustment of a special rod fitted alongside the slave cylinder. On the Triumph 1200 and S model, the adjustment of the clutch is automatic.

The Mini and B.M.C. 1100 and 1300 method of clutch adjustment is depicted in fig. 10:3.

Fig. 10:2   Clutch linkage adjustment on the Vauxhall VX 4/90. (A should equal ⅛ in.)

Hydraulic systems

Faults due to an inefficient hydraulic system can be diagnosed by placing the car on level ground, pressing the clutch pedal, engaging gear and keeping the clutch pedal fully depressed. If the car starts to creep forward (within 30 seconds) the slave-cylinder rubbers and clutch master cylinders require renewing. Figs. 10:4 and 10:5 depict typical master cylinders while figs. 10:6, 7 and 8 depict typical slave cylinders.

After renewing master-cylinder rubbers or slave-cylinder rubbers, the hydraulic system must be bled.

Fig. 10:3   Adjustment of the B.M.C. Mini, 1100 and 1300 clutch

An adjustable stop (arrowed) is provided just forward of the clutch operating lever. To carry out the adjustment pull the operating lever outwards until all free movement is taken up and then check with a feeler gauge that there is a clearance of 0·020 in. (0·5 mm) between the operating lever and the head of the adjustment bolt (early Minis 0·060). Correct if necessary.

If the throwout plunger is removed at any time or if a new clutch plate is fitted, the stop (not fitted to early Minis) on the clutch release shaft must also be reset. To do this, screw the stop and locknut away from the cover boss about two complete turns. Depress the clutch pedal to fully release the clutch, hold it in this position and screw the stop up against the cover boss, release the clutch pedal to fully engage the clutch, screw the stop up one 'flat' of the hexagon and tighten the locknut. Check the clearance at the lever stop screw and adjust if necessary

Chart 10:1

Inspection of the Clutch Assembly

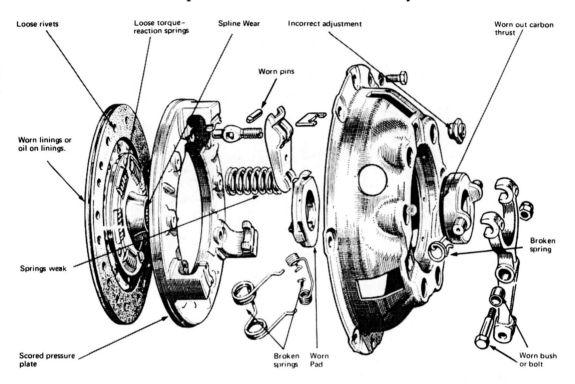

Loose rivets

Loose torque-reaction springs

Spline Wear

Incorrect adjustment

Worn out carbon thrust

Worn pins

Worn linings or oil on linings.

Springs weak

Broken spring

Scored pressure plate

Broken springs

Worn Pad

Worn bush or bolt

Note: If faults with the pressure-plate cover assembly are found a reconditioned assembly is advisable.
(Diaphragm Clutch see chart 10:2)

Bleeding the hydraulic clutch system

An assistant is required to depress the clutch pedal. Also a length of rubber tube to fit over the bleed nipple, a clean glass jar to receive the fluid, and at least $\frac{1}{4}$ pint of clutch hydraulic fluid.

Attach the bleed tube to the slave-cylinder nipple and immerse the end of the tube in a small quantity of fluid. Open the bleed screw half a turn, have the assistant press the clutch pedal and watch for air bubbles in the fluid. When the pedal is fully depressed close the bleed nipple. The pedal should now be allowed to return unassisted to its normal position. The complete cycle should be repeated until the air bubbles cease to appear. The master-cylinder fluid must be topped up after every 3 or 4 strokes of the pedal. On completion of bleeding, top up the master-cylinder reservoir and replace the cap.

Clutch slip

If clutch slip is suspected, chock the front wheels and, with the handbrake fully on, start the engine, press the clutch pedal, place the car in top gear and open the throttle to about 2000 rev/min and let in the clutch slowly. If the engine stalls, the clutch is satisfactory.

Common causes of clutch slip are: binding of the pressure plate release levers; non-return of the clutch pedal owing to the pedal being seized to its fulcrum, and oil on the driven-plate linings, or the linings stripped from the driven-plate.

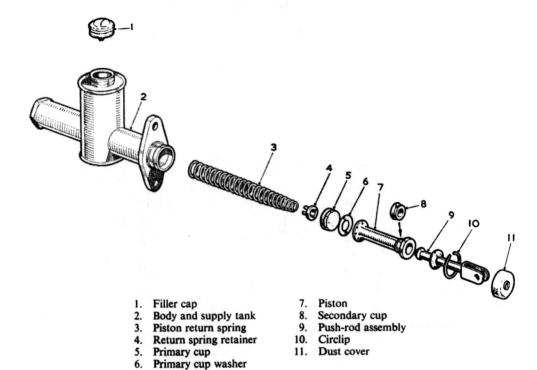

1. Filler cap	7. Piston
2. Body and supply tank	8. Secondary cup
3. Piston return spring	9. Push-rod assembly
4. Return spring retainer	10. Circlip
5. Primary cup	11. Dust cover
6. Primary cup washer	

Fig. 10:4   The clutch master cylinder components

Dragging or spinning clutch

A dragging clutch is noticeable by the gears grating on engagement of first gear; and also when reverse gear is engaged. If the clutch is adjusted and the trouble persists, the clutch will have to be dismantled. A dragging clutch is sometimes due to a distorted driven-plate.

Clutch judder

Judder may be caused by oil on the driven-plate linings, the driven-plate not running true with the flywheel, a disintegrated spigot bush, or the driven-plate binding on the splines.

Clutch fierceness

Fierceness may be due to the friction linings being worn; the driven-plate hub binding on the first-motion-shaft splines, or oil and grease on the linings. Incorrect pressure-plate springs, too, can result in fierceness. If new springs are fitted to a pressure-plate assembly, or an exchange assembly is fitted, it is important to ensure that the springs are of the correct colour; the colour is a code denoting the strength of the springs.

Noisy clutch operation

If a noise is apparent (particularly a high pitched whine or a harsh 'burring' type of sound) when the pedal is depressed about an inch but the noise ceases when the pedal is released, the clutch release bearing (see figs 10:9 and 10:10) probably requires renewing.

If a noise is apparent when the clutch pedal is fully depressed, and the clutch does not operate, it may be owing to collapsed crankshaft thrust washers. If this is suspected, watch the crankshaft pulley whilst the clutch pedal is pressed a few times and the engine running, and observe the end-float on the crankshaft. If excessive float is present (more than 0·005 in.), the sump will have to be removed for renewal of the thrust bearings.

Loud metallic rattles may sometimes be traced to lack of tension on the anti-rattle springs which fit to the toggle arms. A bent release fork may also cause a rattle (when the pedal is pushed) and may be accompanied by clutch drag, fierceness, or juddering.

A graphite-release bearing, if completely worn out, sometimes causes a knock when the pedal is pushed owing to the graphite holder fouling the pressure plate. At a later stage this may result in broken clutch-release plate straps and complete failure of the clutch.

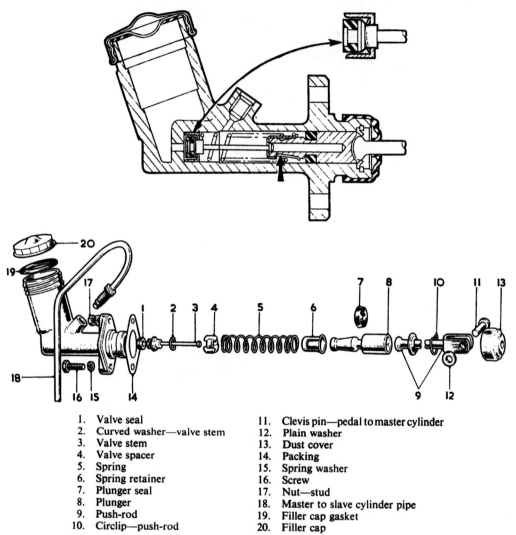

1.	Valve seal	11.	Clevis pin—pedal to master cylinder
2.	Curved washer—valve stem	12.	Plain washer
3.	Valve stem	13.	Dust cover
4.	Valve spacer	14.	Packing
5.	Spring	15.	Spring washer
6.	Spring retainer	16.	Screw
7.	Plunger seal	17.	Nut—stud
8.	Plunger	18.	Master to slave cylinder pipe
9.	Push-rod	19.	Filler cap gasket
10.	Circlip—push-rod	20.	Filler cap

Fig. 10:5   The clutch master cylinder of the B.M.C. 1800. The upper (sectional) drawing shows the correctly assembled unit with detail of the centre valve enlarged; the arrow indicates the thimble leaf. The exploded drawing shows parts in order of assembly.

Removing the clutch assembly

Removal of the clutch may entail removing either the engine or gearbox, depending entirely upon the design of the car. If the engine has to be taken out, lifting equipment will be needed and a decision as to whether the engine and gearbox come out as one unit, or whether the engine can be split from the gearbox and removed separately. On the Austin

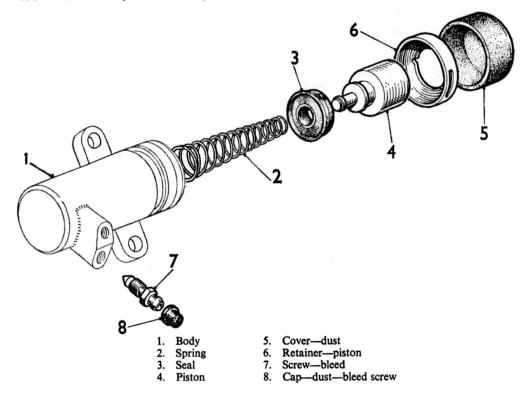

1. Body
2. Spring
3. Seal
4. Piston
5. Cover—dust
6. Retainer—piston
7. Screw—bleed
8. Cap—dust—bleed screw

Fig. 10:6   Assembly details of typical slave cylinders. *Top*: B.M.C. 1800
Bottom: Cortina slave cylinder.

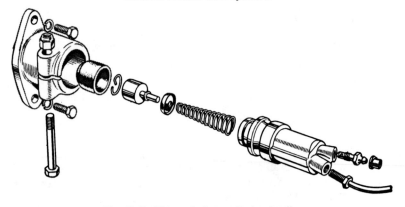

Fig. 10:7   Triumph slave cylinder details

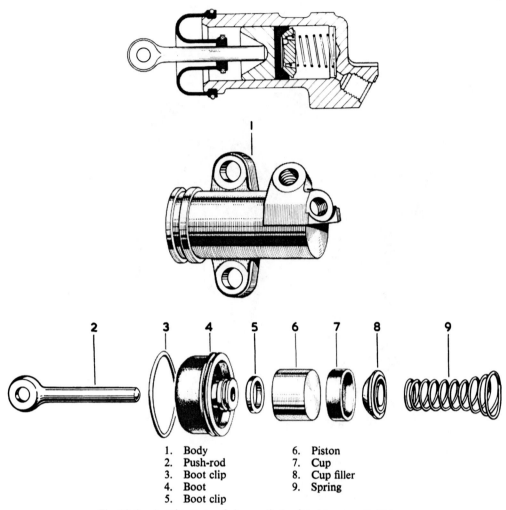

1. Body
2. Push-rod
3. Boot clip
4. Boot
5. Boot clip
6. Piston
7. Cup
8. Cup filler
9. Spring

Fig. 10:8 Another type of slave cylinder fitted to some British cars

Healey 3000, MGB, Hillman and many other cars, the gearbox can be lifted out. On some models this entails the removal of the gearbox tunnel. On other models, if the car can be run on to ramps, or placed over a pit, the speedometer cable, propeller shaft, etc. can be removed and the gearbox lifted out from below the car. In all cases, the rear of the engine must be supported.

On B.M.C. 1100s and Minis, the engine must be supported, the engine mounting disconnected and the clutch cover removed, see fig. 10:10. A socket spanner size $1\frac{1}{2}$ in. AF is required to undo the flywheel-retaining bolt and a special puller will be necessary to release the flywheel/clutch assembly which comes off as a complete unit. (Overhauling, see page 101.)

On the Vauxhall Victor a crossmember has to be removed before the bottom clutch cover plate can be removed and access to the gearbox-retaining bolts obtained.

Several different makes of cars suffer from the clutch thrust pressure-plate-retaining straps breaking. These clutches have now been superseded by a better design (see page 100) so, when fitting a replacement clutch bowl assembly, make sure that it is the modified type without the straps. A broken strap is usually first noticeable as a noise from the clutch casing.

Reassembling the clutch, see page 117.

Inspection of Diaphragm Clutches

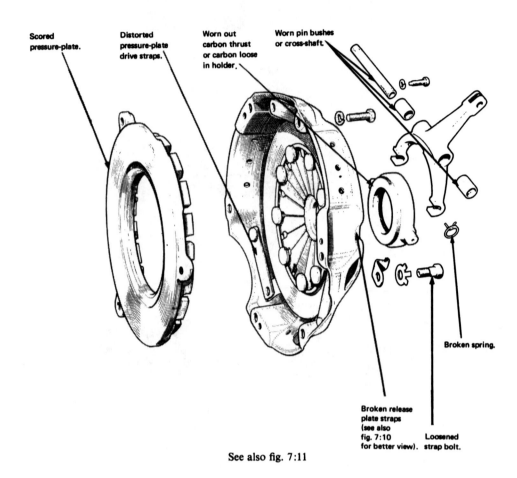

Scored pressure-plate.

Distorted pressure-plate drive straps.

Worn out carbon thrust or carbon loose in holder.

Worn pin bushes or cross-shaft.

Broken spring.

Broken release plate straps (see also fig. 7:10 for better view).

Loosened strap bolt.

See also fig. 7:11

Gearboxes

Without previous experience, stripping and overhauling a gearbox can be a somewhat hazardous, as well as an expensive, task, depending on the type and design of the gearbox. Gearbox parts are expensive, and with a fairly old car it may pay to obtain a second-hand unit or, on a more recent model, a works reconditioned gearbox may be the answer. On the other hand some gearboxes are fairly easy to overhaul, indeed if some specific fault such as jumping out of gear, or faulty synchromesh, is the trouble the task can be successfully tackled, provided the correct sequence of stripping and rebuilding is adhered to.

Diagnosis of gearbox faults (see also chart 10:3)

To diagnose gearbox faults, drive the car in each gear accelerating hard, and then lifting off to allow the car to over-run the engine. A highly pitched whine coming from the gearbox usually denotes worn bearings; a regular light knock, possibly a chipped tooth. Jumping out of gear may be due to weak selector detent springs or it may indicate either a collapsed, or badly worn mainshaft thrust washer. Austin A40s are particularly susceptible to this. A sloppy gear lever can result in difficult gear selection. On some cars, a polythene bush fixed on the base of the gear lever becomes worn, perished or displaced and results in excessive lever movement.

On B.M.C. 1100s, Minis and Mini Coopers, the bushes where the linkage shaft fits through the transmission case (see fig. 7:2) may become worn giving rise to undue movement on the gear lever.

On late Minis fitted with remote control levers, and earlier Mini Coopers and S types, the bolt retaining the primary shaft lever, see fig. 7:2, may become slack resulting in a floppy lever. To gain access to the bolt, remove the rubber dust cover on the rear of the remote control.

Where gear-change levers are attached to the steering column (see fig. 10:11) a floppy lever may be indicative of general wear in the linkage, or if a gear cannot be engaged at all, it may be nothing more than a slack linkage-retaining nut, or a linkage bolt broken off, allowing one of the selector linkage rods to flop about.

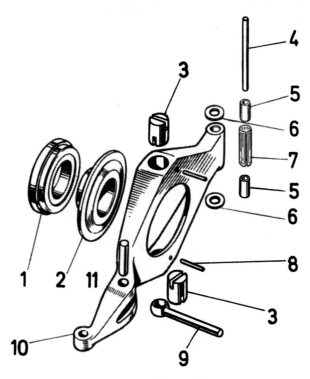

1. Release bearing	7. Tolerance ring
2. Release bearing carrier	8. Lock pin
3. Thrust plugs	9. Push rod
4. Hinge pin	10. Operating lever
5. Bushes	11. Pushrod pin
6. Distance washer	

Fig. 10:9   The clutch release bearing and operating lever assembly fitted to the Herald 1200, 12/50, 13/60 and Spitfire

On high-mileage gearboxes, the cumulative effect of wear on bearings, splines, selectors and thrust washers can result in sloppy gear lever action and jumping out of gear on the over-run. Temporary relief from jumping out of gear may be obtained by fitting stronger selector detent springs. On some cars, where the springs are readily accessible by taking off the gearbox top, it is worth trying new springs. If this does not cure the trouble the

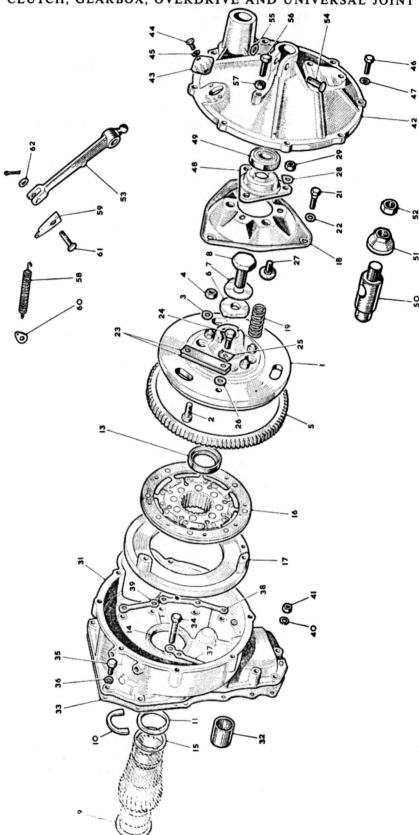

Fig. 10:10   Assembly diagram for early type B.M.C. 1100 and Mini clutches. After 1964 the flywheel oil seal was dispensed with and a different type primary gear assembly was fitted. Early cars may be converted.   Removing clutch assembly, see page 199

KEY TO THE FLYWHEEL AND CLUTCH COMPONENTS

No.	Description	No.	Description
1.	Flywheel assembly.	23.	Driving strap.
2.	Hub screw.	24.	Strap screw.
3.	Lockwasher.	25.	Lockwasher.
4.	Nut.	26.	Washer.
5.	Starter ring.	27.	Thrust plate screw.
6.	Key.	28.	Lockwasher.
7.	Lockwasher.	29.	Nut.
8.	Flywheel screw.	31.	Flywheel housing.
9.	Gear thrust washer (front).	32.	Idler gear bearing.
10.	Gear thrust washer (rear).	33.	Housing joint.
11.	Backing ring.	34.	Housing screw (long).
13.	Flywheel oil seal.	35.	Housing screw.
14.	Housing oil seal.	36.	Washer.
15.	Primary gear retaining washer.	37.	Housing lockwasher.
16.	Clutch driven-plate.	38.	Housing lockwasher.
17.	Pressure plate.	39.	Housing lockwasher.
18.	Pressure spring housing.	40.	Washer.
19.	Pressure spring.	41.	Nut.
21.	Pressure plate screw.	42.	Clutch cover.
22.	Washer.	43.	Cover-plate.

No.	Description
44.	Screw.
45.	Washer.
46.	Cover screw.
47.	Washer.
48.	Clutch thrust plate.
49.	Release bearing.
50.	Throw-out plunger.
51.	Throw-out stop.
52.	Stop locknut.
53.	Clutch operating lever.
54.	Lever pin.
55.	Washer.
56.	Screw.
57.	Locknut.
58.	Lever pull-off spring.
59.	Spring anchor (lever).
60.	Spring anchor (cylinder).
61.	Push-rod pin.
62.	Washer.

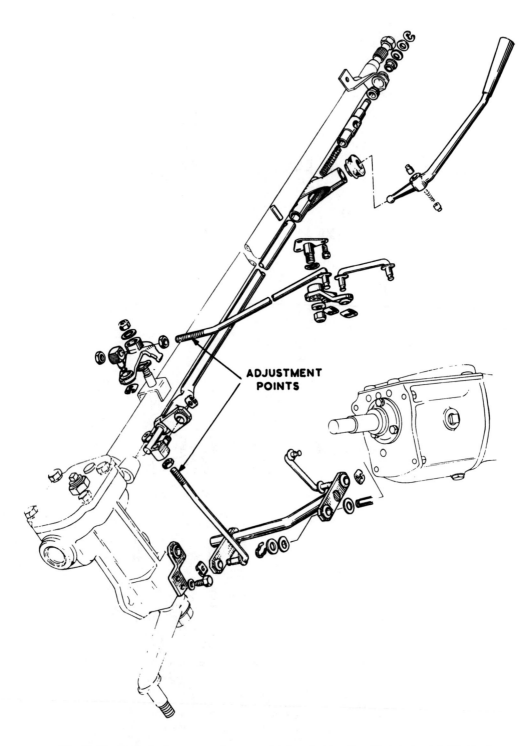

ADJUSTMENT
POINTS

Fig. 10:11   Typical column gear change mechanism showing the adjustment points (Cortina)

Chart 10:3

Diagnosis of Gearbox Faults

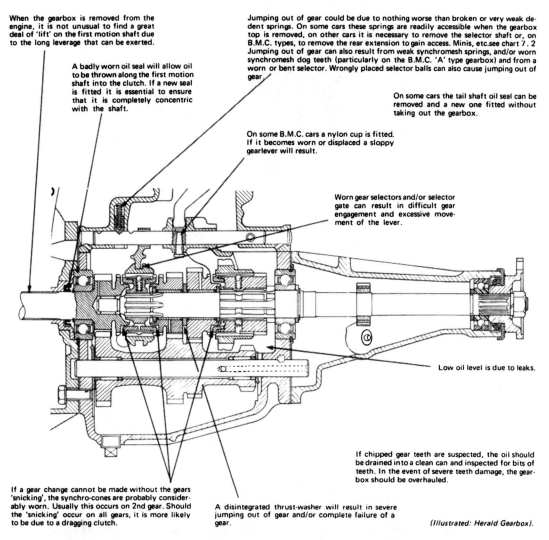

When the gearbox is removed from the engine, it is not unusual to find a great deal of 'lift' on the first motion shaft due to the long leverage that can be exerted.

A badly worn oil seal will allow oil to be thrown along the first motion shaft into the clutch. If a new seal is fitted it is essential to ensure that it is completely concentric with the shaft.

Jumping out of gear could be due to nothing worse than broken or very weak detent springs. On some cars these springs are readily accessible when the gearbox top is removed, on other cars it is necessary to remove the selector shaft or, on B.M.C. types, to remove the rear extension to gain access. Minis, etc. see chart 7.2 Jumping out of gear can also result from weak synchromesh springs, and/or worn synchromesh dog teeth (particularly on the B.M.C. 'A' type gearbox) and from a worn or bent selector. Wrongly placed selector balls can also cause jumping out of gear.

On some cars the tail shaft oil seal can be removed and a new one fitted without taking out the gearbox.

On some B.M.C. cars a nylon cup is fitted. If it becomes worn or displaced a sloppy gearlever will result.

Worn gear selectors and/or selector gate can result in difficult gear engagement and excessive movement of the lever.

Low oil level is due to leaks.

If chipped gear teeth are suspected, the oil should be drained into a clean can and inspected for bits of teeth. In the event of severe teeth damage, the gearbox should be overhauled.

If a gear change cannot be made without the gears 'snicking', the synchro-cones are probably considerably worn. Usually this occurs on 2nd gear. Should the 'snicking' occur on all gears, it is more likely to be due to a dragging clutch.

A disintegrated thrust-washer will result in severe jumping out of gear and/or complete failure of a gear.

(Illustrated: Herald Gearbox).

gearbox will have to be removed from the car for overhaul. Mini and 1100 detent springs are readily accessible by removing two plugs found on the bottom rear side of the transmission case.

Chart 10:3 will help the diagnosis of gearbox troubles. Chart 10:4 shows the location and correct assembly of detent-locking devices in B.M.C. A-type gearboxes and is fairly typical of current practice.

OVERDRIVES

The overdrive fitted to British cars is usually the Borg Warner or the Laycock de Normanville. From a mechanical aspect they are both remarkably trouble-free; unfortunately this cannot be said quite so confidently regarding the electrical control units and external wiring.

Chart 10:4

The Location of Selector Locking Balls, Plungers and Springs (B.M.C. A-Type Gearbox)

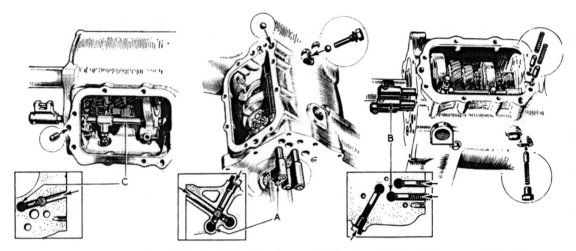

A. First and second gear fork rod in place (gearbox upside-down). B. Third and fourth gear fork rod in place. C. Reverse gear fork rod in place

Laycock de Normanville overdrive

If a Laycock overdrive does not engage, it pays first to investigate the fuse to which the overdrive circuit is connected, and then to ascertain whether the switch operated by the gear lever selector has become loose and so is not making contact. Wires on the fascia panel switch may be detached, or a Lucar connecting terminal may sometimes be found pulled from the solenoid, or gear-lever switch. A guide to operating faults is on page 211.

If a brief inspection fails to indicate the trouble, consult the wiring diagram (fig. 10:12) and methodically check the system. Failure of the overdrive to engage may also be due to the solenoid-operating linkage being out of adjustment and, in the event of trouble, this should always be checked. To do this, on the type of overdrive fitted to the Austin Healey, a hole in the valve setting lever (on the right-hand side of the overdrive unit) must be aligned (using a $\frac{3}{16}$ in. rod) with a hole in the overdrive casing. The adjustment between the operating lever linkage and solenoid, on the Healey, is on the opposite side of the casing to the lever but on smaller units such as fitted to the MGB the operating lever and solenoid are on the same side as each other (see fig. 10:13) and are covered by a small plate.

Most gearbox tunnels are fitted with a plate, or rubber grommet, to allow access to the solenoid, but access to the lever, on some cars, is from underneath the car.

The controls are set correctly when a $\frac{3}{16}$ in. (4·76 mm) diameter rod can be passed through the hole in the lever and into the hole in the casing when the solenoid is energized, *i.e.* with the ignition switched on, top gear engaged, and the fascia switch in the 'overdrive' position (fig. 10:14).

If the solenoid operates, but does not move the setting lever sufficiently to allow the insertion of the rod, the solenoid plunger linkage must be adjusted by screwing the self-locking nut on the plunger in or out as necessary with the plunger pushed into the solenoid as far as it will go. The plunger must be held against rotation with a spanner, a flat being provided on the spindle for the purpose. The adjustment must be carried out so that the operating-lever fork just contacts the nut when the $\frac{3}{16}$ in. rod is positioned.

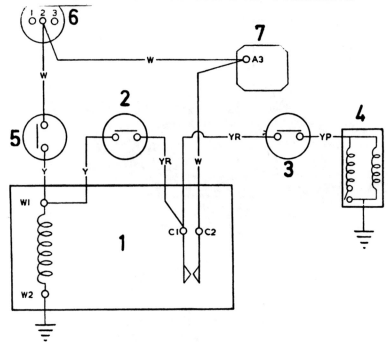

1. Relay
2. Throttle switch
3. Gear switch
4. Solenoid
5. Driver's switch
6. Ignition switch
7. Fuse block

Fig. 10:12 Laycock de Normanville overdrive electrical circuit (single relay)

Removing and refitting an overdrive

There is little difficulty in removing an overdrive from a gearbox but when it is being refitted, trouble in repositioning it may be encountered for it is important not to force the unit. The overdrive will fit quite easily provided the planet carrier and unidirectional clutch are in alignment.

1. Blanking plug
2. Operating lever
3. Solenoid plunger
4. Solenoid

Fig. 10:13 View from right hand side of Laycock de Normanville overdrive unit showing the solenoid cover removed

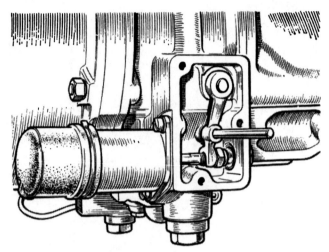

Fig. 10:14   Laycock overdrive. Aligning the operating lever to check the adjustment of the operating plunger

The best way of refitting the unit is to engage a gear and turn the gearbox first-motion-shaft to and fro to assist in starting the shaft on the splines. Take care, however, that it is the lowest part of the oil pump cam that contacts the oil pump plunger otherwise the plunger will be damaged and the unit will not go on.

If difficulty is still encountered, remove the unit and, supporting it in an upright position, insert a dummy shaft to line up the planet carrier with the unidirectional clutch which lies in the back end of the overdrive unit. If the shaft will not go in, use a long thin screwdriver to line up (by eye) the splines in the planet carrier and clutch, turning anticlockwise only.

When this has been done, insert the dummy shaft to complete the alignment procedure, rotating it slightly as it is fed into the unit. Once the shaft is fully home (see fig. 10:15) it may be taken out again and the unit carefully fed onto the gearbox shaft, taking care once again, as the overdrive and gearbox case come together, that the overdrive oil pump engages the cam correctly.

Overdrive relay system

The circuit shown in fig. 10:12 is the single relay system controlled through a manually operated toggle switch, and includes the following components:

(1) *Relay*. An electromagnetic switch used with item (4) to enable an interlocking

Fig. 10:15 Dummy mainshaft inserted into the Laycock overdrive unit to check the alignment of the planet carrier and uni-directional clutch. Note also the oil pump plunger. When refitting the over-drive, care must be taken that the plunger does not become jammed against the eccentric on the mainshaft

safeguard to be incorporated against changing out of overdrive with the throttle closed.
(2) *Throttle switch.* A vacuum-operated switch to over-ride the toggle switch under closed throttle conditions.
(3) *Gear switch.* A small plunger-operated switch allowing overdrive to be engaged only in the two highest forward gear positions.
(4) *Solenoid unit.* An electromagnetic actuator to engage the overdrive mechanism by opening the hydraulic control valve.

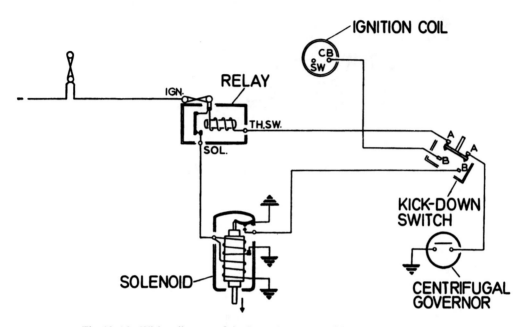

Fig. 10:16   Wiring diagram of the Borg Warner overdrive electrical circuit

Operation

When the driver engages 'overdrive' by closing the toggle switch contacts, current is fed through the ignition switch and fuse unit supply terminal to energize the relay-operating coil. Closure of the relay contacts connects the gear switch and, provided third or top gear is engaged, will energize the solenoid unit and effect a change from direct drive to overdrive.

Change from overdrive to direct drive is by selecting a low gear (when the gear switch contact will open) or by moving the toggle switch to 'NORMAL' with the throttle open (when the vacuum switch will open).

If carried out with the throttle closed (high manifold depression) the vacuum switch will over-ride the toggle switch, delaying the change until the engine takes up the drive.

Borg Warner overdrive

If overdrive fails to engage, or disengage, it is probably due to an electrical fault and, if the fuse and external wiring appear satisfactory, the electrical components should be tested, referring to the wiring diagram (fig. 10:16) and accompanying notes.

The most likely fault is a disconnected wire or, due to infrequent use of the overdrive, corroded switch contacts.

Access to the switch terminals is not easy and an assistant will be required to operate the overdrive whilst the different tests are carried out. It is advisable to have the rear of the car on firm supports so that the overdrive can be operated.

Overdrive fails to release

If the overdrive has failed to release, disconnect the relay at the SOL terminal. If the overdrive then releases the relay must be renewed.

To check the operation of the relay, switch on the ignition and earth the TH SW terminal. If the relay does not 'click' it is faulty and should be renewed.

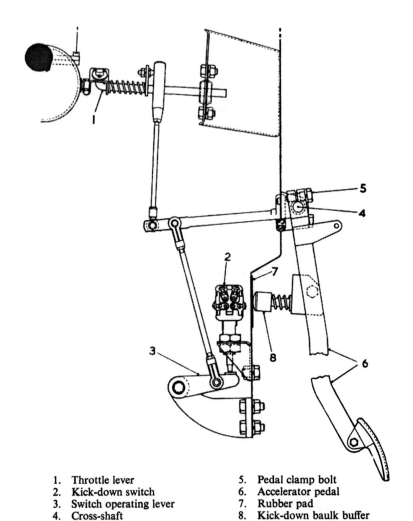

1.	Throttle lever	5.	Pedal clamp bolt
2.	Kick-down switch	6.	Accelerator pedal
3.	Switch operating lever	7.	Rubber pad
4.	Cross-shaft	8.	Kick-down baulk buffer

Fig. 10:17 Accelerator linkage with kick-down baulk (Borg Warner overdrive). Adjustment of the accelerator pedal and Borg Warner overdrive kick-down switch on the Austin Westminster is made by slackening the clamp bolt and repositioning the pedal on the accelerator cross-shaft. The position of the kick-down switch is adjusted by means of the two large clamp nuts securing the switch to its bracket on the bulkhead

Kickdown switch

With the ignition switched on, earth the two terminals of the kickdown switch in turn. If a click does not occur a defective wire or terminal connection is indicated between the throttle switch and relay. If a click occurs when earthing one A terminal but not the other, a faulty kickdown switch is indicated.

Chart 10:5

Guide to Overdrive Operating Faults

Laycock de Normanville overdrive (single-relay system)
(*For Borg Warner overdrive see page* 209)

Overdrive does not engage

Insufficient oil in the unit*.
Solenoid not operating due to fault in electrical system.
Solenoid-operating lever out of adjustment.
Insufficient hydraulic pressure due to pump non-return valve incorrectly seating (probably dirt on ball seat).
Pump not working.
Filter blocked.
Insufficient hydraulic pressure due to sticking or worn relief valve.
Damaged gears, bearings, or moving parts within the unit requiring removal and inspection of the assembly.

Overdrive does not release

(Do not engage reverse gear as damage may be caused to the unit)
Fault in the electrical control system.
Blocked restrictor jet in operating valve.
Solenoid operating lever out of adjustment.
Sticking clutch.
Damaged parts within the unit necessitating removal and overhaul of the assembly.

Clutch slip in overdrive

Insufficient oil in the unit.
Solenoid operating lever out of adjustment.
Insufficient hydraulic pressure due to pump non-return valve incorrectly seating (probably dirt on ball seat).
Insufficient hydraulic pressure due to sticking or worn relief valve.

Clutch slip in reverse or free-wheel condition on overrun

Solenoid operating lever out of adjustment.
Partially blocked restrictor jet in operating valve.
Worn clutch linings.

* On no account should anti-friction additives be used in the oil

If a click occurs as each terminal is earthed in turn with the ignition still switched on, earth the lead on the centrifugal switch that connects to the A terminal of the kickdown switch. If the solenoid fails to operate when the kickdown switch is used, the switch is faulty or out of adjustment.

If the overdrive is failing to release, disconnect one of the A terminals; if the overdrive then releases, renew the switch.

Next, with the engine running, earth each B terminal in turn with the switch fully depressed. If earthing one of the B terminals does not stop the engine, examine the wiring between the kickdown switch and the coil.

Adjusting the kickdown switch

The position of the switch can be altered by slackening the locknut, and the clamping nut, which secures the switch to its bracket. Adjust the switch so that the plunger is fully depressed just before the accelerator pedal touches the floor. Make sure to retighten the clamping nuts and then adjust the throttle rod so that the throttle is fully opened just as the pedal touches the switch plunger.

Checking the centrifugal switch

If the overdrive does not engage, switch on the ignition and earth the terminal that is connected to the kickdown switch. If the solenoid works, the centrifugal switch is faulty and must be renewed.

Checking the solenoid

Switch on the ignition and earth the TH SW terminal of the relay, if the solenoid does not click, check that a sound connection exists at No. 4 terminal on the solenoid. Recheck by earthing the TH SW relay terminal again, if the solenoid still fails to click it is faulty.

Adjusting the control cable

Should a buzzing noise emanate from the solenoid when attempting to engage overdrive, the control cable probably requires adjusting. Disconnect the cable, move the lock-up lever to the full rearwards position, and adjust the cable length so that there is a $\frac{1}{4}$ in. free movement at the dash control before the lever moves.

AUTOMATIC TRANSMISSION

Cars equipped with automatic transmission are usually fitted with an inhibitor (safety) switch to prevent the starter motor from being operated until the selector lever is placed in the Park or Neutral position. It is therefore important, in the event of the starter motor failing to rotate the engine, to ensure that the gear lever is correctly positioned at 'P' or 'N' and that the inhibitor switch circuit is in order.

The inhibitor switch (automatic transmission)

The inhibitor switch on B.M.C. transverse engine cars is located on the gear-change-lever housing, but on the Westminster, Cortina and other cars it will usually be found on the left-hand side of the automatic transmission unit. Should the engine of an automatic-transmission car fail to rotate when the starter switch is used and the gear selection lever is at the 'N' position, the first check should be to ascertain whether the solenoid is operating. If it is not operating, make sure that the wires running to the terminals No. 2 and No. 4 (see fig. 10:18) have not become disconnected. If the wiring is satisfactory, bridge the switch by connecting the No. 2 and No. 4 terminal wires together. Try to start the engine. If the starter solenoid fails to operate, follow the test procedure in chart 10:5. If the solenoid operates, the inhibitor switch is faulty or needs adjusting.

Before an inhibitor switch can be adjusted, it is necessary to disconnect the wires from it.

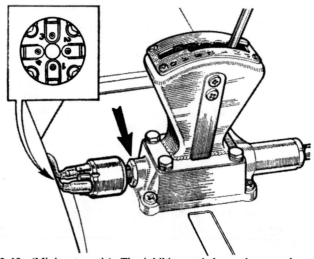

Fig. 10:18 (Mini automatic). The inhibitor switch on the gear-change lever housing. Inset, the 1 and 3, 2 and 4 connections marked on the switch. The switch locking nut is indicated by an arrow

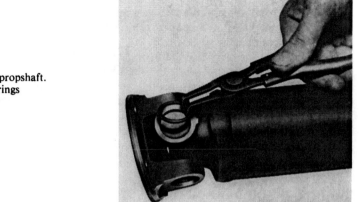

Fig. 10:19 Dismantling the propshaft.
Releasing the circlips or snap rings

Should wires be connected to No. 1 and No. 3 terminal, these will be reverse light wires, and each wire should be marked to ensure that it can be refitted to its own terminal. To adjust the switch, place the gear selector at 'N', slacken the locknut and screw the switch in as far as possible. Connect a test bulb and battery in circuit across No. 2 and No. 4 terminals and screw the switch out until the bulb lights. Screw the switch out another half-turn, tighten the locknut and reconnect the wires to the appropriate switch terminals.

Make sure that the starter motor can be operated only when the gear selector is at 'N' or 'P'. If a reversing light is fitted check that the light is operative only when 'R' is selected. If the switch fails to operate correctly it must be renewed.

Universal joints

A distinct 'clonk' in the transmission, as the initial drive is taken up, may be due to play in a universal joint. If the joints are examined, wear on the thrust faces of the bearings can be ascertained by testing the lift in the joint with the hands, and although lift of more than 0·005 in. is undesirable it does not normally result in any noise and does not call for immediate attention. It is the circumferential movement of the shaft, within the yoke flanges, which

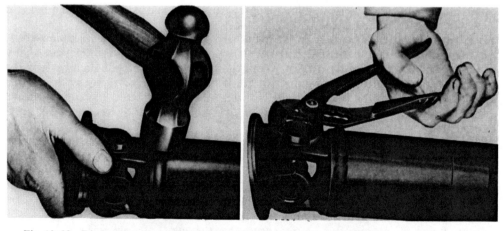

Fig. 10:20 Dismantling the propshaft bearings. *Left*: Tapping the bearing cups partially from the yoke. *Right*: Finally removing the bearing cups

indicates wear on the needle roller bearings or worn yoke flanges and which gives rise to 'knocks'.

On B.M.C. front-wheel-drive vehicles, apart from those with an automatic gearbox, the universal couplings are of a bonded-rubber type and should the couplings develop 'play' a new rubber coupling must be fitted.

The removal of the propeller shaft (or Triumph drive shafts) from the car to fit universal joints is straightforward but when the shaft has been removed, before stripping the joint, if rust is present, give the area a thorough soak with rust solvent and then a little later, thoroughly wash the joint with paraffin to remove any dirt adhering to the joint.

To strip the joint, remove the snap rings, or circlips, by pinching the ends together with a pair of thin-nosed pliers or circlip pliers (see fig. 10:19), whichever is required. In some cases, when the clips have been compressed, a screwdriver will have to be used to prise the clips from the yoke.

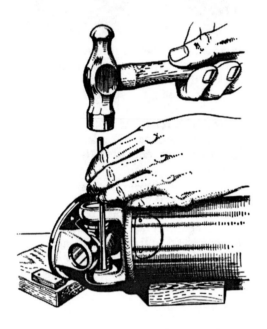

Fig. 10:21 When dismantling a universal joint, if the bearings are very tight they may be tapped out with a small diameter rod from the inside

Should it be difficult to compress the retaining clips, tap the end of the bearing races to release the pressure against them.

It is best to remove the bearings in the mainshaft, prior to those in the yoke flange.

If a lubricator is fitted to the spider, remove it and then tap the radius of the yoke lightly with a copper hammer (see fig. 10:20). When the bearing has emerged slightly, turn the yoke over and take out the bearing (fig. 10:20). If the bearing is too tight, pinch it in the jaws of a vice and pull the yoke from the bearing. Alternatively, tap the bearing from the inner side using a small diameter bar (see fig. 10:21). When the bearing has been taken from the yoke, remove the other bearing.

Next, remove the spider trunnion (complete with the flange) from the shaft, rest the assembly on wood blocks, and tap the lug of the flange yoke to remove the bearing. Repeat the operation on the other bearing.

When a universal joint has been dismantled, it must be reassembled using a new spider and bearing assembly. If wear has taken place in any of the yoke cross holes, rendering them oval, the parts sustaining the wear must be renewed.

Before fitting the new universal joint, ensure that the yoke holes are perfectly clean and completely free from burrs and if any corrosion is present remove it with fine emery cloth.

Assemble the needle rollers in the bearing races with a suitable grease, such as Duckhams L.B. 10, or Castrolease LM, so as to retain the needle rollers in position during assembly of the joint.

Insert the spider into the flange yoke, making sure the rubber seals remain in position. If the spider is of the type fitted with a lubricator boss, make sure the spider is fitted with the boss facing away from the yoke flange.

Employing a soft-nosed drift about $\frac{1}{32}$ in. smaller in diameter than the hole in the yoke, press or tap one of the bearings into position, locating the spider into the bearing carefully, and taking care not to disturb the needles from their position. Push the bearing through the yoke hole until it is well past the retaining circlip groove. Insert the retaining spring clip into its groove, start the opposite side bearing onto the spider and then push the bearing into position in the yoke (see fig. 10:22).

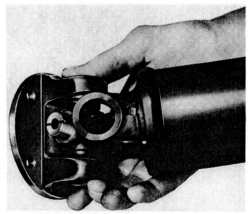

Fig. 10:22 Refitting the propshaft bearings. Pushing the bearings into the yoke

When the bearing reaches its correct position, the other bearing will be pushed against its retaining circlip. The drift may then be removed and the circlip fitted to retain the bearing which has just been fitted.

If difficulty is encountered in pushing any of the bearings into place it may be due to a needle dropping from its position so preventing the bearing from going fully 'home'. If undue force is exerted the bearing will crack.

If in doubt, when assembling a universal joint, always remove the bearing and make sure that the needles are still correctly placed.

When the flange yoke bearings have been fitted, insert the spider into the yoke holes of the shaft and using the same procedure as with the flange yoke, fit the bearings and insert the retaining clips.

If the splined drive shaft is disconnected from the main tubular shaft, it is essential that it is reconnected properly (see fig. 7:38).

When refitting the propeller shaft to the car, ensure that the driving flanges are absolutely clean and free from burrs. Failure to ensure that these conditions are met may result in excessive propeller shaft vibration.

Renewing and adjusting front wheel bearings

If the front hubs are removed during a steering overhaul, or if a harshness develops when the car is cornering, the wheel bearings should be checked. If they show signs of wear or pitting, they must be renewed.

To renew the wheel bearings, the brake drums must be removed. On rear-drive vehicles, if disc brakes are fitted, before the hubs can be taken off, the two bolts holding the brake calliper to the swivel axle must be removed and the calliper released. To avoid the need to disconnect the brake pipes, tie the calliper into a suitable position ensuring that the pipes do not take any strain.

Hubs are retained by a nut and split pin which are covered by a grease-retaining cap. With wire-wheel hubs, a little manipulation may be necessary to remove the split pin which locks the bearing-retaining nut; this involves working the pin through the access hole in the hub.

When the retaining nuts have been removed, the hubs may be taken off.

The first step towards renewing the bearings is to remove the oil seals and the outer and inner bearings complete with the races, which are a tight fit in the hub. The bearings come out easily but the races will have to be removed with a drift and a hammer. New races may be fitted in the same manner.

When the new races are repositioned, pack the bearings with an h m p grease. Subsequently fit the inner bearing and the oil seal, position the collar (distance piece) and insert the outer bearing and assemble the hub onto the axle.

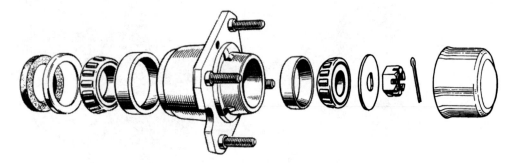

Fig. 10:23   Assembly view of drum-brake hub-bearing details (Herald)

Taper roller bearings

If the bearings are of the taper roller type, the bearings must be correctly adjusted; to do this proceed as follows: fit the washer and retaining nut and tighten the nut until the bearings bind. This will put the races fully against the locating flange in the hub. Remove the nut, washer and outer bearing.

The bearings may have to be adjusted by means of shims (available in several different thicknesses) to obtain an end-float of 0·002 to 0·004 in. The best way of adjusting the bearings is to insert sufficient shims against the outer side of the distance collar to produce excessive end-float, noting the thickness of the shims used and tighten the retaining nut to 60 lb ft. Measure the amount of movement with a dial gauge, remove the retaining nut, the washer and the bearing, and reduce the thickness of shims accordingly, so as to produce the correct end-float.

Re-check the measurement with a dial gauge; if all is well, tighten the hub nut to the correct torque setting and then turn the nut to the next nearest split pin slot which aligns with the hole in the stube axle.

Bearings that are not of a taper roller type do not require adjustment by shims, but it is essential to fit the races and bearings correctly, to repack the hub with grease and to refit the distance piece and oil seal (see fig. 8:21).

Some taper wheel bearings do not have a distance collar interposed between the outer and inner bearing; with this type, when the hub has been repositioned on the stub axle, fit the flat washer and the nut, and with a box spanner, or socket, fitted to an 8 in. bar, fully tighten the nut, and then slacken it a half-turn (see fig. 10:24) and insert the split pin. Make sure the hub turns freely. If it does not, the hub must be removed, the cause located and the fault rectified.

Chart 10:6

Common Rear End Faults

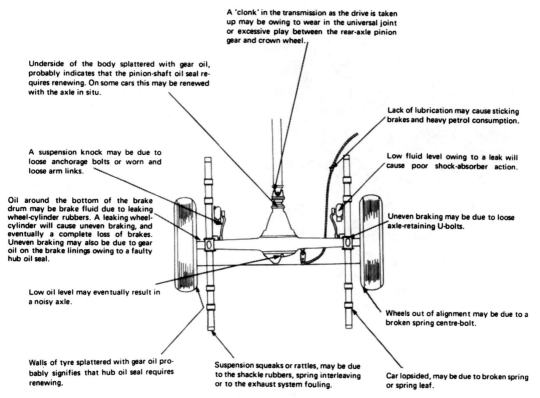

A 'clonk' in the transmission as the drive is taken up may be owing to wear in the universal joint or excessive play between the rear-axle pinion gear and crown wheel.

Underside of the body splattered with gear oil, probably indicates that the pinion-shaft oil seal requires renewing. On some cars this may be renewed with the axle in situ.

Lack of lubrication may cause sticking brakes and heavy petrol consumption.

A suspension knock may be due to loose anchorage bolts or worn and loose arm links.

Low fluid level owing to a leak will cause poor shock-absorber action.

Oil around the bottom of the brake drum may be brake fluid due to leaking wheel-cylinder rubbers. A leaking wheel-cylinder will cause uneven braking, and eventually a complete loss of brakes. Uneven braking may also be due to gear oil on the brake linings owing to a faulty hub oil seal.

Uneven braking may be due to loose axle-retaining U-bolts.

Low oil level may eventually result in a noisy axle.

Wheels out of alignment may be due to a broken spring centre-bolt.

Walls of tyre splattered with gear oil probably signifies that hub oil seal requires renewing.

Suspension squeaks or rattles, may be due to the shackle rubbers, spring interleaving or to the exhaust system fouling.

Car lopsided, may be due to broken spring or spring leaf.

After inserting the split pin and locking it in position, fit the grease-retaining cap. Refit the brake drum or other braking system components.

Rear axles (see chart 10:7)

Rear axle noises can be deceptive and many cases of suspected rear axle trouble can be traced to other faults. Knocks and vibrations may be due to a propellor shaft bearing.

Noise which has culminated in complete lack of drive is most likely to be due to stripped crown wheel or pinion teeth. Lack of any appreciable noise culminating in lack of drive is most likely to be due to a broken axle shaft or, on some cars, to a broken differential shaft pin.

Rear axle noises can prove expensive to correct, and it is usually possible to purchase a second-hand assembly for considerably less than the price of a new crown wheel and pinion, (but see figs. 10:25 and 26).

Rear axle trouble (see also chart 10:7) is usually attributable to one of the following causes:

(A) *Noise on over-run*

1. Excessive clearance between pinion gear and crown wheel.
2. The spacer, fitted between the pinion inner and outer bearing, severely worn resulting in the pinion tending to wind in and out of mesh as the car is accelerating or over-running.
3. Loose pinion bearing lock ring or nut.
4. Worn bearings (probably more pronounced when pulling hard).
5. Pinion and crown wheel teeth breaking up.
6. No oil.

Fig. 10:24 Adjusting front wheel bearings which do not have a spacer interposed between the outer and inner bearing. The bearing should be tightened and then the nut 'backed off' to allow a running clearance

(B) *Noise when pulling hard*

    1. Inadequate clearance between pinion and crown wheel.
    2. Worn bearings.
    3. Crown wheel teeth breaking up.
    4. No oil.

(C) *Noise when cornering*

    1. Differential thrusts collapsed.
    2. Hub bearings worn.
    3. Differential gears disintegrating.

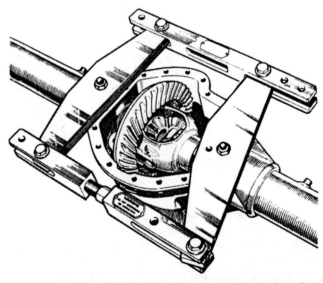

Fig. 10:25 On some types of axles such as fitted to Vanguards and some late B.M.C. vehicles, a stretcher tool has to be used to remove the crown wheel and differential assembly. Unless the proper tools are available and one is familiar with the work, it should not be attempted

Chart 10:7

Rear Axle Faults

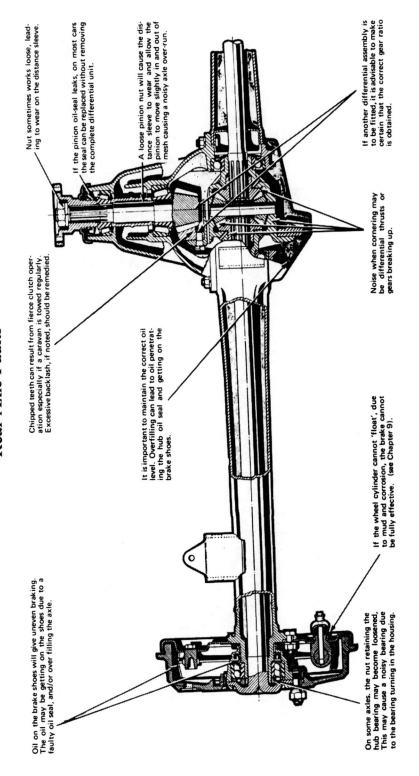

Nut sometimes works loose, lead-ing to wear on the distance sleeve.

If the pinion oil-seal leaks, on most cars the seal can be replaced without removing the complete differential unit.

A loose pinion nut will cause the dis-tance sleeve to wear and allow the pinion to move slightly in and out of mesh causing a noisy axle over-run.

If another differential assembly is to be fitted, it is advisable to make certain that the correct gear ratio is obtained.

Chipped teeth can result from fierce clutch oper-ation especially if a caravan is towed regularly. Excessive backlash, if noted, should be remedied.

It is important to maintain the correct oil level. Overfilling can lead to oil penetrat-ing the hub oil seal and getting on the brake shoes.

Noise when cornering may be differential thrusts or gears breaking up.

Oil on the brake shoes will give uneven braking. The oil may be getting on the shoes due to a faulty oil seal, and/or over filling the axle.

If the wheel cylinder cannot 'float', due to mud and corrosion, the brake cannot be fully effective. (see Chapter 9).

On some axles, the nut retaining the hub bearing may become loosened. This may cause a noisy bearing due to the bearing turning in the housing.

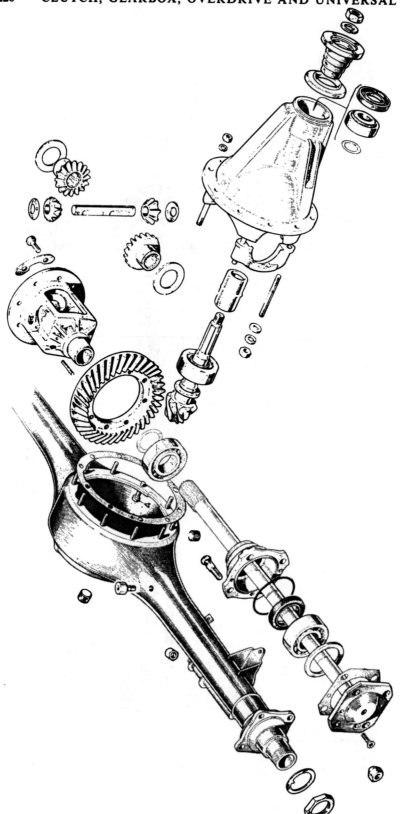

Fig. 10:26   Although this type of crown wheel and pinion unit is fairly easy to remove from the axle case, it is not advisable to completely dismantle the unit for, on reassembling, unless the pinion and crown wheel are meshed together correctly and the bearing pre-load adjusted, the result could prove disastrous. As with gearbox overhauls, rear axles are best left to the expert for although the correct backlash and pre-load figures can be ascertained from the manufacturer's manual, special tools are usually required. Rear axle noise, see chart 10:6

Fig. 10:27   Before a crown wheel and pinion unit can be removed from the axle
casing the axle shafts must be withdrawn

(D) *Lubricant leakage*

1. Leakage from the hub indicates (a) the hub bearing oil seals require renewing, see figs. 10:26, 28 and 29, (b) the axle has been overfilled.
2. Leakage from the pinion indicates that the pinion oil seal requires renewing.
3. Leakage from around the banjo cover bolts will either be due to loose bolts or a faulty gasket.
4. General leakage due to a blocked breather.

Fig. 10:28   Using a universal hub puller to remove the rear hub bearing assembly
(B.M.C. car) to renew a hub oil seal

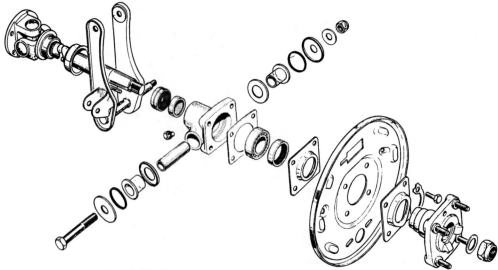

Fig. 10:29   Outer axle shaft and hub assembly details (Herald)

DRIVE SHAFTS (B.M.C. FRONT-WHEEL-DRIVE CARS)

In the event of a knock from the constant velocity joint, the drive shaft should be removed and a new bell joint fitted. The knock is due to wear of the ball cage in the joint, but as there are several variations in the size of the ball cages and a special measuring gauge is needed to service the joint, a complete new bell joint must be fitted. Also, besides the ball cages, there have been variations in the number of splines and diameter of the drive shafts so, before starting on the job, consult pages 226 to 230.

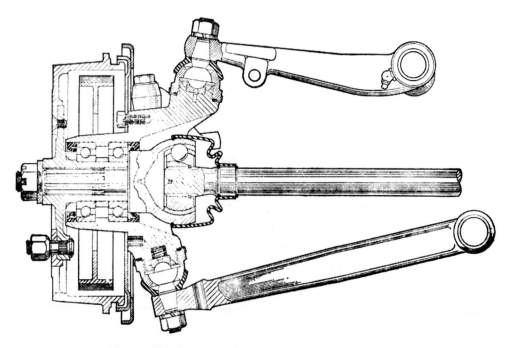

Fig. 10:30   A section through the Mini front suspension assembly

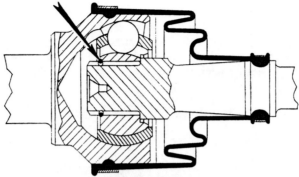

Fig. 10:31   A constant velocity joint: note the snap ring securing the drive shaft
(see also page 227)

Removing drive shafts

On B.M.C. 1800 cars, place a 1in. packing between the upper arm and lower bump rubber. Jack up the car, remove the appropriate road wheel, disconnect the steering rod ball joint from the steering arm and then take the split pin from the drive shaft nut and remove the nut. Tap the end of the shaft with a hide mallet and remove the outer cone.

Next, disconnect the drive shaft at the flexible inner coupling by removing the 4 nuts from the coupling U-bolts. Mark the driving flange and flexible joint to identify them for refitting in their original position. Support the drive flange and disc assembly on a suitable block and remove the nut and washer retaining the upper swivel hub ball pin.

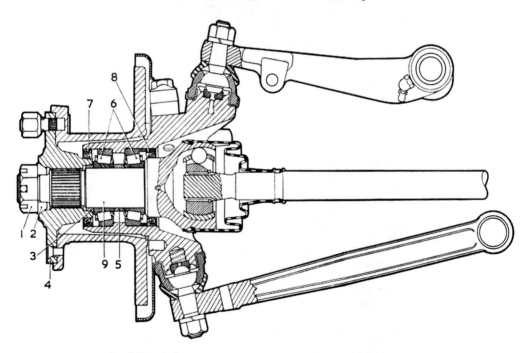

1. Drive shaft nut	5. Bearing distance piece
2. Outer tapered collar	6. Taper-roller bearings
3. Inner tapered collar	7. Outer oil seal
(early models only)	8. Inner oil seal
4. Hub and disc assembly	9. Drive shaft

Fig. 10:32   The front hub assembly, Mini Cooper 'S'

Release the upper arm from the swivel hub and allow the hub to pivot outwards, disengage the inner drive coupling and pull the shaft out of the hub assembly (see fig. 10:33). It will be necessary to manipulate the shaft to gain adequate clearance.

When removing a Mini drive shaft, jack up under the lower suspension support arm and insert a ½ in. packing piece between the top support arm and rebound stop (see fig. 10:34) and remove the jack.

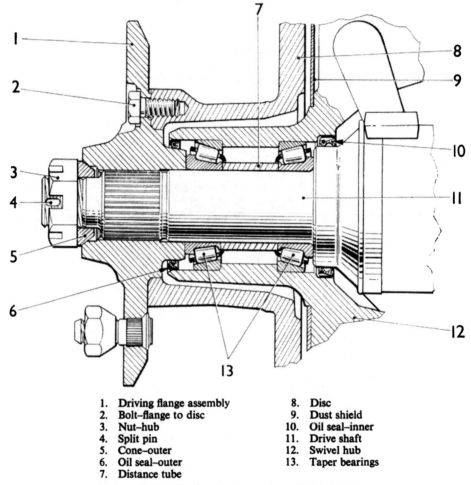

1. Driving flange assembly	8. Disc
2. Bolt–flange to disc	9. Dust shield
3. Nut–hub	10. Oil seal–inner
4. Split pin	11. Drive shaft
5. Cone–outer	12. Swivel hub
6. Oil seal–outer	13. Taper bearings
7. Distance tube	

Fig. 10:33   The front hub assembly of B.M.C. 1800

Next, take the split pin from the drive shaft nut and remove the nut. Jack up the vehicle and place a suitable support under the front sub-frame and remove the appropriate wheel. On the 1100 and Mini fitted with disc brakes, remove the brake calliper retaining bolts and support the calliper so that the brake hose does not take any strain.

Disconnect the ball end from the steering lever and disconnect the drive shaft at the inner flexible joint. Release the upper suspension arm from the ball pin and remove the lower arm pivot pin.

Turn the hub on the inner lock and force downwards (on drum brakes, care must be taken not to strain the brake hose), manipulate the drive shaft into a suitable position and tap it through the hub and remove.

Although refitting the drive shaft is a reversal of the removal process, on cars that have a spacer fitted between the outer and inner hub bearing, the spacer must be aligned correctly before the shaft can be pushed through the hub.

Overhauling a drive shaft

To remove the bell joint from the drive shaft, remove the wire or clip securing the rubber boot to the bell joint and slide the boot back along the shaft. Hold the shaft vertically (see fig. 10:35) give the hub of the inner race a sharp blow with a soft-faced mallet. This will contract the spring retaining ring which is located in a groove in the extreme end of the shaft (see fig. 10:36) and allow the joint to be tapped from the shaft.

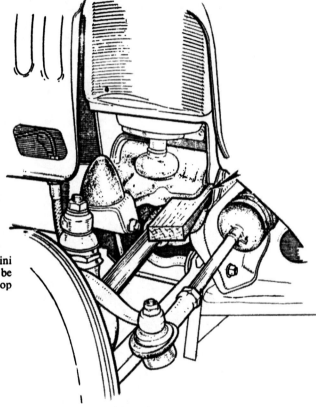

Fig. 10:34 When removing a Mini drive shaft, a block of wood should be inserted between the suspension top support arm and rebound rubber

Before fitting the new bell joint, inspect the rubber boot on the shaft and renew it if perforated or perished. It is also important to renew the round-sectioned spring on the end of the shaft.

Lubricate the new bell joint with Duckhams MB grease (a special pack is available, part number AKF 1457, and the complete pack should be used) and smear a little of the grease on the inside of the boot.

To refit the bell joint, hold the drive shaft in a vice and locate the inner race on the shaft. Press the bell joint assembly against the spring ring and locate the ring centrally and contract it into the chamfer of the inner race with screwdrivers, see fig. 10:37 (an assistant is necessary). When the ring is suitably positioned, give the end of the bell assembly a sharp blow with a soft-faced mallet. This will close the ring and allow the assembly to slide onto the shaft. Check that the inner race is fully home against the circlip and slip the rubber boot into position securing it with two turns of 18 swg soft iron wire. Make sure to bend the ends of the wire away from the direction of rotation (see fig. 10:38) and refit the shaft to the car.

It should be noted that the earliest types of Mini drive shafts had sliding joints that were fitted with grease nipples. If an early type drive shaft has to be replaced by the later type that has a sliding joint pre-packed with grease, the lower pivot pin must be changed and perhaps the hub driving flange, etc.

Complete drive shafts are available for B.M.C. cars as service replacements. Parts are also available as per A.B.C. in fig. 10:39.

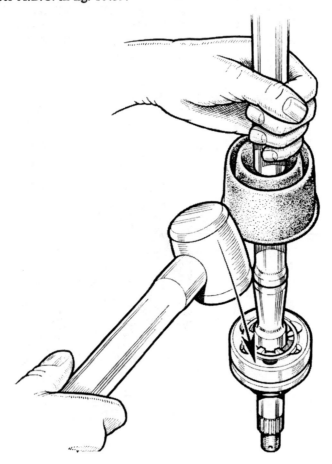

Fig. 10:35   Releasing the shaft from the inner race of the joint

If drive-shaft knock occurs it should be noted that there have been several changes made to Mini drive shafts since their inception. Parts are not available for some of the earlier models and later type shafts are not interchangeable for the early type unless the drive flanges, etc. are also changed. The accompanying information and identification should be noted.

Part number of complete drive-shaft assembly			Identification
1st type	RH	21A 204	Shaft and constant-velocity joint bell-
	LH	21A 205	housing with 18 splines.
2nd type	RH	21A 204	Shaft 19 splines; constant-velocity
	LH	21A 205	joint bell-housing 18 splines.
3rd type	RH	21A 261	Shaft and constant velocity joint bell-
	LH	21A 262	housing with 19 splines.

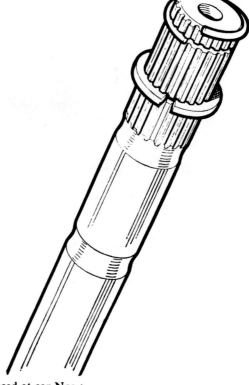

Fig. 10:36 The splined bell joint end of
the drive shaft showing the circlip and
the round-section spring ring. The latter
must be renewed when a new constant
velocity joint is fitted

Mini range, 3rd type drive shafts were introduced at car Nos.:

Austin	Mini	26591	Morris	Mini Minor	24832
	Mini Countryman	31526		Mini Traveller	34814
	Mini van/pick-up	26236		Mini van/pick-up	27536

Mini Cooper, Wolseley Hornet and Riley Elf models were fitted with 21A 261/2 drive shafts from the commencement of production.

21A 204–5 (1st and 2nd types) are no longer available complete or in breakdown form and if worn they must be replaced by 3rd type assemblies. In this event the drive flange hub bearings and distance rings must also be changed. The new shaft 21A 261-2 must also be fitted if the drive

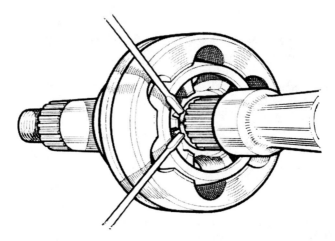

Fig. 10:37 Centralize the spring ring before tapping the shaft onto the splines

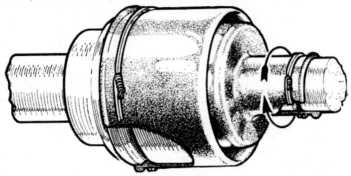

Fig. 10:38   The method of securing the rubber boot with wire, the ends of which must be bent away from the direction of rotation (lh shown)

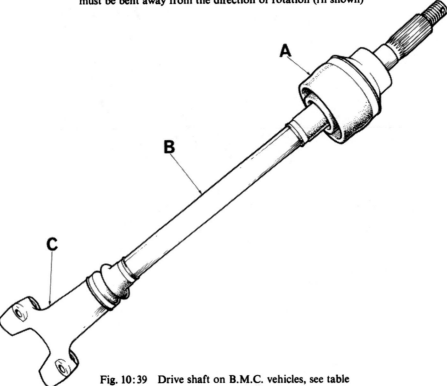

Fig. 10:39   Drive shaft on B.M.C. vehicles, see table

	Description	Mini Part No.	Cooper S Part No.	1100 Part No.
(A)	Bell and joint assembly	17H 8600[1]	27H 4750	27H 4750[2]
	Boot, joint, shaft to bell-housing	21A 636	21A 1695	21A 636
	Boot, joint, shaft to sleeve.	21A 963	21A 963	21A 963
(C)	Sleeve assembly	21A 1696	21A 1696	21A 1696
	Circlip, inner, shaft to inner race, retaining	17H 8596	17H 8596	17H 8596
	Circlip, outer, shaft to inner race, retaining	17H 8597	17H 8597	17H 8597
(B)	Shaft—RH	27H 4775	27H 4775	17H 8598
	Shaft—LH	27H 4776	27H 4776	17H 8599

[1] 3rd type drive shafts complete are parts Nos. 21A 261 (RH) and 21A 262 (LH).

[2] Prior to car Nos. 20094 (Morris) and 10768 (MG) it will also be necessary to fit new driving flange, Part No. BTA 367.

flange hub bearings and distance rings have to be replaced, or if it should be decided to fit bell joint 17H 8600 (see also below).

Type of drive shaft	Parts available for servicing		Remarks
1st type 21A 204–5	Boot	21A 636	In the event of drive shaft knock 21A 261/2 must be
	Boot	21A 963	fitted together with drive flange 21A 231,* Ring-
	Sleeve	21A 430	distance-bearing 21A 233*, Bearing–inner 21A 234
	Circlip	17H 8597	Bearing–outer 21A 234*.
	Circlip	17H 8596	
2nd type 21A 204–5	As above plus		
	Shaft	27H 4775	The shaft (B) (see fig. 10:39) alone may be exchanged,
	Shaft	27H 4776	but if a bell joint is required the above additional
	Bell joint	17H 8600	parts must be fitted (see asterisk).
3rd type 21A 261–2	As previously listed.		

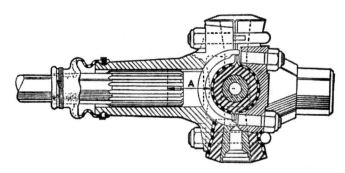

Fig. 10:40   A section through the sliding joint, showing (A) the depth to which the shaft joint flange should be repacked with grease when a new rubber boot is fitted

When fitting a replacement drive assembly (of the type fitted with a rubber boot on the sliding joint—see fig. 10:40) to the left-hand side of an early model Mini a modified lower arm inner pivot pin must be fitted to ensure sufficient clearance for the rubber boot. See fig. 10:41 for the dimension of the modified pivot pin.

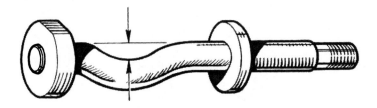

Fig. 10:41   Mini lower arm pivot pin. To accommodate the rubber boot of later drive shaft assemblies the measurement at the position indicated must be 0·312 in.
(7·9 mm.)

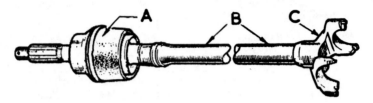

Fig. 10:42   Drive shaft B.M.C. '1800'

B.M.C. 1800 drive shafts

Bell joint assemblies 'A' or shaft 'B' (see fig. 10:42) may be replaced separately. Later cars have modified drive shafts with sliding splines on the inner yoke C and modifications in the transmission case.

Service bell joint kits contain the correct type and the required amount of lubricant to pack the joint.

CHAPTER 11

Checking and Overhauling the Fuel System

If petrol pump trouble is suspected, check the flow from the pump by disconnecting the carburetter inlet pipe. If a mechanical pump is fitted (see fig. 11:1), rotate the engine (see page 233) using the starter motor, but if an SU electric pump is fitted (see figs. 11:2 and 11:3) activate the pump by switching on the ignition. If the fuel flow proves inadequate, or spasmodic, either the pump is faulty or a blockage or air leak is occurring. If the pump is electrically operated, the current supply may be faulty. Fault finding: see also Chart 1:6 at the end of chapter 1.

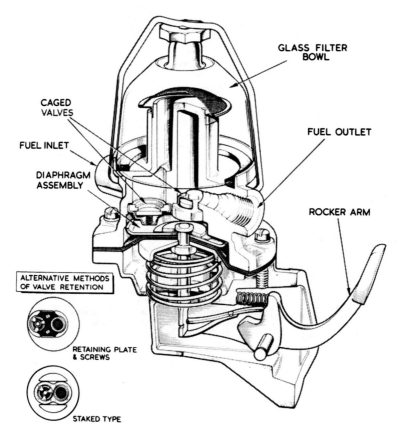

GLASS FILTER BOWL

CAGED VALVES

FUEL OUTLET

FUEL INLET

DIAPHRAGM ASSEMBLY

ROCKER ARM

ALTERNATIVE METHODS OF VALVE RETENTION

RETAINING PLATE & SCREWS

STAKED TYPE

Fig. 11:1 The AC 'FG' type petrol pump. AC petrol pumps are easy enough to overhaul. Particulars of the various type pumps and repair kits are given in the Appendix. See also assembly diagram Fig. 11:4

Testing a mechanical pump

Check the unions and nuts on the suction side of the pump for tightness, and also the set-crews or nuts which hold the pump to the crankcase. Inspect the underside of the pump using

Chart 11:1

Checking the Fuel System

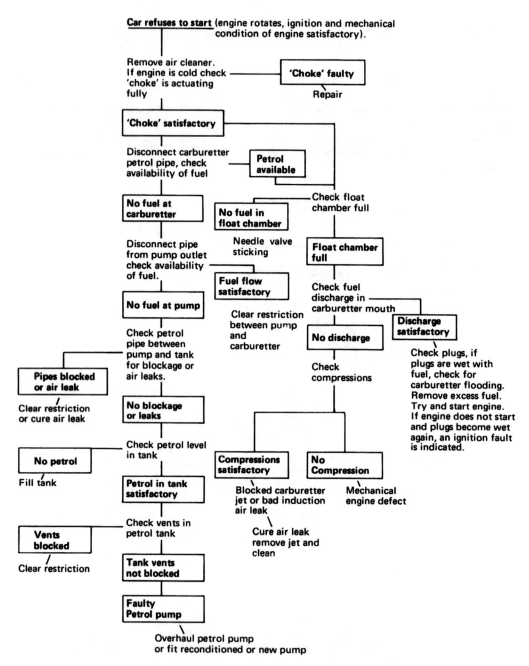

Difficult starting, stalling etc., see chart 1:6 and page 9. Fuel pump troubles chart 11:2

Chart 11:2

Fuel Pump Troubles

Mechanical Pump

Insufficient fuel.	Air leaks at filter bowl or suction side of pump.
	Air leak at diaphragm flange.
	Leaking valves.
	Perforated or incorrectly flexed diaphragm.
	Worn or broken pump linkage.
	Vapour lock in fuel line.
	Blocked filter.
Delivery pressure too high.	Carburetter needle valve not seating correctly.
	Incorrect needle valve.
	Diaphragm not flexed correctly.
	Incorrect diaphragm spring.
	Pump mounting gasket too thin.
	Pump body fitting flange refaced (too much metal removed).

Electrical Pump

Insufficient fuel.	Air leaks on intake side of pump.
	Air leak at diaphragm flange.
	Leaking valves.
	Perforated or incorrectly flexed diaphragm.
	Vapour leak in fuel line.
	Faulty electrical feed.
	Bad earth.
	Contact points burnt.
	Corroded throw-over mechanism.
	Incorrectly adjusted throw-over mechanism.
	Excessive diaphragm tension.
	Blocked filter.
Pump beats rapidly.	Carburetter flooding.
	Air leak in suction line.
	Fuel leak.
	Vapour lock.
Non-operation.	Electrical supply fault, or bad earth.
	Diaphragm pressure too great.
	Throw-over mechanism incorrectly adjusted.
	Faulty valves.
	Faulty coil winding.
	Faulty internal pump connection.
	Points dirty.
	Corroded throw-over mechanism.

a mirror; if fuel is dripping from the drain hole in the pump body, a cracked, or perforated, diaphragm is indicated and the pump should be removed and overhauled. If there is an oil leak between the crankcase and the pump body, remove the pump and check to see whether the fitting face is bowed. If it is, a new bottom pump body must be fitted or the flange refaced.

If the fuel flow is spasmodic, after the engine has rotated a few times, remove the petrol cap from the tank. If this cures the trouble the tank breather is blocked and the blockage should be located and rectified.

An AC pump can be tested when it has been removed from the car by operating the rocker arm and blocking the outlet with the finger, but if the valves are dry immerse the inlet in a can of petrol and operate the rocker arm. Petrol should be ejected from the pump after

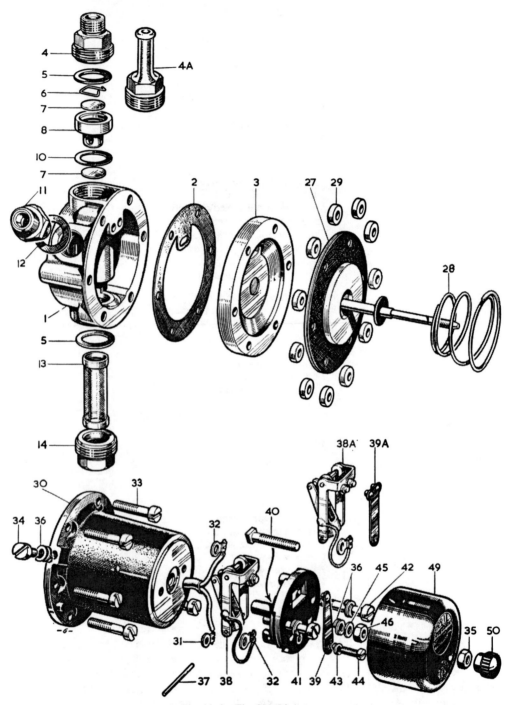

Fig. 11:2　The SU 'L' type pump

the rocker arm has been operated four or five times. The pump should then be emptied, the arm operated again and, when the rocker is fully depressed, the outlet blocked with a finger. The air drawn into the pump should be held in compression for a brief period. If difficulty is experienced in retaining the compression, the diaphragm may be faulty or, if a new diaphragm has been fitted or the pump does not pick up petrol fairly quickly, the valves are probably not seating adequately and the top half of the pump must be stripped and the trouble located.

When refitting the pump to the engine, make sure to place the rocker arm so that it rides correctly on the camshaft eccentric otherwise difficulty will be encountered in aligning the holes in the petrol pump flange with those of the crankcase.

KEY TO SU 'L' TYPE COMPONENTS (fig. 11:2)

BODY GROUP

1. Body—aluminium
2. Joint—washer
3. Plate—body
4. Outlet union
4A. Outlet union (AUA66)
5. Washer—filter and outlet union
6. Spring clip—valve
7. Valve disc
8. Valve cage
10. Washer—valve cage
11. Double-ended inlet union
12. Washer—inlet union
13. Filter
14. Filter plug

DIAPHRAGM GROUP

27. Diaphragm sub-assembly
28. Spring—diaphragm
29. Roller

COIL GROUP

30. Coil housing sub-assembly—12-volt (Spec. AUA25 and AUA58)
31. Terminal tag—5 B.A. (all Specs.)
32. Terminal tag—2 B.A. (all Specs.)
30. Coil housing sub-assembly—6-volt (Spec. AUA26)
33. Screw—2 B.A.—coil housing to body
34. Earthing stud—long—2 B.A.

35. Terminal nut—2 B.A. (Specs. AUA25 and AUA26)
    Terminal nut—2 B.A. (Spec. AUA58)
36. Spring washer—double coil—2 B.A

ROCKER AND PEDESTAL GROUP

37. Spindle—contact breaker—inner and outer rocker
38. Rocker and blade sub-assembly—single contact (Specs. AUA25 and AUA26) including:
32. Terminal tag—2 B.A.
39. Spring blade—single contact
38A. Rocker and blade sub-assembly—double contact (Spec. AUA58) including:
32. Terminal tag—2 B.A.
39A. Spring blade—double contact (Spec. AUA58)
40. Terminal screw—2 B.A.
41. Pedestal sub-assembly
42. Pedestal screw—2 B.A.
43. Spring washer—5 B.A.
44. Screw—5 B.A.—spring blade
45. Lead washer
46. Terminal nut—2 B.A.—recessed
49. End cover—black—12-volt (Specs. AUA25 and AUA58)
    End cover—brown—6-volt (Spec. AUA26)
50. Terminal knob

If a pump develops excess pressure when it is on the car, poor idling, a heavy petrol consumption and a flooding carburetter may result. The excess pressure may be due to a faulty rocker arm or, if the pump fitting flange has been refaced, too much metal may have been removed. Too low a pressure may be caused by too thick a gasket between the petrol pump and crankcase.

When a pump has been overhauled and refitted to an engine, if the carburetter floods, the carburetter needle should be renewed. If this does not cure the flooding, try fitting an extra gasket between the crankcase and the pump.

Testing electrical petrol pumps

Sometimes inadequate fuel delivery from an electric pump is due to nothing more than a poor electrical connection. If stranded by the roadside, prior to carrying out any tests, give the pump a good thump with the hand. Sometimes the pump will start operating and continue for a short period, but though this procedure may suffice as a 'get home' repair, give the pump a thorough examination at the earliest opportunity.

During hard frosts, any moisture in the petrol tank is likely to freeze and cause a blockage in the petrol outlet pipe; this is by no means uncommon. During frosty weather, petrol tank breather pipes, too, may become blocked by ice and cause faulty fuel feed.

To check the electrical supply to the pump, disconnect the feed wire from the pump terminal, connect a test lamp between the lead and earth, and switch on the ignition; if the bulb lights, current is available. The lead should not be shorted against the side of the pump, for if a spark does occur and fuel vapour is present, the spark may cause a fire. If the test bulb does not light, the lack of current may be owing to a flat battery, a faulty ignition switch or wiring, possibly a snap connector pulled from its socket. If the pump is rubber mounted and a spark cannot be obtained, the fault may be a broken pump earth lead.

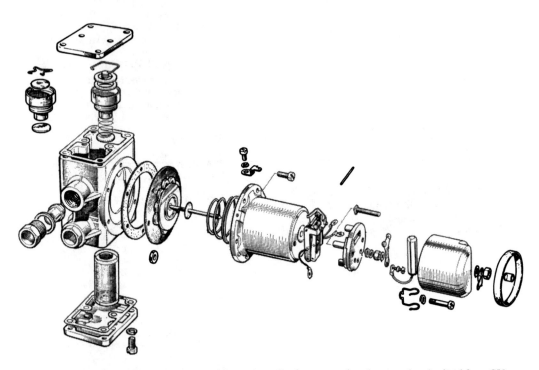

Fig. 11:3   The SU 'LCS' type pump. New points, diaphragms and gaskets can be obtained from SU stockists. The SU company are also able to supply detailed leaflets concerning the overhaul of each type of pump. See Appendix 7.

It is usually possible to fit new contact points to an SU petrol pump whilst the pump remains *in situ*. To do this disconnect the pump, remove the pump cover, take off the fixed 'springy' point, take out the throw-over mechanism pin and remove the two 'pedestal' screws.

Next, unscrew the throw-over points from the armature, counting the number of turns it takes to release them. Fit the new points, screwing them onto the armature the same number of turns. Refit the the pedestal screws making sure to position the electrical wires and tags correctly. Fit the throw-over mechanism pin and fit the 'springy' point. Refit the cap, reconnect the pump and test

If the bulb lights, remove the bakelite cover (see figs. 11:2 and 11:3), reconnect the feed wire to the terminal and inspect the points. If they are closed, non-operation is probably due to dirty contact points and they should be checked by touching a bare piece of wire across them. If the pump then performs a stroke, the points need cleaning. Insert a folded piece of fine emery paper between them, pinch the points lightly together and move the emery paper to and fro. If this cures the trouble, it will be a temporary cure only and the points should be renewed at the earliest opportunity, or a replacement pump fitted.

Chart 11:3

Mechanical Pump Inspection Guide

Part	*Possible Fault and Remedy*
Metal filter bowl.	Distortion around the central hole. Retaining bolt fibre washer disintegrated.
Glass filter bowl.	Chipped or cracked glass.
Filter gauze.	Breaks in the mesh.
Pump body (upper casting).	Cracks caused by overtightening of the filter bowl screw. Stripped or partially stripped threads, excessive corrosion in sediment chamber. Renew if necessary.
Diaphragm.	Renew.
Diaphragm spring.	Corrosion. Renew. Check that the new spring is of the correct strength.
Oil seal washer.	Renew.
Pump body lower casting.	Distortion of flanges. Worn rocker pin holes. Renew if necessary.
Rocker arm spring.	Renew.
Rocker arm.	Wear on contact face. Arm bent.
Inner rocker arm (link).	Broken spring. Worn slots that engage the diaphragm pull rod. Renew if necessary.
Rocker arm pin.	Wear. Renew if necessary.
Valves.	Renew.
Valve springs.	Renew.
Valve retainer plate.	Distortion. Renew if necessary.
Valve seats.	Wear. Fit new body if the seats in the body are unsatisfactory.

Renew all gaskets.

If the pump is mounted underneath the car, it is important to ensure that the rubber protective sleeve which fits over the bakelite cap is maintained in a good condition otherwise the corrosive effect of moisture may prove troublesome. A thorough coating of 'Damp-Start' lacquer, after the petrol pump has been serviced, is an added safe-guard.

If starting difficulty is encountered, or severe spitting-back through the carburetter, a simple test is to disconnect the carburetter petrol pipe and place the end of the pipe in a glass jar. Switch on the ignition to operate the pump. If bubbles appear when the petrol is filling the jar there is an air leak and all unions and joints must be checked to ensure they are completely airtight. If the pump is overheating and operates sluggishly, it may be due to a blockage on the suction side but should the pump 'tick' and not deliver fuel, dirt may have lodged under one of the valves. On many SU pumps it is possible to unscrew the delivery union and lift out the valve case to give the valves a thorough clean. When refitting the valve case make sure to refit the thin fibre washer below the cage and the thick one under the outlet union.

If the fuel flow to the carburetter is satisfactory, the cause of severe spitting back may be a sticking carburetter needle valve. If the flow is normal at first but decreases rapidly and the pump labours, check that the petrol tank vent is not blocked. If this is not the trouble, check for any restriction on the inlet side of the pump such as a blocked pipe or filter. If a poor flow of petrol is obtained, but the pump operates rapidly, an air leak on the suction side is indicated. If the pump fails to tick when the carburetter petrol pipe is disconnected there is probably an obstruction in the suction line.

Servicing a mechanical pump

Remove the cover from the top of the pump, lift out the filter, and using a compressed air line, or cycle pump, blow any deposit from the filter and pump sediment chamber. When this has been done, refit the filter, place a new filter-bowl gasket in position, refit the bowl and tighten the retaining setscrew or thumb screw. On pumps where the bowl is retained by a setscrew, a fibre washer is fitted between the setscrew head and filter bowl and must be replaced so that the bowl remains completely airtight.

Overhauling a mechanical pump

Special kits for overhauling AC pumps are available. Before stripping the pump, lightly scribe a line across the diaphragm flanges to facilitate correct reassembly. Remove the screws around the diaphragm and separate the top body from the bottom. Release the diaphragm from the inner rocker arm by pushing on the diaphragm and turning it through 90°.

Should the pump fitting face be faulty, the body will have to be dismantled by removing the rocker-arm retaining fulcrum pin to release the inner and outer rocker arms, packing washers and return spring. When this has been done, the body must be refaced, taking off the minimum amount of metal, for the thickness of the flange affects the working stroke of the pump.

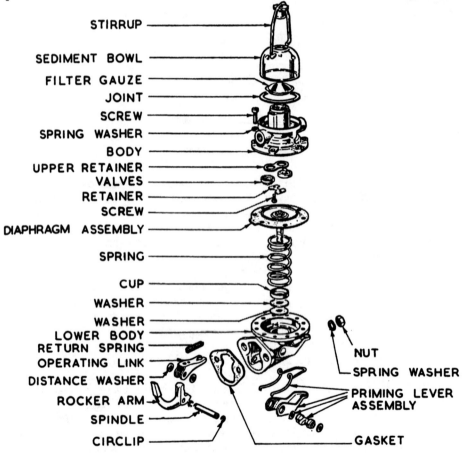

STIRRUP
SEDIMENT BOWL
FILTER GAUZE
JOINT
SCREW
SPRING WASHER
BODY
UPPER RETAINER
VALVES
RETAINER
SCREW
DIAPHRAGM ASSEMBLY
SPRING
CUP
WASHER
WASHER
LOWER BODY
RETURN SPRING
OPERATING LINK
DISTANCE WASHER
ROCKER ARM
SPINDLE
CIRCLIP

NUT
SPRING WASHER
PRIMING LEVER ASSEMBLY
GASKET

Fig. 11:4   Assembly diagram of AC 'FG' type fuel pump

From the top body, release the valves and springs by removing the setscrews which hold the retaining plate. When the pump has been completely stripped, clean and inspect the parts, referring to chart 11:3.

Reassembly of the upper pump body is a reversal of the dismantling procedure. Special points to watch are that the valves and springs are assembled correctly. With the pump body inverted, the outlet valve should be balanced on its spring, whereas the inlet valve spring must rest on the valve. The valve always faces the direction of fuel flow. With the valves and springs in position, carefully refit the retainer plate in position and insert and tighten the screws.

To reassemble the rocker arms, packing washers and spring into the lower housing, a pilot pin is a great help. It should be longer and of slightly smaller diameter than the standard pin and tapered for a short distance at one end.

Insert the pilot pin through one of the holes in the pump body, engage the packing washer, inner rocker arm and outer rocker arm, insert the return spring and the other packing washer. If the retainer clips have been removed from the rocker arm fulcrum pin, 'spring' one of the clips into a groove on the pin and carefully insert the pin, pushing out the pilot. Fit the other spring clip to the fulcrum pin.

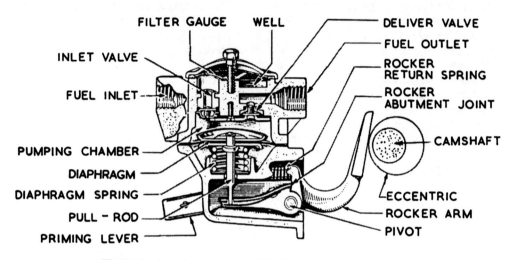

Fig. 11:5 A section through an AC 'Y' type mechanical fuel pump

Before fitting the diaphragm, if the pump is of the type which contains an oil seal, this should be fitted. Next place the diaphragm spring in position and (see figs. 11:4 and 11:5) set the diaphragm assembly on the spring with the notches on the diaphragm rod in alignment with the rocker arm. Centre the upper end of the spring in the diaphragm protection washer and press downwards, turning the diaphragm a quarter turn to the left so that the flattened, notched end of the pull rod engages with the slot in the rocker link and locks the diaphragm into position.

The top and lower pump bodies may now be fitted together. Make sure the faces are clean and coat them with a thin layer of shellac. Align the scribe marks, fit the two parts together and lightly flex the diaphragm with the rocker arm; insert but do not tighten the screws.

Next press the rocker arm hard towards the pump so as to flex the diaphragm and then carefully tighten the screws by diametrical selection. A poorly flexed diaphragm is restricted in movement and can reduce the quantity of fuel delivered by the pump. When the screws have been tightened, if the edges of the diaphragm protrude beyond the pump flanges, the flexing of the diaphragm has been incorrect and must be rectified.

The filter should now be fitted, followed by the cover gasket and cover, and (see fig. 11:5) the retaining screw inserted with its fibre washer in place. The cover on some types of pump is retained by a thumb screw but in either case make sure to tighten carefully so as not to strip the threads.

Overhauling an electric pump

Although service exchange pumps are reasonably priced, overhauling an electric pump is not difficult and, if care is exercised when setting the diaphragm and points, there is no reason why it should not be tackled.

The SU 'L' type pump (see fig. 11:2), in its 6 volt and 12 volt form, is one of the most widely used electric pumps. In its low-pressure form, the pump is usually mounted in the region of the engine at approximately carburetter level with the delivery pipe as short as possible. The high-pressure pump is usually mounted over the rear fuel tank. The early HP 'L' type was fitted with a coil housing $\frac{9}{16}$ in. longer than the low-pressure model, but later pumps were the same external length, the appearance of the pump is therefore not a reliable method of identification, but on the later low-pressure 'L' type, which normally has a single-point contact blade, the earth screw is a 2 BA size whereas on the HP type it is 4 BA.

Strip the pump by reference to fig. 11:2 but note carefully where the electrical connecting tags are fitted so as to ensure correct refitment. Remove any gummy deposit from the parts by soaking them in methylated spirits (ethyl alcohol) followed by a good brushing.

When they are clean, inspect the parts, checking the pump body for cracks and damaged joint facings. Make sure all of the threads on the various parts are not stripped or damaged. If the valve seats, in the pump body or in the outlet valve cage, are pitted or corroded, the part should be discarded. Make sure to check all electric wires for faulty insulation and ascertain that none of the connector tags are loose on their wires.

Examine the steel pin which secures the rocker mechanism and check the pedestal for cracks. If the pin is bent or worn, obtain a new pin. It should be noted that this component is specially hardened and must not be replaced by anything other than the correct part. When overhauling a pump, always renew the diaphragm, all of the fibre washers, gaskets, sealing rings and contact points.

  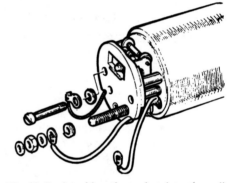

Fig. 11:6 Fitting the rocker assembly to the pedestal, (inset) the correct position of the centre toggle spring

Fig. 11:7 Attaching the pedestal to the coil housing. The correct assembly of components on the terminal stud is important

When the new parts are to hand, invert the pedestal and fit the rocker assembly. Make sure that the rockers are perfectly free to swing, position the centre toggle and spring correctly (see fig. 11:6) so that, with the inner rocker spindle in tension against the rear of the contact point, the centre toggle is above the spindle on which the white rollers run.

Fit the square-headed terminal bolt, the spring washer, the correct terminal tag (see fig. 11:7) and the lead washer. Fit the rocker pedestal to the coil housing but take care when tightening the screws, for the pedestal is easily cracked.

Before fitting the diaphragm, place the armature spring into the coil housing with the large diameter towards the coil. Fit the impact washer to the armature (see fig. 11:8) insert the spindle into the hole in the coil and screw it into the threaded trunnion in the centre of the rocker assembly. Screw the diaphragm inwards until the rocker 'throw over' will not operate. Insert the centralizing rollers (see fig. 11:9). On late type rocker mechanisms, fit the contact blade and adjust the finger settings (see fig. 11:10) and then remove the contact blade. On earlier assemblies, check the gaps between the rollers and the coil housing, setting the contact blade to obtain the correct reading (see fig. 11:11). The contact blade should be in light contact only, as too much tension restricts the travel of the rocker mechanism. Next,

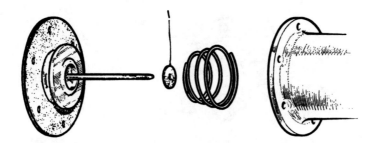

Fig. 11:8 Fitting the diaphragm to the coil housing. Note the impact washer (1)

Fig. 11:9 Inserting the diaphragm centralizing rollers

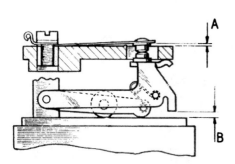
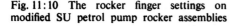

Fig. 11:10 The rocker finger settings on modified SU petrol pump rocker assemblies

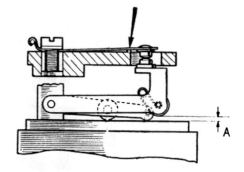

Fig. 11:11 The contact gap setting on earlier-type SU petrol pump rocker assemblies

slowly unscrew the diaphragm about one-sixth of a turn or until the rocker 'throw-over' can be operated and then take it another two-thirds of a complete turn (four holes) and align the holes in the periphery of the diaphragm with the holes in the pump body.

It is not necessary to stretch the diaphragm but a wedge may be inserted under the trunnion (see fig. 11:10) and helps hold the diaphragm in alignment whilst the screws are inserted.

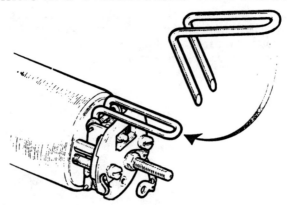

Fig. 11:12   Fitting the roller retaining fork

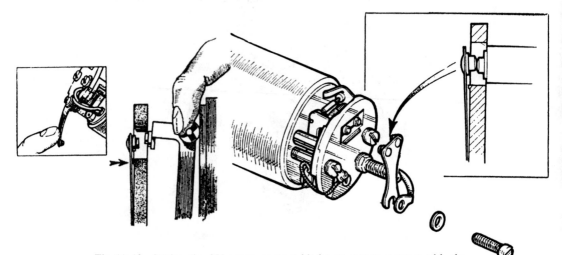

Fig. 11:13   Setting the SU pump contact blades to ensure contact with the pedestal ridge. (*Left*) Early type (*Right*) Later type.

If a fuel pump has been overhauled, it must be primed when it is refitted to the vehicle. Disconnect the fuel pipe at the carburetter and switch on the ignition until a flow is obtained, then switch off, reconnect and tighten the pipe

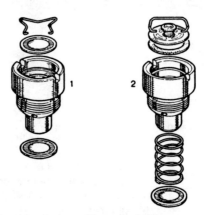

Fig. 11:14   Petrol pump valve assembly on the LCS type pump. (1) Earlier type. (2) Later type

Fig. 11:15 Fitting an armature guide plate (used in place of rollers on some pumps)

If the bottom body is clean, fit the new disc valve to its cage, ensuring that the smooth face is towards the valve seat. Insert the circlip and fit the other valve disc cage and washers.

Next, reassemble the pump body to the coil housing. Place the sandwich-plate gasket in position, fit the sandwich plate, concave face to the diaphragm, place the diaphragm gasket into position and offer the coil housing to the pump body and sandwich plate, taking care to align the holes and making sure that the filter plug is correctly positioned (see fig. 11:2). Make certain that the centralizing rollers are not displaced as they would cut into the diaphragm as the screws are tightened and render the diaphragm useless. Insert the retaining screws, and tighten them by diametrical selection, subsequently removing the wedge.

On pumps fitted with a plastic armature guide plate in place of rollers, when the diaphragm is correctly set, fit the guide plate, flat face towards the diaphragm, by turning back the diaphragm edge (see fig 11:15) and inserting an end lobe into the recess between the armature and the coil housing. Follow this process until all four lobes are approximately in position, then press each lobe firmly home, finishing with the two end ones to prevent distortion.

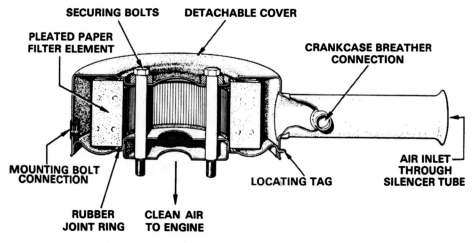

Fig. 11:16 Typical construction of a dry-element carburetter air filter. A blocked filter can cause a severe loss of power and excessive fuel consumption. In the event of such troubles, the filter should always be examined and renewed if 'choked'. This type of filter element cannot be successfully cleaned

Refit the contact blade to the pedestal and make sure that the points are aligned correctly (see fig. 11:13).

When the pump has been assembled, connect it to a battery for testing. If the pump functions correctly, mount it 3 feet above a container of petrol. When switched on, the pump should prime itself. The fuel flow rate may be timed but unless faulty delivery at high rev/min is thought to be occurring, this is not necessary if the pump has been assembled carefully. Fuel flow rates, see Appendix 7.

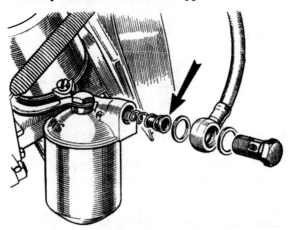

Fig. 11:17 A partially blocked fuel filter can result in weak fuel mixures resulting in loss of performance and valve burning. Cleaning should be carried out every 5,000–6,000 miles

As an added test, the pump outlet should be gradually restricted; this should slow the pump operation and when the outlet is completely blocked the pump should cease to 'tick' for at least twelve seconds. If it continues ticking, a valve is probably leaking and the valves (see fig. 11:4) should be removed and the seats examined for the presence of foreign matter and possible flaws.

Refit the bakelite end-cover, seal the joint with adhesive tape and, if the pump is fitted in a situation where moisture is prevalent, fit a rubber seal over the end of the pump.

CHAPTER 12

Carburetters

Complaints of unsatisfactory carburation are not uncommon, but many complaints* prove to be due to faults other than the carburetter; indeed it is fairly safe to say, inadequate fuel delivery, dirty fuel, and air leaks are more usual than a faulty carburetter.

High petrol consumption and difficult starting may be due to poor fuel, low compression or faulty ignition and before making tuning adjustments to the carburetter or overhauling it, make sure that the mechanical condition of the engine and the ignition system is put in order.

The most widely used carburetters on British cars are Zenith, Solex, SU and Stromberg CD. The SU and Stromberg are of the constant-vacuum, variable-choke type, whereas the Zenith and Solex are fixed-choke, multi-jet carburetters. Although faults in the fuel system and carburetter may be detected by reference to charts 1:6 and 11:1, different types of carburetter tend to suffer from specific troubles, SU carburetters tend to suffer from sticking pistons if they are not serviced regularly and the older models (see fig. 12:1) are susceptible to jet-gland leakage.

On Zenith carburetters (see chart 12:1), the economy-device diaphragm may need occasional renewing, and on older cars, air leaks due to a worn throttle spindle can result in erratic slow running that cannot be cured by normal adjustment. Blocked jets on fixed-choke carburetters, and partial blockage of the nylon jet-feed tube on the modern HS type SUs are some of the more common carburetter faults.

SU carburetters

Having a variable choke, SU carburetters must be maintained in good working order, and should be cleaned regularly. Sticking pistons can give erratic running, stalling at idling speed and lack of power combined with heavy fuel consumption.

The old 'H' type SU carburetter (see fig. 12:1) in wide use prior to 1960 is largely superseded by the HS type (see fig. 12:2). The principal difference between the two carburetters is the jet unit and float chamber. In the HS type, the cork glands and sealing washers have been dispensed with, the jet being retained in a single bearing bush, the fuel being fed through a small-diameter nylon tube which runs from the base of the float chamber to the jet head. The tube is attached to the float chamber by a nut and nipple and may easily be detached (with the jet) for renewal.

Although the basic method of overhaul is the same with all types of SU carburetters, on the HD type (see figs. 12:3 and 12:4) the jet glands are sealed by a diaphragm and the jet adjustment, like that of the HS8 type (see fig. 12:5) is by a screw which operates upon a lever that moves the jet. On the HS8 type however, the jet bearing is not sealed.

Overhauling an SU carburetter

Strip the carburetter by removing the float chamber, the suction chamber and piston. Disconnect the jet spring, and remove the linkage from the jet head and remove the jet and bearing.

If an H type carburetter is being stripped, carefully arrange the upper jet bearing, the washers, bottom jet bearing, gland spring etc. in the order of removal so that they may be refitted correctly, see fig. 12:6.

Clean the parts and inspect for wear and damage. Check the piston-plate (if fitted) screws for tightness and ascertain that the carburetter induction flange is not bowed. If an asbestos-

---

* Emission controlled cars: see also chapter 14.

type insulating distance piece was fitted between the carburetter and induction manifold, a warped carburetter flange due to overtightening the carburetter retaining nuts is not unusual.

Should the carburetter flange be distorted, carefully true the face using a fine file. Make sure to fit a new insulating washer before refitting the carburetter to the manifold.

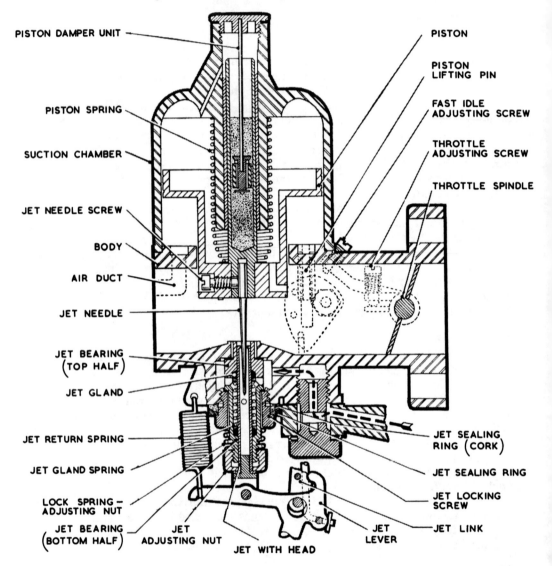

Fig. 12:1 Diagrammatic view of an H type carburetter

Other points to watch are: stripped threads on float-chamber bolts, and wear in the throttle bore caused by the throttle disc biting into the bore. A shallow groove of 0·001 in. is insignificant, but if the groove is more pronounced a new carburetter body is advisable.

If the throttle spindle is severely worn, it is advisable to fit a new spindle. To do this, remove the throttle disc and withdraw the throttle spindle. Knock out the pin on the throttle arm and withdraw the arm from the spindle.

Fit the new spindle by inserting it into the carburetter body. Position the throttle disc and

Chart 12:1

Fault Diagnosis of Multi-Jet Carburetters

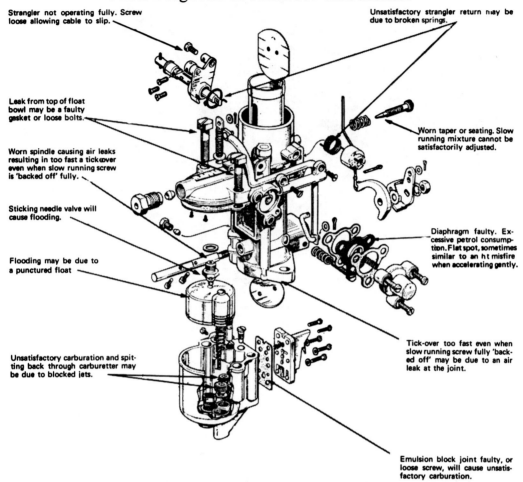

Strangler not operating fully. Screw loose allowing cable to slip.

Unsatisfactory strangler return may be due to broken springs.

Leak from top of float bowl may be a faulty gasket or loose bolts.

Worn taper or seating. Slow running mixture cannot be satisfactorily adjusted.

Worn spindle causing air leaks resulting in too fast a tickover even when slow running screw is 'backed off' fully.

Sticking needle valve will cause flooding.

Diaphragm faulty. Excessive petrol consumption. Flat spot, sometimes similar to an ht misfire when accelerating gently.

Flooding may be due to a punctured float

Tick-over too fast even when slow running screw fully 'backed off' may be due to an air leak at the joint.

Unsatisfactory carburation and spitting back through carburetter may be due to blocked jets.

Emulsion block joint faulty, or loose screw, will cause unsatisfactory carburation.

It should be noted that some cars are extremely sensitive to engine operating temperature. For instance, the removal of a thermostat from the cooling system may result in faulty carburation

fit two new throttle-disc retaining screws, but before tightening the screws, actuate the spindle a few times to centralize the disc in the carburetter bore. Finally tighten the retaining screws and open the end of the screws to lock them in position.

Reposition the arm so that the slow-running screw is hard against the abutment on the body, release the screw a few threads, hold the throttle disc in the closed position and drill a $\frac{1}{32}$ in. hole through the spindle in alignment with the holes in the arm. An assistant will be required to hold the spindle and arm in position whilst the spindle is being drilled.

Fit a taper-retaining pin and carefully peen the end over with a small ball peen hammer.

If the spindle holes in the body of the pump are badly worn and a new spindle will not take up the excess clearance, the spindle holes can be drilled out and bushes fitted. Fitting bushes, however, is best entrusted to a machine shop for it is imperative that the holes are correctly aligned and the bushes fit tightly into the carburetter body.

Rather than fit bushes, it is better to fit either a new body, or a good second-hand one. If another body is to be used, it is advisable to preassemble the carburetter, for a slight

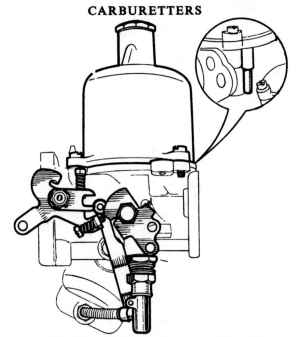

Fig. 12:2　The HS type SU carburetter. Inset, piston lifting pin

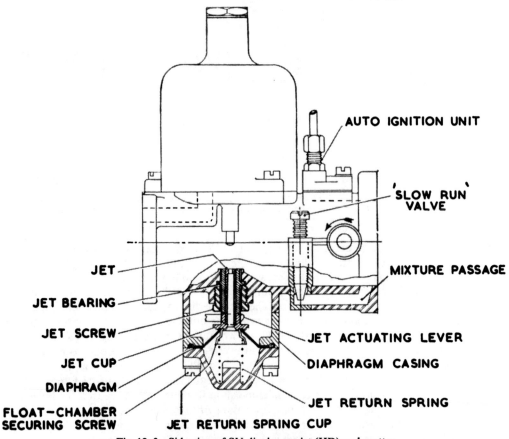

AUTO IGNITION UNIT

'SLOW RUN' VALVE

JET

JET BEARING

JET SCREW

JET CUP

DIAPHRAGM

FLOAT−CHAMBER SECURING SCREW

JET RETURN SPRING CUP

MIXTURE PASSAGE

JET ACTUATING LEVER

DIAPHRAGM CASING

JET RETURN SPRING

Fig. 12:3　Side view of SU diaphragm-jet (HD) carburetter

easing with a fine file may be necessary around the part of the body where the suction chamber locates.

When the carburetter has been preassembled, check that the piston moves satisfactorily and then remove the suction chamber and piston and clean thoroughly to remove any filings and grit.

When finally rebuilding the carburetter, reassemble the jet unit using all new washers, corks and glands etc. as necessary. If the needle has been removed from the piston, insert the needle, tighten the retaining screw and fit and centralize the jet assembly as per page 250.

Insert the float, check the fuel level (see page 254) position the float chamber and gasket, and refit the lid, tightening carefully.

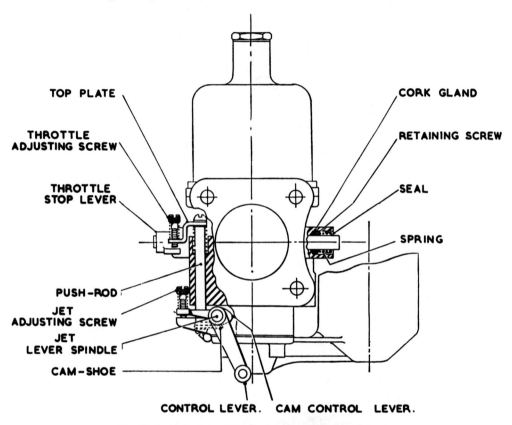

Fig. 12:4   End view of SU diaphragm-jet (HD) carburetter

When refitting the float chamber to the carburetter, renew the mounting rubbers or washers and make sure to tighten the retaining nut or bolt securely. If the bowl is rubber mounted, it is essential to ensure that each rubber locates properly as the float-chamber retaining nut is tightened.

When the carburetter has been rebuilt, make sure that the piston drops satisfactorily from the fully raised position and that the throttle spindle operates properly. Fill the damper orifice with oil (see fig. 12:8) and refit the carburetter to the engine.

Changing needles

The type of carburetter needle fitted as standard is arrived at by extensive tests and unless the engine specification has been substantially changed it is not advisable to fit a needle with a different profile from the standard recommendation. Manufacturers, however, normally suggest an alternative weaker needle from standard (see Appendix 6) and sometimes a slightly richer needle.

If, due to unsatisfactory performance, a needle has to be changed, remove the piston damper, detach the suction chamber from the carburetter, remove the piston and loosen the needle clamping screw. The identification letters, or numerals, are on the shank of the needle but a magnifying glass may be required to distinguish them.

When fitting a needle, note that the needle position in the piston is with the shank flush with the bottom of the piston (see fig. 12:9). In some cases when a new needle has been fitted it may be necessary to centralize the jet. If the piston tends to stick when it is lifted and released it is essential (see jet centring).

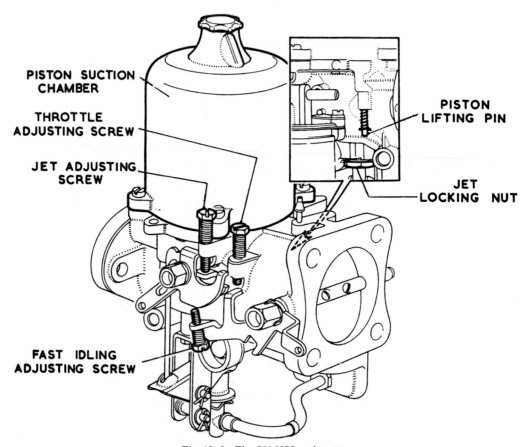

Fig. 12:5   The SU HS8 carburetter

When the suction chamber has been replaced and the jet recentred, make sure to refill the damper orifice (see fig. 12:16) with a thin oil and refit the damper.

Jet centring (see fig. 12:10)
When the carburetter piston is lifted and released it should return freely on to the carburetter bridge when the jet is in the fully 'up' position. If the piston sticks, or falls freely only when the jet is lowered, the piston needs cleaning, or the needle is bent, or the jet requires re-centring.

On HS8 carburetters, before the jet can be recentred, it is necessary to remove the jet anchor pin, jet fork pivot pin, and to remove the fork bracket and allow the linkage to swing to one side (see fig. 12:11).

On carburetters other than the HS8 type, it is usually sufficient to disconnect the jet by removing the bottom clevis pin and on some cars even this is not essential.

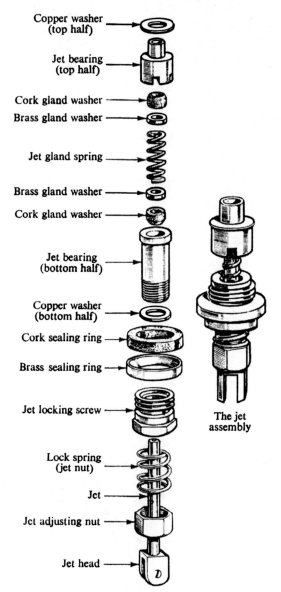

Copper washer
(top half)

Jet bearing
(top half)

Cork gland washer

Brass gland washer

Jet gland spring

Brass gland washer

Cork gland washer

Jet bearing
(bottom half)

Copper washer
(bottom half)

Cork sealing ring

Brass sealing ring

Jet locking screw

Lock spring
(jet nut)

Jet

Jet adjusting nut

Jet head

The jet
assembly

Fig. 12:6   The jet components of an SU H type carburetter

During a carburetter overhaul, when the jet unit is being recentred, the jet must be placed in its uppermost position or, in the case of the HS8, pressed fully home. This must be done temporarily to enable the needle to go right home into the jet and so centralize it.

On HD carburetters before the jet can be recentred, the four float-chamber holding-up bolts must be taken out to release the float chamber and sandwich piece, together with the jet and the diaphragm assembly.

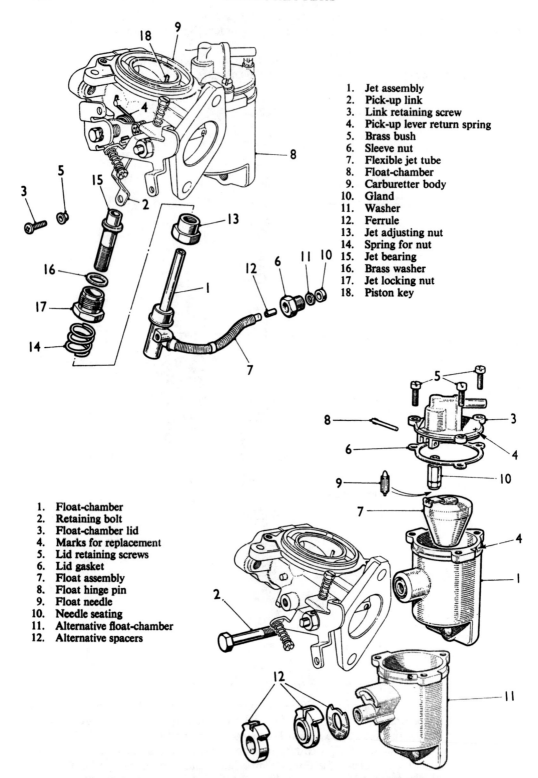

1. Jet assembly
2. Pick-up link
3. Link retaining screw
4. Pick-up lever return spring
5. Brass bush
6. Sleeve nut
7. Flexible jet tube
8. Float-chamber
9. Carburetter body
10. Gland
11. Washer
12. Ferrule
13. Jet adjusting nut
14. Spring for nut
15. Jet bearing
16. Brass washer
17. Jet locking nut
18. Piston key

1. Float-chamber
2. Retaining bolt
3. Float-chamber lid
4. Marks for replacement
5. Lid retaining screws
6. Lid gasket
7. Float assembly
8. Float hinge pin
9. Float needle
10. Needle seating
11. Alternative float-chamber
12. Alternative spacers

Fig. 12:7   Jet arrangements and float chamber assembly of SU HS type carburetter

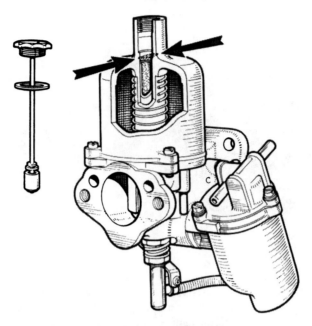

Fig. 12:8   Servicing SU carburetters (see also fig. 12:17). Piston dampers on SU carburetters should be topped up with the recommended engine oil until the level is ⅓ in. (13 mm) above the top of the hollow piston rod. On dust-proofed carburetters, identified by a transverse hole drilled in the neck of the suction chambers and no vent hole in the damper cap, the oil level should be ⅓ in. (13 mm) below the top of the hollow piston

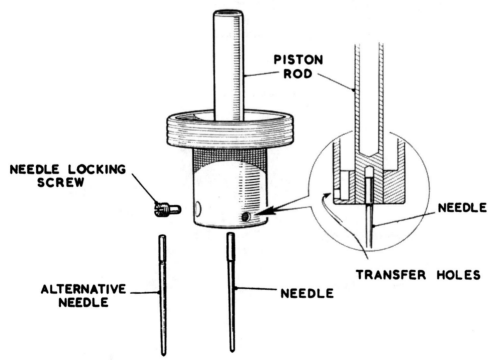

Fig. 12:9   SU and Stromberg carburetter needles must be fitted correctly and securely locked into position with the locking screws. Illustrated, SU

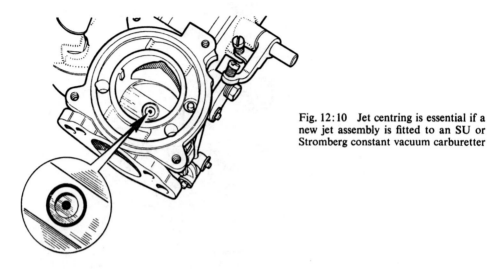

Fig. 12:10   Jet centring is essential if a new jet assembly is fitted to an SU or Stromberg constant vacuum carburetter

Next, separate the sandwich piece from the float-chamber arm to release the jet diaphragm, and position a ring spanner on the jet clamping nut. Replace the jet into the jet bearing, lining up the four holes in the diaphragm with the tapped holes in the body.

Slacken the jet locking nut sufficiently for the jet bearing to move freely.

Hold the jet in the 'up' position and, with the damper cap removed, press on top of the piston rod with a pencil to hold the piston onto the bridge. Tighten the jet locking nut, making sure that the jet is still in its correct position.

Holding the jet fully 'up', lift the piston and check that it falls freely onto the jet bridge and that the jet moves freely in the bearing without the needle binding in the jet.

Reassemble the sandwich piece and float chamber, making sure that the jet and diaphragm are kept to their same angular positions relative to the body (see fig. 12:12 and also fig. 12:13).

Setting the fuel level

The fuel level of H type carburetters may be raised, or lowered, by bending the float hinged-

Fig. 12:11   On HS8 carburetters before the jet can be re-centred it is necessary to remove the jet anchor pin, jet fork pivot pin, and to remove the fork bracket and allow the linkage to swing to one side. Arrowed: jet locking nut

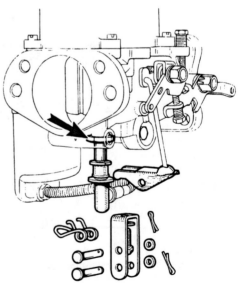

lever upwards, or downwards and to correct the level it should be bent so as to touch a test bar placed across the diameter of the float-chamber lid. The diameter of the test bar for the $1\frac{7}{8}$ in. and $2\frac{1}{8}$ in. (outside diameter) float chambers should be $\frac{7}{16}$ in. (see fig. 12:14).

On the HS carburetter (nylon float) the fuel level may be adjusted by inverting the float-chamber top (so that the needle valve is held in the shut off position by the weight of the float) and setting the gap between the float lever and rim of the float-chamber lid to $\frac{1}{8}$ in. (see fig. 12:15). If the measurement has to be reset, bend the lever with a pair of thin-nosed pliers.

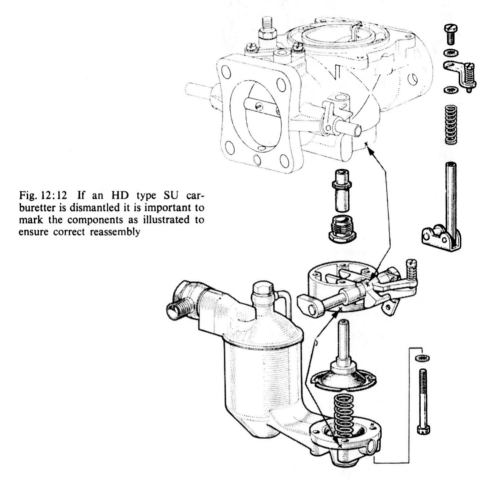

Fig. 12:12 If an HD type SU carburetter is dismantled it is important to mark the components as illustrated to ensure correct reassembly

The fuel level of the SU carburetter is fairly critical and the recommended size of the test bars is such that the petrol level will be maintained in the jet not more than $\frac{3}{8}$ in. below the top of the jet bridge.

Piston dampers

Some piston dampers do not have a hole drilled in the damper cap. The non-drilled type must only be used in conjunction with a suction chamber that has a transverse hole in the neck of the suction chamber. The hole is drilled at an angle and runs from just below the inner thread (for the damper cap) into the main chamber.

If the wrong combination of damper and suction chamber is used the carburetter will not function correctly.

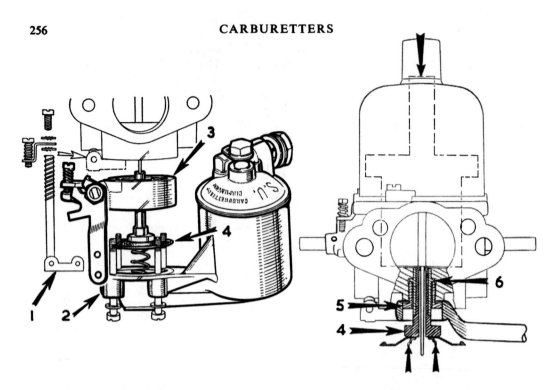

Fig. 12:13   Jet centring on the HD type SU carburetter. To do this, mark the position of the jet housing and float-chamber in relation to the carburetter body. Remove the plate retaining screw and withdraw the cam rod assembly (1). Unscrew and remove the float chamber securing screws, remove the float chamber (2) and the jet housing (3) and release the jet assembly (4).

Next slacken the jet locking nut (5), using a ring spanner, until the jet bearing is just free to move. Remove the piston damper, hold the jet (4) in the 'fully up' position, and apply light pressure to the top of the piston rod. Tighten the jet locking nut (5).

Check that the jet can move in the bearing freely and reassemble, ensuring that the jet and diaphragm are kept to the same angular position and that the beaded edge of the diaphragm is locked in the housing groove. Refill the piston damper with oil

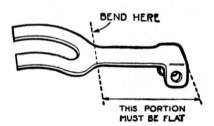

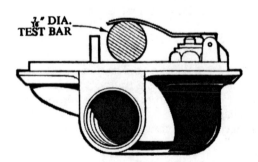

Fig. 12:14   The correct setting of the float lever on $1\frac{7}{8}$ in. and $2\frac{1}{4}$ in. SU float chambers that have metal floats

The type of damper with a hole in the cap must be used with a suction chamber that does not have a hole in the neck. Before 1955 identification of the caps was obvious for the damper with a cap vent hole had a circular knurled-brass cap, whereas the type without the hole had a hexagonal cap. From 1956 both types were given a hexagonal cap for the knurled type tended to come unscrewed.

Late-type caps are made in a black moulded plastic material and have a part number stamped on the top. On $1\frac{1}{4}$ in. carburetters 8103 denotes that a 0·040 in. free lift is provided whereas AUC8114 has a 0·070 in. free lift. A 0·040 in. damper allows the carburetter piston to lift 0·040 in. before the damper becomes effective. The 0·070 in. damper produces an initially weaker mixture than the 0·040 in. damper for the initial piston lift is higher, thereby causing less depression over the jet which in turn decreases the amount of fuel discharged.

On $1\frac{1}{4}$ in. carburetters produced before 1955 the suction-chamber piston spring is smaller in coil diameter at one end. The small coil should be cut off and the skid washer which fits between the spring and piston discarded to allow more piston lift, which is beneficial to top end performance. On carburetters produced after 1955 this modification was adopted as standard.

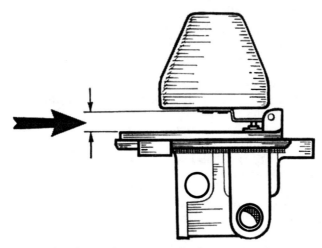

Fig. 12:15   The correct setting for SU nylon floats
With the needle valve held in the shut-off position by the weight of the float only, there should be a $\frac{1}{8}$ to $\frac{3}{16}$ in. (3·2 to 4·8 mm) gap between the float lever and the rim of the float chamber lid. The float lever may be set by bending at the crank

Tuning SU carburetters

Before any carburetter tuning is carried out, it is essential that the ignition system is in order, the ignition timing correct, and the tappet clearances properly adjusted. It is also advisable to check all carburetter and inlet manifold flanges for air leaks and to retighten manifold retaining nuts.

If any sign of petrol leakage is apparent from around the jet nut or base of the carburetters, or if the top or bottom of the float chamber is damp with fuel, the carburetter should be carefully examined and the cause of the leak ascertained. Unless the carburetter is fairly new, when the leakage is likely to be due to an incorrect float level or a faulty needle valve, it is advisable to remove the carburetter for a complete strip and overhaul.

Tuning single carburetters

Three adjustments are possible on the SU: (1) slow running; (2) idling mixture; (3) cold idling speed. Before any adjustments are carried out, set the float level, warm the engine to normal operating temperature and remove the air cleaner. Set the slow running to about

Fig. 12:16 Replenishing damper chambers (see also Fig. 12:8). It is essential to use the correct grade of oil

800 rev/min and adjust the jet nut one flat at a time until the engine runs smoothly. There is a special spanner available from the SU carburetter company for adjusting the jet nut, but the nut can usually be moved satisfactorily with the fingers.

After each adjustment, check that the jet head is in firm contact with the base of the jet nut, and lift the carburetter piston by the lifting pin and adjust accordingly (see page 261).

When the final adjustment has been carried out, make sure that the jet release spring has not been displaced.

To adjust single HD carburetters (see also page 262), warm the engine until normal operating temperature is reached, remove the air cleaner, close the throttle, and screw the by-pass idle valve on to its seating and then unscrew it $3\frac{1}{2}$ turns.

Set the float level. Remove the piston and suction chamber and turn the mixture adjusting screw until the jet is flush with the bridge of the carburetter. Refit the piston and suction chamber and check that the piston falls freely. Lower the jet by turning the adjusting screw $2\frac{1}{2}$ turns and top up the piston damper orifice with oil (see fig. 12:16).

Restart the engine and adjust the by-pass idle screw to obtain a tickover of 800 rev/min. Next, turn the jet adjusting screw until the fastest idling speed, consistent with smooth running, is obtained. If the engine increases in speed, slow it down on the by-pass screw.

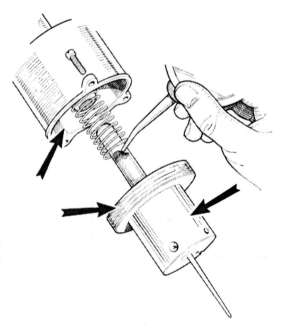

Fig. 12:17 Cleaning SU carburetters. Every 6000 miles carefully remove the piston/suction on chamber unit. Using a petrol-moistened cloth, clean the inside bore of the suction chamber and the two diameters of the piston (see arrows). Lightly oil the piston rod only and reassemble (see also fig. 12:8)

Check the mixture strength by means of the lifting pin which is on the side of the carburetter and carry out any further adjustments necessary (see page 261).

The suction chamber and piston

The suction chamber and piston should be taken off every 6000 miles for the removal of varnish and carbon deposits (see fig. 12:17). On multiple carburetter installations be sure to keep the same dashpot and piston to the appropriate carburetter. Also, when refitting the suction chambers make sure that each one is fitted exactly in the same position. With carburetter dashpots that are secured by two retaining screws it is easy to refit the chamber in a different position with the result that the piston needle may not be central in the jet.

In the event of burrs on the piston perimeter, the rubbing surface must be located and removed with a fine file. Any high spots in the suction chamber should be removed by using a scraper. The manufacturers recommend a clearance of between 0·002 and 0·003 in. between the bore of the suction chamber and piston, and that the piston 'drop' should be timed. Unless considerable wear has taken place 'timing' the drop is not essential.

Fig. 12:18 Removing the air cleaners on a Spitfire prior to setting the carburetter mixture, slow running, etc

If a new needle is fitted (see fig. 12:9) make sure to tighten the clamping screw securely, to replace the suction chamber piston spring and to refit the dashpot piston in a completely dry condition with just a few drops of thin oil on the piston guide rod. Take care not to bend the carburetter needle when locating the piston onto its slide and tighten the dashpot-retaining screws evenly. When reassembly is complete, ensure that the piston is free and drops back from a raised position satisfactorily. Before replacing the damper, refill the damper orifice with a thin oil. If the piston fails to drop, see page 250.

Suction chambers with a transverse drilling (running from just below the damper thread to the main chamber) must only be fitted with dampers that do not have a hole in the damper cap (see page 255).

Tuning twin HS type SU carburetter installations

Warm the engine, switch off, and remove the air cleaners (see fig. 12:18). Slacken off the coupling-lever pinch bolts (see fig. 12:19) on the carburetter interconnecting spindle so that each carburetter can be operated separately.

If the installation is not fitted with coupling levers, slackening a pinch bolt on the interconnecting clamp serves the same purpose.

When each carburetter throttle can be operated independently, disconnect the mixture control cable and slacken the two pinch bolts to free the jet actuating lever.

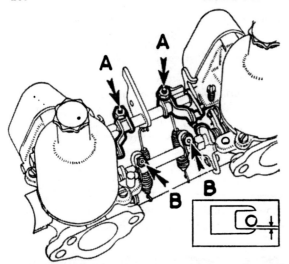

Fig. 12:19 When setting twin carburetter installations, to allow each
carburetter to be operated independently,
slacken the clamp bolts. When retighten-
ing, set the throttle interconnection
coupling levers (A) so that the link pin
is 0·008 in. away from the lower edge
of the fork (see inset). Tighten the
clamp bolts. With both jet levers at their
lowest position, set the jet interconnection lever clamp bolts (B) so that both
jets commence to move simultaneously

Some installations have a jet interconnecting rod which must be disconnected by removing
a clevis pin but, on any type of installation, provided each carburetter adjustment can be
carried out without altering the adjustment of the other carburetter, difficulty should not be
encountered.

Screw the mixture-adjusting nut on each carburetter to the topmost position and then
unscrew them 12 flats of the hexagon. Undo the throttle-adjusting screws until they are just
off the stops and then screw them up 1 turn. Re-start the engine and make any further
adjustments necessary to obtain a suitable working tickover speed (800 rev/min).

To synchronize the throttle openings, it is best to use a balancing meter. Alternatively,
listen to the hiss in the intake (see fig. 12:20) using a length of rubber hose, placing one end
of the hose against the ear and the other at the intake. Try each carburetter in turn and adjust
the screws (see also fig. 12:20) until the same intensity of hiss is obtained at each carburetter.

When the synchronization is satisfactory, adjust the mixture strength by moving each jet-
adjusting nut up or down the same amount until an even idling speed is obtained (upwards
for weakening and downwards for richening) which gives the fastest idling speed consistent
with even firing. When this has been found it may be necessary to lower the idling speed
slightly by unscrewing the throttle-adjusting screws on each carburetter an equal amount,
subsequently rechecking the hiss from the intakes.

Fig. 12:20 Listening to volume of hiss
at carburetter intakes; (1) throttle stop
screw

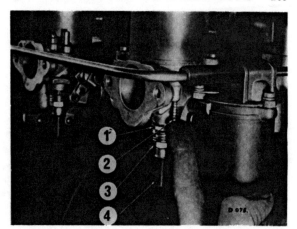

Fig. 12:21   Using piston lifting pin to check mixture strength.
1.   Jet bearing
2.   Piston lifting pin
3.   Jet adjusting nut
4.   Jet

A weak idling mixture gives a splashy irregular type of misfire with a colourless exhaust, whilst a rich idling mixture produces a 'rhythmical' regular misfire with a blackish exhaust.

When the mixture is approximately correct on both carburetters, lift the piston of the rear one, using the lifting pin (see fig. 12:21) on the side of the body. This will produce:

(a) uneven firing and a decrease in speed due to excessive weakness

(b) an increase in speed denoting richness or

(c) a momentarily slight increase in speed denoting a correct mixture.

Adjust the rear carburetter accordingly and make the same check and necessary adjustment to the front carburetter. After each adjustment, ensure that the jet is firmly against the adjusting nut.

Recheck and readjust the rear carburetter. The final position for the adjusting nuts is seldom precisely the same, indeed a difference of two or three flats is quite normal although the mixture is sensitive to even one or two flats on the hexagon.

When the mixture is correct, if the carburetter installation is fitted with interconnecting rod levers, tighten the clamp bolts and reconnect the 'choke' cable, adjust the clearance

Fig. 12:22   To tune an SU HS8 carburetter, start the engine and adjust the throttle adjusting screw (4) to give desired idling as indicated by the glow of the ignition warning light.
Turn the jet adjusting screw (2) up to weaken or down to richen until the fastest idling speed consistent with even running is obtained. Readjust the throttle adjusting screw (4) to give correct idling if necessary

between the lever pins and the forked lost-motion levers attached to each carburetter spindle. To do this, insert a 0·006 in. feeler gauge between the throttle shaft stop and screw. Move the throttle shaft levers downwards until their pins rest lightly upon the fork of the lost motion lever and tighten the levers at this position. Remove the feeler gauge and ensure that the lever pins have clearance in the forks (see fig. 12:19).

If lost-motion levers are not fitted, it is only necessary to retighten the interconnecting clamp and reconnect the choke connecting rod by refitting the clevis pin. Take care, when tightening the interconnecting clamp, that the throttle is not moved.

Reconnect the choke cable, checking that the jet adjusting levers (if fitted) are working correctly and return fully when the cable is pushed fully in. Next pull out the 'choke' control and when the linkage is about to move the carburetter jets, adjust the choke/fast-idle adjusting screws to give an engine speed of about 1000 rev/min when hot. Make certain, when the 'choke' is pushed in, that there is a clearance between the fast-idle cam and the abutment screw.

Tuning HD carburetters

HD carburetters are made in three sizes, $1\frac{1}{2}$ in., $1\frac{3}{4}$ in. and 2 in. and are sometimes fitted with an auxiliary, electrically or thermostatically operated, cold-starting attachment in place of a manual enrichment control.

HD carburetters are tuned in much the same manner as other SUs but on the HD carburetter, the jet is adjusted by turning a screw which acts upon a lever that lifts or lowers the jet according to the direction in which the screw is rotated. The adjustment of the idling mixture on the HD carburetter is also different from the HS type for, on the HD, the mixture is conducted through a passageway and regulated by a screw valve, instead of being regulated by a partially open throttle disc.

On multi-carburetter installations, when the engine has reached operating temperature, set the float levels and remove the air cleaners. Slacken one of the bolts on the throttle spindle connector, close the throttles and screw the by-pass idle screws on to their seatings and then open them $3\frac{1}{2}$ turns. Disconnect the 'choke' linkage, remove the suction chamber and set the jets of each carburetter so that they are flush with the carburetter bridge. Refit the suction chamber and check that the pistons fall freely from a raised position. Turn each jet adjusting screw $2\frac{1}{2}$ turns so as to lower the jet slightly.

Restart the engine and adjust each carburetter using the by-pass idle screw until the hiss from each carburetter is equal and the engine speed about 800 rev/min. Next, adjust the mixture by screwing both jet-adjusting screws up or down by the same amount until the fastest idling speed, consistent with even firing, is obtained. As the mixture is adjusted, it may be necessary to cut down the engine speed by screwing down the by-pass idle screw. Check the mixture on each carburetter by lifting the piston with its lifting pin and readjust accordingly (see page 261).

When the adjustments have been completed, make sure to retighten the throttle spindle connecting-link clamp bolt and reconnect the 'choke' linkage.

THE SU AUXILIARY ENRICHMENT (THERMO) CARBURETTER

The auxiliary enrichment (thermo) carburetter, fitted to many of the Jaguar range of cars is used to provide automatically differing degrees of mixture enrichment at:
    (a) Starting.
    (b) Idling and light cruising conditions.
    (c) Full throttle conditions.

Control

The unit may be controlled by either:

(a) A thermostatically operated switch housed in the cylinder head coolant jacket and set to bring the apparatus into operation below 35°C (95°F).

(b) A manually operated switch, which is generally provided with a warning light.

Operation

The auxiliary carburetter is a separate unit attached to the main carburetter. When fitted to H-type carburetters the construction of the main carburetter jet assembly differs from normal in the method of mixture adjustment.

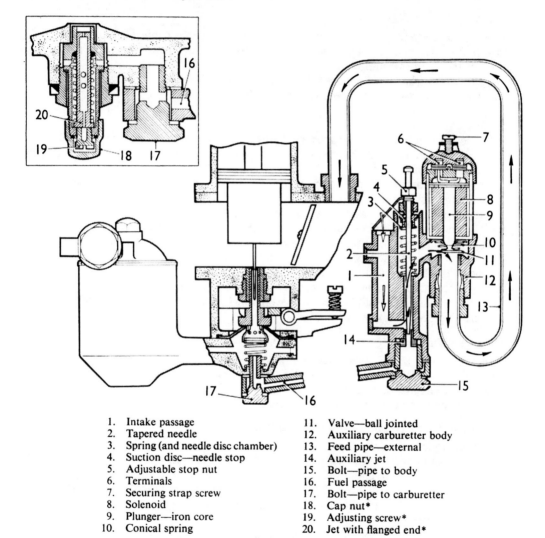

1.	Intake passage	11.	Valve—ball jointed
2.	Tapered needle	12.	Auxiliary carburetter body
3.	Spring (and needle disc chamber)	13.	Feed pipe—external
4.	Suction disc—needle stop	14.	Auxiliary jet
5.	Adjustable stop nut	15.	Bolt—pipe to body
6.	Terminals	16.	Fuel passage
7.	Securing strap screw	17.	Bolt—pipe to carburetter
8.	Solenoid	18.	Cap nut*
9.	Plunger—iron core	19.	Adjusting screw*
10.	Conical spring	20.	Jet with flanged end*

* 'H'-type jet assembly used with auxiliary carburetter

Fig. 12:23 The SU auxiliary enrichment (Thermo) carburetter with 'HD'-type carburetter and inset 'H' type jet assembly.

The device consists of a solenoid operated valve and a fuel metering needle which draws its fuel from the base of the auxiliary jet supplied from the main carburetter.

When the device is operated, air is drawn from the atmosphere through the air intake into a chamber and is mixed with fuel as it passes the jet. The mixture then passes upwards past the shank of the needle, through a passage, and so past the aperture provided between the valve and its seating. From here it passes directly to the main induction manifold through the external feed pipe as shown.

Solenoid and valve

The device is brought into action by energizing the solenoid. The iron core is thus raised carrying with it the ball-jointed disc valve against the load of the conical spring, thereby opening the aperture between valve and seating.

Valve seating

A cup washer is fitted against the solenoid face to centralize the conical spring. Any leakage between the valve and its seating would allow the device to operate and affect the idling setting of the main carburetter(s).

If switched on while the engine is idling the solenoid will not lift the valve until the manifold depression is reduced by opening the throttle momentarily.

Fuel level

The fuel level in the auxiliary carburetter is controlled by the main carburetter float chamber. It can be seen from fig. 12:23 that this results in a reservoir of fuel remaining in the well of the auxiliary carburetter.

When starting with the device in operation, this fuel is drawn into the induction manifold to provide the rich mixture necessary for instant cold starting.

When in operation the needle disc chamber is in direct communication with the inlet manifold and the depression, dependent on throttle opening, varies the position of the needle by exerting a downward force upon the suction disc and needle assembly. Thus:

(a) At idling the relatively high depression will draw the needle into the jet until the head abuts against the adjustable stop.

(b) At larger throttle openings a reduced depression is communicated to the needle disc chamber and the spring will tend to overcome the downward movement of the needle, thus increasing mixture strength.

Tuning and adjustment

As both the main and auxiliary carburetters operate when starting from cold, the main carburetter(s) must be tuned correctly before attempting any adjustment to the auxiliary carburetter. Reference should be made to the appropriate Service Sheets and to the mixture instructions given below for H-type carburetters.

Mixture adjustment H-type with auxiliary carburetter

The procedure for mixture adjustment is the same as for normal H-type carburetters except that a jet-adjusting screw is used in place of the normal jet-adjusting nut. Proceed as follows:

(1) Remove the cap nut.

(2) Adjust the jet as required, by turning the slotted screw up to weaken or down to enrich the mixture. The slight leakage of fuel through the jet during this operation can be ignored.

(3) Replace the cap nut with its sealing washer.

Auxiliary carburetter

Tuning of the auxiliary carburetter is confined to adjustment of the stop nut which limits the downward movement of the needle, and is carried out with the engine at normal running temperature and the main carburetter(s) tuned.

Proceed as follows:

(1) Switch on the auxiliary carburetter:

(a) Where the thermostat has automatically broken the circuit, energize the solenoid by short-circuiting the thermostatic switch to earth, or if this is inaccessible earth appropriate terminal of the auxiliary carburetter with a separate wire.

    (b) Where a manual switch is fitted, switch on.

(2) Open the throttle momentarily to allow the valve to lift.

(3) Adjust the stop nut:

    (a) Initially clockwise (to weaken) until the engine begins to run erratically.

    (b) Then anticlockwise (to enrich) through the phase where the engine speed has risen markedly and up to the point where over-richness results in the engine speed dropping to between 800 and 1,000 rev/min. with the exhaust gases noticeably black in colour.

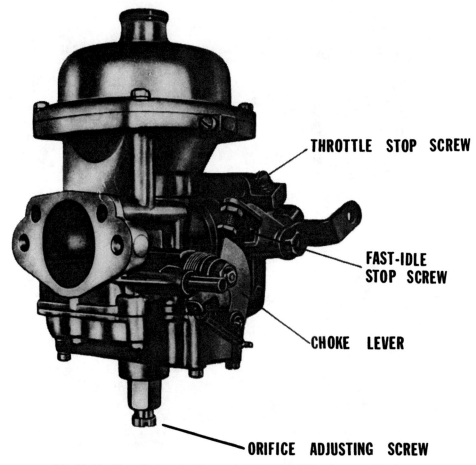

THROTTLE STOP SCREW

FAST-IDLE STOP SCREW

CHOKE LEVER

ORIFICE ADJUSTING SCREW

Fig. 12:24 The adjustment points on the Stromberg CD carburetter. The Stromberg operates on the same principle as the SU carburetter

Zenith Stromberg

The constant depression, or constant vacuum, Stromberg (see fig. 12:24) embodies many excellent features, and is probaby the best of this type carburetter; the concentric float chamber (see fig. 12:25) and central jet orifice (see fig. 12:26) ensuring a steep flooding angle and lack of surge problems. In principle, it is similar to the SU, the mixture strength being governed by the profile of the jet needle (see fig. 12:27).

Adjusting single Stromberg installations

First warm the engine to operating temperature, switch off and remove the air cleaner and take out the damper. With a pencil, push the air valve onto the bridge in the throttle bore. Screw up the jet-adjusting screw until it is possible to feel the jet coming into contact with the underside of the air valve, and then unscrew the jet-adjusting screw 3 turns.

Start the engine and adjust the slow-running screw to obtain an idling speed of approximately 700 rev/min. With a long screwdriver, lift the air valve a small amount ($\frac{1}{32}$ in.). If the engine speed rises, the mixture is too rich. If the engine stops it is too weak. Adjust the jet nut accordingly to get a smooth firing engine making any necessary adjustments to the throttle stop screw (referring to fig. 12:24). When the carburetter is correctly adjusted, the engine speed will remain constant, or fall slightly, when the air valve is lifted the stipulated amount.

Fig. 12:25   The float chamber removed from a Stromberg CD carburetter

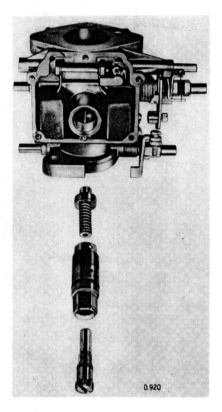

Fig. 12:26   The jet bushing screw and adjusting screw on the Stromberg CD

Adjusting twin Stromberg carburetters

Loosen the clamp bolt on the carburetter throttle interconnecting spindle, unscrew the throttle stop screws until the throttles close completely, and then screw in the stopscrews so that the ends just contact the abutment and then turn the screw another $1\frac{1}{2}$ turns.

Make sure that the fast-idle screw is clear of its cam and that the cold start levers are fully off their stops when the dashboard control is pushed fully 'home'.

Adjust the jet-adjusting screws (as in tuning single Strombergs) by unscrewing them $1\frac{1}{2}$ turns from the point where the jet orifice comes into contact with the base of the air valve. Start the engine and, using a throttle synchronizer, or by listening at each carburetter intake,

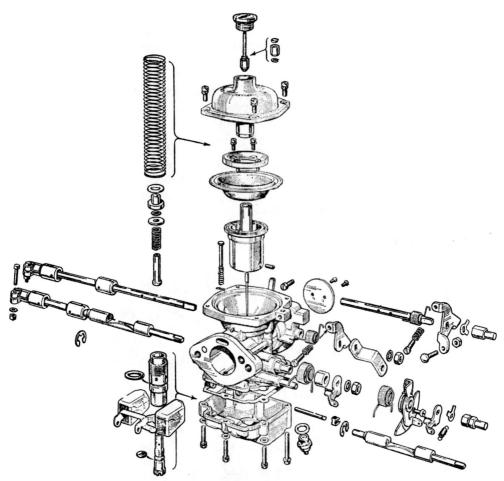

Fig. 12:27   A diagrammatic view of the Stromberg constant vacuum carburetter. Diaphragms, gasket sets, 'O' rings and needle valves are available in separate kit packs

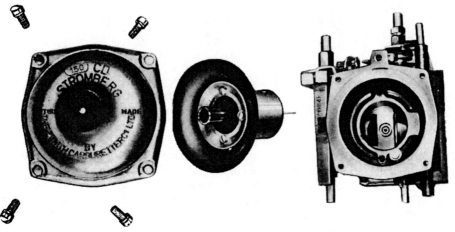

Fig. 12:28   Top view of the Stromberg with cover and air valve assembly removed

set the throttle-adjusting screws so that the same degree of 'hiss' is obtained from each carburetter, and then retighten the clamp bolts on the interconnecting spindle.

Make the final mixture adjustments with the jet-adjusting screws, turning each a similar amount until the engine is running smoothly (see tuning single Stromberg).

Jet centring

If the jet assembly is disturbed, it must be recentred otherwise the air valve will not fall freely, and the correct mixture will not be supplied. To recentre the jet, screw up the orifice adjuster until the orifice is just above the throttle tube intake bridge. Slacken the hexagonal nut above the orifice-adjusting screw half a turn to release the orifice bush.

Lift the air valve and allow it to fall; if the valve does not fall freely the first time, remove the damper and push the valve down with a pencil and repeat the lifting and dropping process. When the air valve falls freely tighten the jet assembly, checking that the air valve can still fall freely. If it does not, the recentring operation must be repeated; if difficulty persists, the unit must be stripped and cleaned.

When the air valve falls correctly and the jet is recentred, readjust the idling mixture and slow running speed.

Fig. 12:29 Diaphragm location on the Stromberg

Servicing Stromberg CD carburetters

A sticking air valve may be due to deposits of gum or carbon and the air valve should be removed for cleaning. Remove the screws which retain the top cover and lift the cover and diaphragm from the main body of the carburetter (see fig. 12:28).

Wipe the air valve and its bore with a rag moistened with paraffin or petrol. Clean, dry rag only should be used on the diaphragm but should the diaphragm become moistened and expand it must be allowed to dry before it can be fitted correctly to the retainer. If the diaphragm is perished, it must be renewed and if the jet needle is bent or worn, a new one of the correct specification must be fitted. The needle may be released from the piston by slackening the retaining screw and as with SU needles it must be fitted with the shoulder flush with the lower face of the air valve, and the locking screw tightened securely.

The diaphragm is secured to the air valve by a ring and screws. If a new diaphragm is fitted it is essential to locate the bead correctly and fully tighten the screws. Location for the bead and locating tab on the periphery of the diaphragm is by a location channel at the top of the main body (see fig. 12:29).

When refitting the air valve, place a few drops of oil on the guide rod.

If leakage from the jet assembly or base of a Stromberg CD carburetter occurs the float chamber should be removed, the gasket renewed and the 'O' ring (see fig. 12:26) on the orifice adjusting screw and the ring situated between the jet assembly and the float chamber renewed. On completion of the work, refill the damper orifice with SAE 20 oil. To check the float level see fig. 12:30.

Stromberg CDSE carburetters, see chapter 14.

Fig. 12:30 To check the float level on a Stromberg CD carburetter fitted to the Herald, remove the unit from the engine and take off the float chamber. Invert the carburetter. Check that the highest point of the float, when the needle is against its seating, is 18 mm above the face of the main body (see A); reset the level by bending the tag that contacts the end of the needle.

A thin fibre washer placed under the needle-valve seat will lower the fuel level. The float level plays an important part in correct carburation

MULTI-JET CARBURETTERS

Based on the fundamental principle of having fixed-sized choke and jets, this type of carburetter, of necessity, has many circuits or passages, incorporating jets and air bleeds that come into operation at various engine speeds and depressions to govern the amount of fuel discharged. One of the most popular fixed-choke carburetters on older British cars is the Zenith 26VME carburetter (see fig. 12:31), as fitted to Hillman and Austin A35, etc. followed by the Solex carburetter, as fitted to some of the Vauxhall and Triumph models.

The principal enemy of all carburetters is dirt and water and, with fixed-choke carburetters, the cleanliness of the fuel system is of paramount importance; it is essential with this type of carburetter that the sediment chambers in the petrol pump and in the carburetter float bowl are regularly cleaned, and that the various circuits, jets and air bleeds are blown through with air pressure at least every 6000 miles.

While there are numerous types of Zenith and Solex carburetters, all of which differ in various technical aspects, from the practical point of servicing and overhaul, the principle is much the same whichever the model.

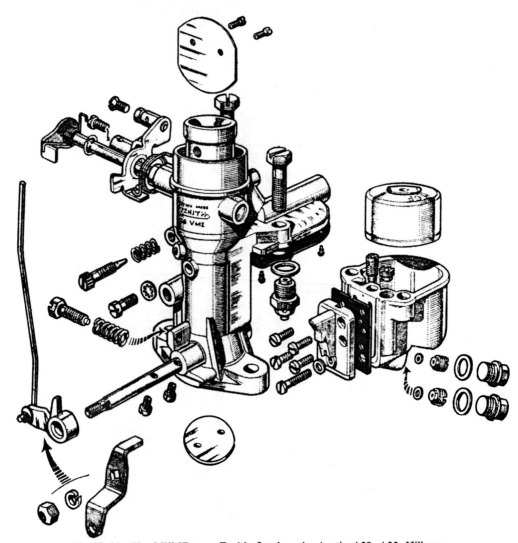

Fig. 12:31 The 26VME type Zenith fitted to the Austin A30, A35, Hillman Husky, 1959/60 A40 and many other cars. Note that an accelerator pump and economy device is not fitted to this carburetter. If the main jet or compensating jet is removed take care that the small washers beneath the jets are not inadvertently lost

Zenith carburetters

Overhauling a Zenith carburetter is principally a matter of cleaning and renewing worn parts. On some Zeniths such as are fitted to some of the Vauxhall Vivas it is necessary to renew all of the gaskets, the 'O' type sealing ring and to fit a new economy device diaphragm, and throttle spindle gasket.

Some Zeniths (see fig. 12:32) do not have economy devices and sealing rings, and though the stripping and reassembly procedure of this type of carburetter is slightly different, provided reference is made to the diagram of the appropriate model, no difficulty will ensue.

Unless the interconnecting linkage is worn, or the throttle spindle requires renewing, there is normally little point is dismantling the linkage mechanism. A special point to watch on some models is that the 'O' type sealing rings are fitted correctly and to renew the fibre washer between the main jet and the carburetter body.

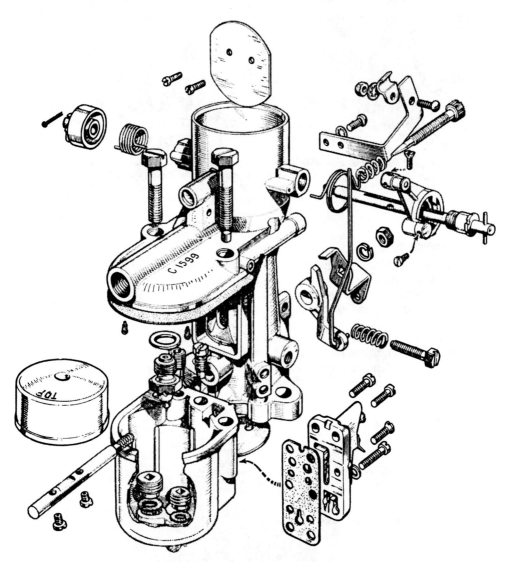

Fig. 12:32   The 30VM8 Zenith as fitted to Hillman series I, II and III (1956–59) and Husky (1959). If another carburetter is to be fitted it is important to refer to the number on top of the float chamber. A 30 VM8 with an identical number must be obtained otherwise the jet setting will not be correct

It is also advisable to renew the needle valve assembly, to fit a new washer beneath it and on some models, when refitting the top of the carburetter to the main body, it is essential to lift the choke lever and position it on to the top of the strangler cam.

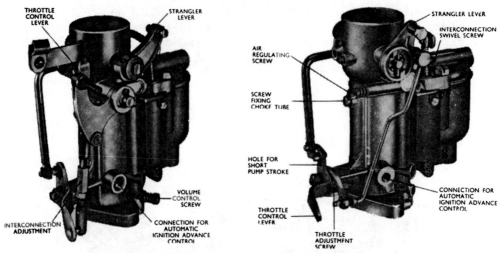

Fig. 12:33 The Zenith Series 30 VIG carburetter; (*Left*) with slow running mixture volume control, (*right*) with slow-running air-regulating screw which regulates the amount of air admitted to the slow running circuit. In carburetters fitted with volume control screws, the quality of the mixture is constant and the screw regulates the volume only

Cleaning and servicing

Normal servicing of Zenith carburetters entails removing the jets and blowing through the circuits with a high-pressure air line in the reverse direction of fuel flow. With carburetters fitted with a detachable float bowl (see figs. 12:33, 34 and 35), undo the bolts which secure the float chamber to the main body of the carburetter, remove the chamber and take out the float.

On some Zeniths, one of the float-bowl retaining bolts has a square on the end of the bolt which can be fitted into the main and the compensating jets to remove them from the float chamber. On other Zeniths, a screwdriver may be used, for the jets are slotted for the purpose.

When overhauling a Zenith it is advisable to remove the emulsion block. Apart from VN

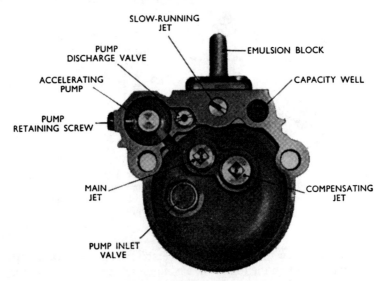

Fig. 12:34 Zenith 30 VIG carburetter float bowl (with float removed)

series carburetters (see fig. 12:36) and W types (see fig. 12:38; the emulsion block is held in position by three or by five screws which, when removed, allow the block to be pulled from the float chamber

On the VN type, the emulsion block is positioned inside the float chamber (see fig. 12:37) and must be removed to gain access to the main and compensating jet, positioned at the base of the emulsion block. The block is retained by two screws.

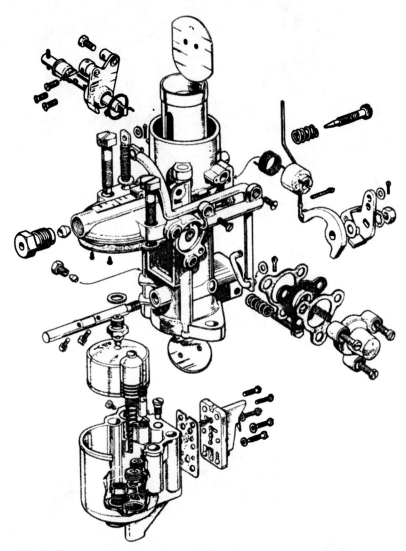

Fig. 12:35 Assembly details of Zenith 30 VIG as fitted to the Hillman Minx Series IIIA and many B.M.C. Austin models equipped with 'B' type engines

The old JS type carburetter as fitted to the Austin A35 (up to 1955) does not possess a compensating jet, compensation being obtained by a different method to that normally adopted by Zeniths.

When the emulsion block has been removed, take out all of the jets, air bleeds, non-return valves, accelerator pump piston and the capacity tube if fitted, and lay them carefully on a sheet of clean paper.

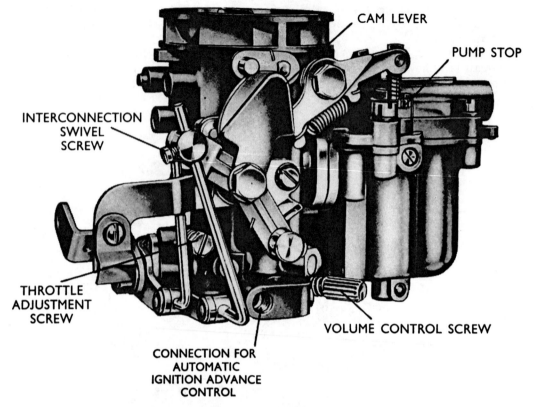

Fig. 12:36   The series VN Zenith carburetter as fitted to the Hillman, Ford Classic, Capri, Corsair and Cortina Super, Vauxhall Velox and Cresta and numerous others

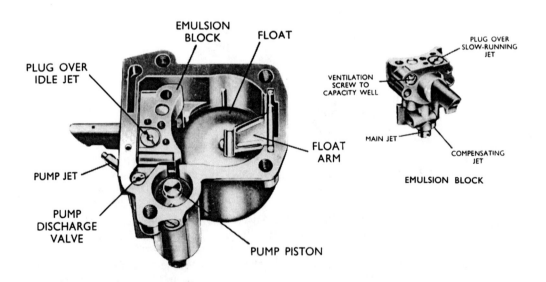

Fig. 12:37   Zenith series VN float chamber and emulsion block

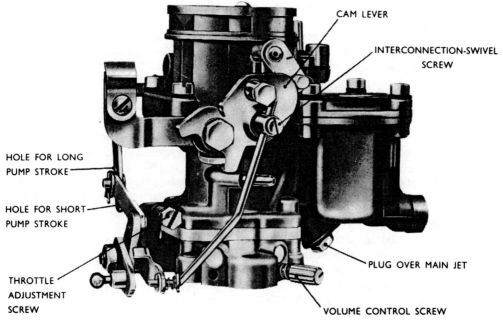

Fig. 12:38   The Zenith series WIA and WIP carburetter

If a high-pressure air line is available, blow through the jets and air bleeds, etc. before they are refitted, but on no account push wire through them. This can only cause detrimental burrs and abrasions.

When the emulsion block is refitted, it is essential to use a new gasket.

On the W type (see fig. 12:39), and IV Series, the float chamber is not separately detachable and the top of the carburetter has to be removed to gain access to the jets, although on

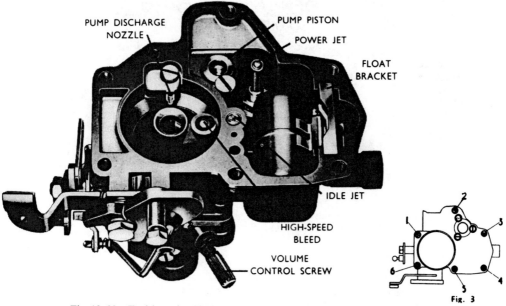

Fig. 12:39   Zenith series WIA carburetter with top removed. *Inset*: correct tightening sequences when refitting carburetter top

the W type (see fig. 12:38) the main jet is accessible by removing a brass plug on the side of the carburetter.

Although the VNT Zenith carburetter is fitted with a thermostatically operated, fully automatic butterfly type strangler the 30 VIG, VN Zenith and other types have manually operated stranglers, sometimes described as automatic or semi-automatic.

Semi-automatic 'choke' or strangler·

On this type, the strangler flap, secured to the spindle with two screws, has a small opening covered by a spring diaphragm. When the strangler is operated the butterfly flap and diaphragm both remain closed, but after the initial firing period, as the engine speed increases, the added depression causes the diaphragm to open, air is admitted and the desired weakening effect obtained. As with the plain butterfly type, the strangler must be returned as soon as possible after starting, but it is trouble free in operation except that on occasions the operating lever sticks owing to contact with the base of the air cleaner or air-cleaner retaining clip.

Fully automatic strangler

When the operating control is pulled, the strangler flap is held closed by means of a light spring. The added depression when the engine fires caused the flap to open. As with the semi-automatic type strangler a buzzing sound sometimes occurs due to the flap pulsating as the depression varies and, as with all strangler devices, the manual control should be pushed fully home as soon after starting the engine as possible.

Starting troubles sometimes occur with this type of strangler due to the spring breaking.

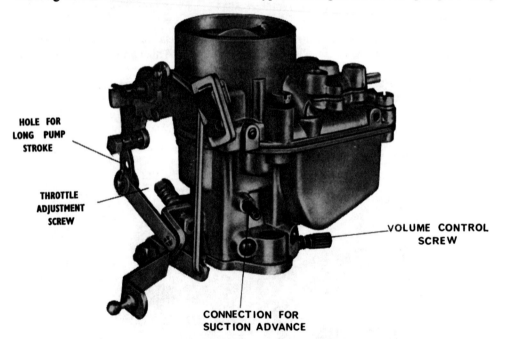

Fig. 12:40   The economy device on the series IV Zenith carburetter (as fitted to the Singer Gazelle, etc.) is situated above the float chamber

Zenith series IV

The series IV carburetters (see fig. 12:40), as fitted to 1965–1967 Hillman Minx, Singer Gazelle and Vauxhall Victor, differ from most of the other Zeniths as they consist of three separate assemblies (see fig. 12:41). The carburetter is a dual-float type with an economy device and accelerator pump, see also figs. 12:42 and 12:43.

To gain access to the float chamber and emulsion block, take out the four screws on the top of the carburetter, disconnect the pump linkage and lift off the carburetter top. The main jet and compensating jet are fitted at the base of the emulsion block, the jet fitted at an angle being the main jet.

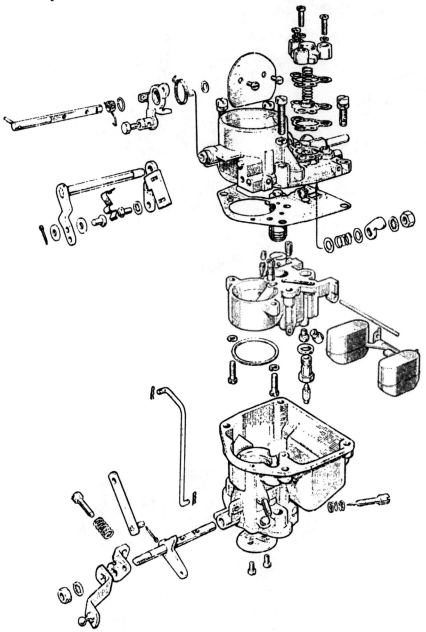

Fig. 12:41   Assembly diagram of Zenith IV type carburetter. Gasket sets, needle packs and economy device kits are available from Zenith agents

The jets are cadmium plated, have special flow characteristics and are not interchangeable with the plain brass jets of other Zenith carburetters.

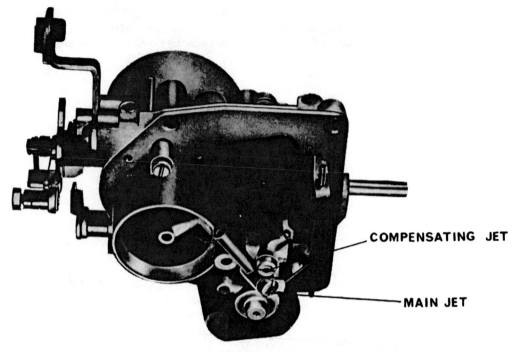

Fig. 12:42   Zenith series IV carburetter emulsion block, 'Main' and 'Compensating' jets

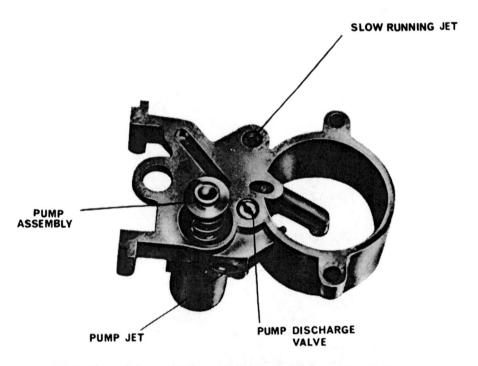

Fig. 12:43   Zenith series IV. Top view of the emulsion block showing the slow running jet, accelerator pump, pump discharge valve and jet

To remove the emulsion block, pull out the float hinge pin, remove the float assembly and take out the fuel inlet needle. Remove the needle seating (taking care not to lose the aluminium sealing washer) and the two screws which hold the emulsion block in position, and lift off the block, retaining the pump assembly which is released on removal of the emulsion block. The gasket between the block and the top cover can now be removed.

A rubber 'O' ring seal is fitted between the top of the choke tube and the bore of the emulsion block and it is essential that it is renewed when the carburetter is reassembled. Care must be taken to ensure that it is positioned correctly.

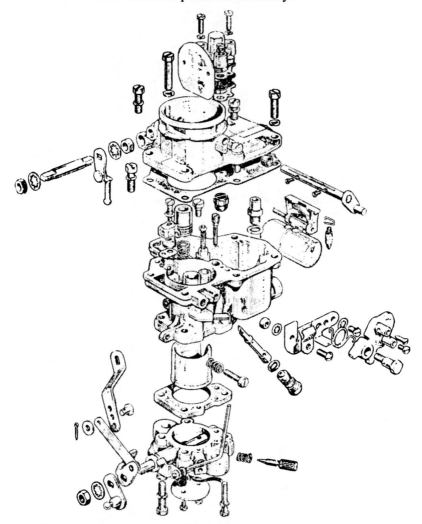

Fig. 12:44   Assembly details of WIA and WIP series Zenith carburetters

When reassembling the carburetter, make sure the top cover gasket aligns with the drilled holes in the underside of the cover, and the short cam-lever on the pump spindle fits into the recess to line up with the pump assembly in the emulsion block.

Zenith series WIA and WIP (see figs. 12:38 and 12:39)

There are several variations in the W type range of carburetters, see assembly diagram figs. 12:44 and 12:45, the 42 WIAT for instance is fitted with a thermostatically controlled strangler.

Most of the W types are fitted with a mixture volume control, but the 36 WIA-3 and the 36 WIP-3 (see fig. 12:46) is fitted with an air regulation screw.

On the W type, to gain access to the float chamber, accelerator pump, etc., remove the six screws retaining the top cover and lift the cover from the body.

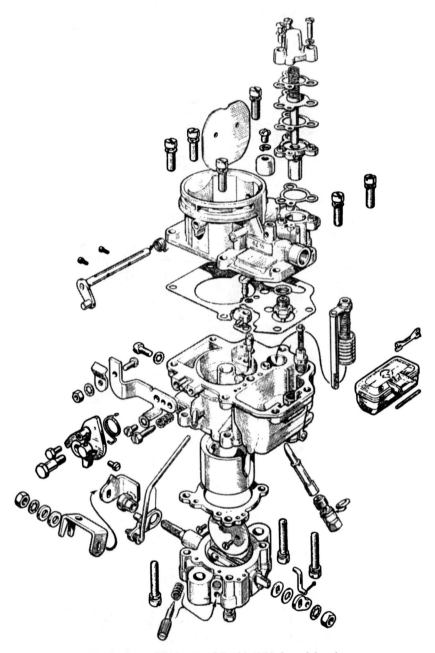

Fig. 12:45   Assembly details of Zenith WIA-2 model carburetters

The main jet is accessible externally from beneath the float chamber and the jet holder may be removed with a spanner. It is fitted with a flexible sealing ring and care must be taken not to damage it when handling the jet. Care must also be taken to ensure that the discharge nozzle does not drop out and sustain damage.

If the main discharge tube drops out, or is removed, when it is refitted it must be positioned correctly with the sharpened tip at the extreme top. If for any reason the discharge valve is removed, it is important that the pump piston is not depressed otherwise the ball underneath it may be blown out and lost.

When refitting the top cover to W type carburetters, tighten the screws in the correct sequence (see fig. 12:39) ensuring that the gasket remains in position and, in the case of the 34 mm and 36 mm sizes, it is important to ensure that the pump lever under the cover engages correctly on top of the accelerator piston and that the needle valve enters its seating.

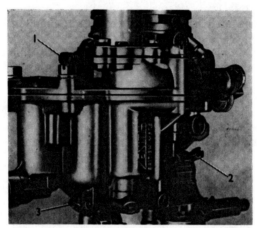

Fig. 12:46   The 36 WIP-3 Zenith fitted to the Vauxhall VX 4/90. (1) Mixture control screw. (2) Throttle stop screw. (3) Main jet

30 VIG carburetters (see fig. 12:33)

If faulty carburation occurs, the carburetter should be stripped, and all the circuits blown through with compressed air and the size of choke tube, jets and air bleeds checked to ascertain whether they conform to the setting specification for the engine concerned. The jets are clearly marked (see page 282). The economy device diaphragm (if fitted) and gaskets should be renewed.

Emulsion block

When a new emulsion block gasket is fitted, it is essential that it is replaced completely dry without sealing compound. On carburetters where the emulsion block is attached to the outside of the float chamber (see fig. 12:35), if it becomes necessary to remove the block, it is essential when it is refitted that the aluminium washers are fitted to the lower screws. Moreover, when the lowest screw has been tightened, carefully centre-punch the metal at the side of the screwhead to lock it.

Main and compensating jets

The mixture at all speeds above idling is supplied through the main jet and the compensating jet (see fig. 12:34), which both feed the beak of the emulsion block. The main jet influences the power and speed at medium to wide throttle openings, its output being directly related to the depression existing in the 'waist' of the choke tube (see fig 12:35) into which the beak of the emulsion block protrudes.

The choke tube controls the weight of charge inspired by the engine, usually the size of the tube is the smallest that enables maximum power to be developed. The flow from the compensating jet is complementary to that of the main jet, and the compensator is in operation at all speeds above idling speed. On the 30 VIG Zenith, the jet is ventilated to the atmosphere through the top of the capacity well, which provides a reserve of fuel during acceleration. In some carburetters, the well is open at the top, on others it is completely closed or vented by a drilling in a tubular headed screw. Variation in the size of the drilling affects the mixture strength, but in a different manner to changes in sizes of the compensating and main jet. A decrease in hole size enriches the mixture when the economy valve is open, but has little influence when the valve is closed, when the extent of ventilation is through a small permanent air bleed at the side of the economy device.

As the compensating jet is vented to the atmosphere, the flow from the jet is not affected by depression increases to the same degree as the main jet. Variation in the size of the compensating jet therefore has less influence on the mixture strength affecting acceleration and low-speed pulling than alterations to the main jet.

Carburetter jets are calibrated in units of hundredths of a millimetre and are normally available in steps of five units, a higher number denoting a larger calibration. Half-size main and compensating jets can be obtained, a jet marked 82 being approximately halfway between 80 and 85. For economy tuning they are often useful.

Choke tube sizes are marked in the intake side and indicate the diameter of the 'waist'. They are obtainable in stepped sizes of 1 millimetre only.

On Zenith carburetters, compensation for altitude is only necessary at heights above 5,000 feet.

Slow-running jets

The slow-running jet is a calibrated jet which supplies a measured quantity of fuel to the slow-running hole on the engine side of the throttle, and to the progression hole or holes at the edge of the throttle butterfly.

Unless the engine specification is considerably changed, it is extremely rare that any change in jet size is necessary. Should the engine fail to idle, the slow-running jet and circuit should be thoroughly blown through with compressed air and the idling adjusted as per page 283.

Pump jet

The pump jet is a small calibrated jet which fits into the emulsion block (VIG type) immediately behind the beak and meters the amount of fuel injected into the main air stream when the accelerator pump piston is activated.

To gain access to the jet on the VIG, the emulsion block must be removed. Care must be exercised when removing or replacing the jet as the threads are easily damaged. The jets are available in steps of ten units (50–90 inclusive). Sizes below 50 are not supplied, as they become blocked very easily. If it is desired to decrease the amount of fuel discharged from the jet, a modified pump inlet valve should be obtained from the Zenith Carburetter Company. This provides a small leak, to enable some of the fuel to return to the float chamber when the pump is operated.

Needle valve and seating

The diameter of the seating hole is stamped in millimetres on one side of the hexagonal body. The correct size depends upon the pressure in the fuel line, the cubic capacity of the engine and its power output. Unless any of these are substantially changed, if a new needle valve assembly has to be fitted it should be the same size as that removed.

The petrol level

The petrol level of Zenith carburetters is set at the factory. If flooding occurs, the float and needle valve must be examined and the pump pressure checked.

Zenith carburetter adjustment

There are three principal adjustments on Zenith carburetters:

1. Regulation of the idling mixture by an air regulating screw, or by a volume control screw (see fig. 12:33).
2. Slow running by means of a throttle stop screw.
3. Strangler throttle interconnection to open the throttle slightly when the strangler is operated.

The strength of the slow-running mixture influences the initial acceleration from low speeds only. If a flat spot occurs when the throttle is first opened, or the engine stalls in traffic, try slightly enrichening the slow-running mixture by turning the mixture regulating screw a little. The slow-running screw may have to be turned a little to prevent the engine from 'hunting'.

To completely reset the idling mixture and slow running, proceed as follows: Warm the engine until the correct operating temperature is reached, and turn the slow-running screw so the engine will run at about 600 rev/min. Screw the mixture screw in fully and then out one-half turn at a time until the engine runs smoothly. If the idling speed increases, readjust the idling screw.

A rich idling mixture results in an irregular engine beat and the exhaust smelling sweet and showing puffs of black smoke.

An unduly weak mixture results in the engine tending to 'miss a beat' and the exhaust will have 'splashy' note and be comparatively odourless.

A tendency to richness is better than the mixture being too weak. The former will give immediate response when the throttle is snapped open whereas a weak mixture will cause hesitation and possibly explosions in the silencer when the throttle is closed at high speeds. On high-mileage cars an idling speed on the high side is advisable as this will enable the car to accelerate more cleanly from low speeds with less likelihood of a plug 'oiling' during long idling periods.

On the old 26 and 30VM types of carburetter the best position for the mixture control from the point of a good 'pick up' (when the short type of adjusting valve screw is used) is normally within 1 turn of the fully home position. If the carburetter is fitted with the elongated-taper screw-valve, the adjustment should be within 3 turns of the fully home position. If the adjustment does not come within this range, an incorrect slow-running jet is indicated, or the wrong type of adjusting screw has been inadvertently fitted to the carburetter.

On the VN type of carburetter the screw is turned inwards to weaken the mixture and unscrewed to richen it.

When the carburetter is fitted with an air-regulating type of idling adjustment, the screw controls the air to the slow-running jet by governing the depression acting upon the jet. As the air is reduced, the depression increases and a greater volume of fuel issues from the jet. The action of the screw is directly opposite to that used in the volume control system. Volume control screws are fitted immediately above the main carburetter flange, whereas the air-regulating type are fitted about an inch below the strangler butterfly-spindle level.

Strangler/fast-idle adjustment

If the adjustment on the interconnecting rod between the choke and throttle is disturbed at any time, it must be reset to give an idling speed sufficiently high when the strangler is pulled, but which will not hold the throttle open when it is not in use. Adjustment is best carried out by trial and error.

If the fuel pump pressure is too high for the needle valve it will not be possible to obtain a smooth tickover, and petrol fumes may become noticeable in the car when it is running down-hill.

Although fitting a new needle-valve assembly will probably cure the trouble, the fuel pump pressure should be checked and be reset if necessary.

With difficult starting when cold, if a Zenith carburetter is fitted, it is essential to ascertain that the strangler is operating fully and to ensure that there is a ready supply of petrol in the float chamber. If a lack of fuel is found, check the needle valve to ensure that it has not 'gummed up'. If this fails to locate the trouble, the rest of the fuel system must be examined.

Difficult starting when the engine is hot may be due to flooding. To check this, remove the air cleaner and observe the discharge beak in the venturi. If petrol drips when the engine has been idling (seconds after switching off) check that the needle valve assembly is of the correct size; make sure that it is fully tight, check that the float is not perished and is fitted the correct way up. If flooding persists, try fitting a new needle valve; if this does not cure the trouble, the flooding is probably due to excess fuel pump pressure.

Erratic slow running

Check that the float chamber gasket is in good condition, and that the retaining bolts are tight. Make sure that the slow-running outlet hole and progression hole or holes in the throttle barrel are clear. Check that the carburetter flange is not bowed and admitting air and that the inlet manifold is free from air leaks. Check that the volume screw (or air regulating screw) taper is not faulty or the seating damaged.

High fuel consumption

High fuel consumption, if due to a carburetter fault, may be owing to the carburetter flooding or the strangler flap not returning fully, a faulty emulsion block gasket, slack emulsion block screws or a faulty economy device (see also below). The ball-valve assembly (see fig. 12:35) must also be removed to check that the ball is free, and if necessary cleaned. On the VN carburetter, the valve upper seating screws into the top face of the emulsion block, but on the VIG carburetter it screws into the top face of the float chamber. When the screw is removed, the ball can be taken out; if gum is present, the complete area should be thoroughly washed with methylated spirit. On replacement, the ball should be completely free. Any tendency for the ball to stick to the upper seating may result in petrol continually issuing from the pump jet.

Accelerator pump

In cases of poor acceleration, if an accelerator pump if fitted, check that fuel is ejected from the 'beak' when the accelerator is pushed. If fuel is not ejected, the pump is faulty. Remove the float bowl and completely strip and wash thoroughly with methylated spirit, giving particular attention to the non-return valve in the base of the float chamber.

If the accelerator pump link is in the summer hole, transfer it to the outer hole. With 30 VIG-11 carburetters as fitted to the Vauxhall Wyvern 1956 to early 1957 (4 ES engine) and the 1955 and early 1956 Velox and Cresta, the pump block must be turned to provide the maximum stroke. This is a square block fitted to the spring-loaded pump rod on the top of the float chamber. It has a vertical lug on one corner and may be arranged in either of two positions. When the lug is placed immediately below the end of the pump control lever, the short, summer stroke results. To change to the longer, winter stroke, the block must be lifted about a quarter inch and revolved through 180° so that the lug is adjacent to the air intake.

To change to the winter setting on 36WIP-3 carburetters, see fig. 12:47.

Poor acceleration

In all cases of poor acceleration, the jets and jet circuits must be thoroughly cleaned, paying particular attention to the compensating jet and main jet. The accelerator pump should be checked to ensure that it ejects a regular unbroken stream of fuel into the venturi. The economy device should also be removed and thoroughly cleaned.

Economy device

The economy device is a diaphragm-operated valve which is mounted on some models on

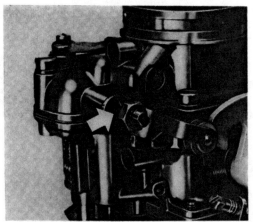

Fig. 12:47 The 36 WIP-3 Zenith showing the
accelerator pump arm in the summer position.
The arrow indicates the slot for the winter
setting

the side of the carburetter but on the W types, as fitted to 1956 and 1957 Sunbeam Rapiers
and 1959 Sunbeam Alpines, and the 42 VIS Zenith as fitted to the 1949–1954 Austin A70
and the Westminster up to 1959, it is fitted on the top of the float chamber. One side of
the diaphragm is spring-loaded, and is directly influenced by the engine depression. If the
diaphragm becomes perforated, or leaks occur past the gaskets which are fitted each side of
the diaphragm, the device cannot function and petrol consumption increases. In all cases of
high petrol consumption, the device should be dismantled and can be removed from the
carburetter by taking out the three retaining screws (early types had four screws).

A replacement diaphragm, gaskets and spring should be fitted; these are available in kit
form from most Zenith agents.

When reassembling the economy unit, ensure that the spring locates squarely in the recess
of the metal cup in the centre of the diaphragm and tighten the screws evenly and securely.

The choke tube

Engines which are power tuned may require a larger choke tube and correspondingly larger
jets. If modifications have been carried out on the power unit and the carburetter settings
are unsatisfactory, the Zenith Carburetter Company will be able to suggest satisfactory
settings.

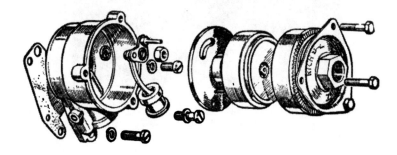

Fig. 12:48 Diagrammatic view of the automatic choke operating mechanism as
fitted to the Zenith series VNT carburetters

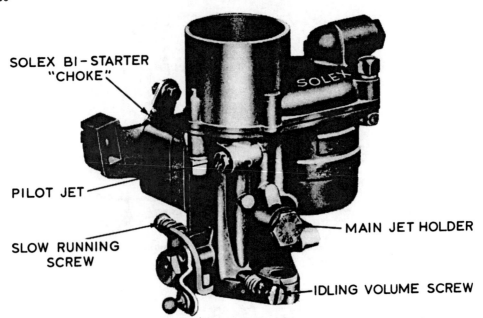

SOLEX BI-STARTER "CHOKE"

PILOT JET

SLOW RUNNING SCREW

MAIN JET HOLDER

IDLING VOLUME SCREW

Fig. 12:49   The Solex B26 ZIC-2 carburetter

Fig. 12:50   The Solex bi-starter device as fitted to a B26 ZIC-2 series carburetter.
Solex carburetters are seldom fitted with the strangler type of enrichening device for one of the characteristics of Solex carburetters is the bi-starter. This in effect is a simple auxiliary carburetter fitted to the side of the main carburetter which is brought into operation by pulling the dashboard knob. The system allows the air and petrol fed to the engine to be calibrated independently of the normal running jets in the main carburetter. The Zero starter system fitted to some Solex carburetters is also very similar

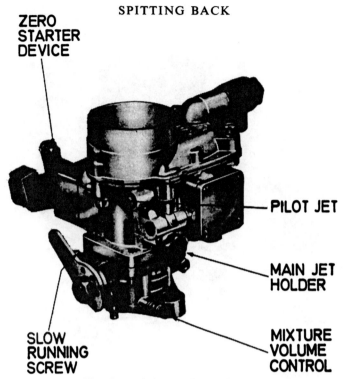

ZERO
STARTER
DEVICE

PILOT JET

MAIN JET
HOLDER

SLOW
RUNNING
SCREW

MIXTURE
VOLUME
CONTROL

Fig. 12:51   Solex 32 BI0-2 carburetter

Spitting back

Spitting back in a carburetter when accelerating from a low speed, if due to a carburetter fault, may be caused by a blockage in the compensating jet or possibly too small a compensating jet. If the spitting back occurs only at irregular intervals, however, and the car has little power and top speed, it may be due to the main jet being partially blocked or too small.

Banging and popping in the silencer may result from unburnt fuel igniting in the silencer*,

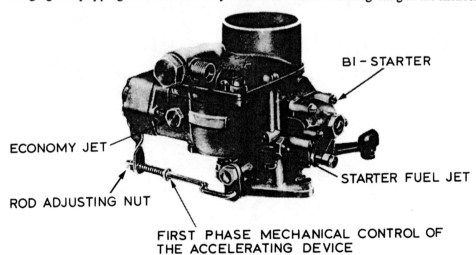

BI - STARTER

ECONOMY JET

STARTER FUEL JET

ROD ADJUSTING NUT

FIRST PHASE MECHANICAL CONTROL OF
THE ACCELERATING DEVICE

Fig. 12:52   The Solex 32 PBL-2 carburetter with two phase accelerating and
economy device

* Exhaust emission controlled cars, see also chapter 14

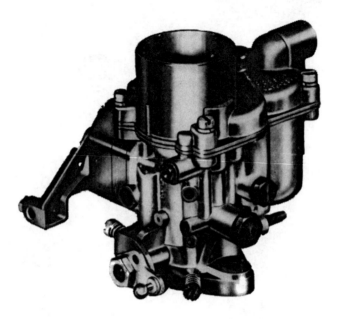

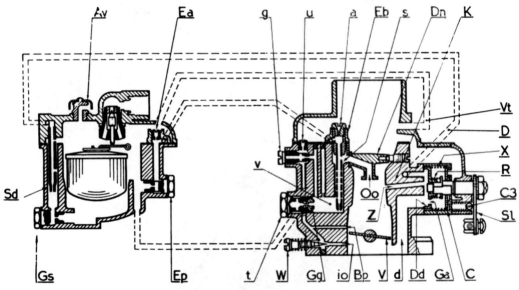

Av	Air vent	Ep	Econostat petrol jet	Sl	Starter lever
a	Correction jet	Ga	Starter air jet	s	Emulsion tube
Bp	By-pass orifice	Gg	Main jet	t	Main jet holder
C	Starter air valve	Gs	Starter petrol jet	u	Pilot air bleed
C3	Starter lever locating ball	g	Pilot jet	V	Throttle
D	Starter petrol channel	io	Idling orifice	Vt	Float chamber vent
Dd	Starter valve	K	Choke tube	v	Reserve well
Dn	Discharge nozzle	Oo	Spraying orifice	W	Volume control screw
d	Starter outlet port	R	Starter valve spring	X	Starter air valve spring
Ea	'Econostat' air bleed	Sd	Starter dip tube	Z	Quick drive-away channel
Eb	'Econostat' body				

Fig. 12:53   The Solex B30 ZIC-2 carburetter

due to an exhaust leak, but where popping back occurs when over-running downhill it may be due to an incorrect slow-running mixture or blocked slow-running jet.

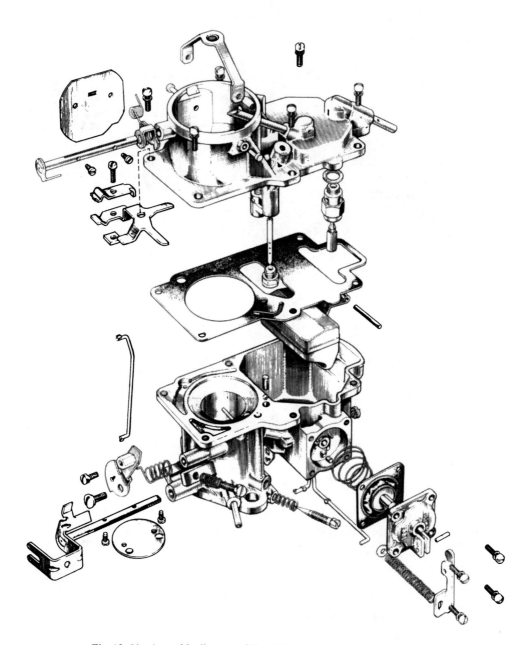

Fig. 12:54   Assembly diagram of Ford Cortina manual choke carburetter

Thermostatically operated automatic choke (Hillman Imp) see fig. 12:48

This is mounted at the side of the carburetter and is influenced by exhaust manifold temperature. The thermostat is adjusted when the carburetter is built and normally does not need resetting. If starting troubles occur, ensure that the air cleaner clip does not foul the strangler

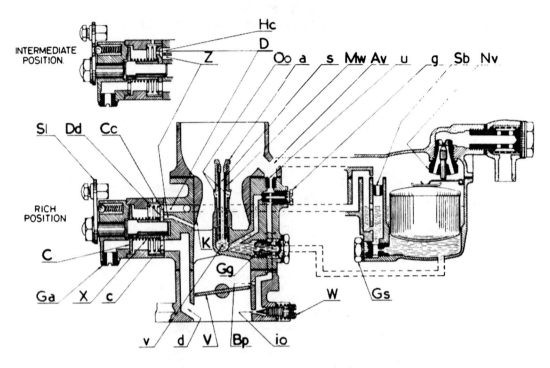

s	Emulsion tube	Bp	By-pass orifice
u	Pilot air bleed	Mw	Spraying well
a	Correction jet	g	Pilot jet
D	Starter channel	W	(Idling) Volume control screw
Ga	Starter air jet	K	Choke tube (Venturi)
Gs	Starter petrol jet	Gg	Main jet
Sb	Starter air bleed	Nv	Needle valve
Ce	Rich position orifice	c	Starter mixture outlet orifice
Hc	Starter progression hole	d	Starter outlet channel
Dd	Starter petrol valve	Av	Float chamber vent
Sl	Starter lever	C	Starter air valve
Z	Starter progression channel	X	Air valve spring
V	Throttle valve	Oo	Spraying orifices
io	Idling orifice	v	Reserve well

Fig. 12:55   The circuit and jets of the Solex 32 B10-2 carburetter

The idling mixture, when the engine is warm, is supplied through (io) and through (Bp) when the throttle is first opened but before it opens enough for the main spraying orifices (Oo) to discharge. Petrol for the spraying orifices is supplied from the reserve well (v), metered by the main jet (Gg) and pilot jet (g), air bleed (u) providing pre-emulsion. When idling, emulsifying air is also drawn in through (Bp), the volume of all this mixture being controlled by a screw (W). On leaving (io) the emulsion is mixed with the air passing round (V), this latter being held slightly open by an adjustment screw on a bracket at the end of the throttle spindle. As the throttle is opened, engine suction is directed to the hole (Bp) which discharges the richer mixture required to balance with the increased volume of air passing (V).

The Main Spraying Circuit

As the throttle is opened further, air speed through the venturi or choke tube (K) rises and petrol, drawn from the float chamber through the main jet (Gg) into the spraying well (Mw) via the reserve well (v), is pre-emulsified by the holes in the emulsion tube (s), drawing air from the correction jet (a) and sprays from orifices (Oo). As the engine speed rises, the bigger hole in the bottom of the emulsion tube is rapidly uncovered, and balancing air to weaken off the mixture is increasingly supplied in the amount required through (a).

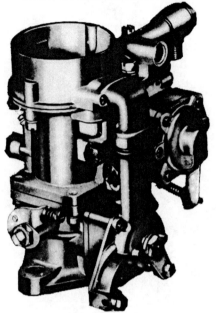

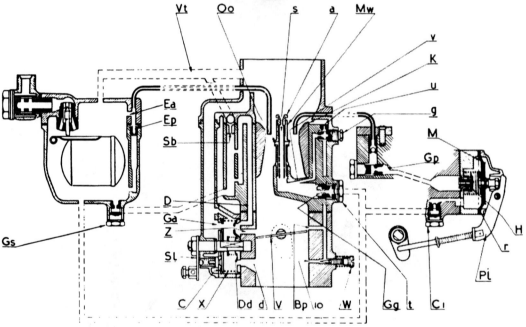

a	Correction jet	Gp	Pump jet	Sl	Starter lever
Bp	By-pass orifice	Gs	Starter petrol jet	Sb	Starter air bleed
C	Starter air valve	g	Pilot jet	s	Emulsion tube
Cl	Pump non-return valve	H	Pump valve	t	Main jet holder
D	Starter petrol channel	io	Idling orifice	u	Pilot air bleed
Dd	Starter valve	K	Choke tube	V	Throttle
d	Starter outlet channel	M	Pump membrane	Vt	Float chamber vent
Ea	'Econostat' air bleed	Mw	Main spraying well	v	Main reserve well
Ep	'Econostat' petrol jet	Oo	Spraying orifices	W	Volume control screw
Ga	Starter airjet	Pl	Pump lever	X	Starter air valve spring
Gg	Main jet	r	Pump spring	Z	Quick drive-away channel

Fig. 12:56   Solex B40 PA10-5 carburetter

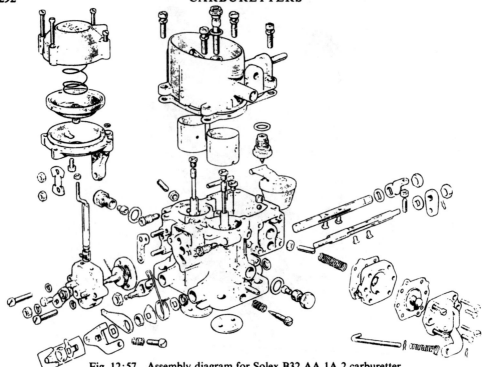

Fig. 12:57   Assembly diagram for Solex B32 AA 1A 2 carburetter

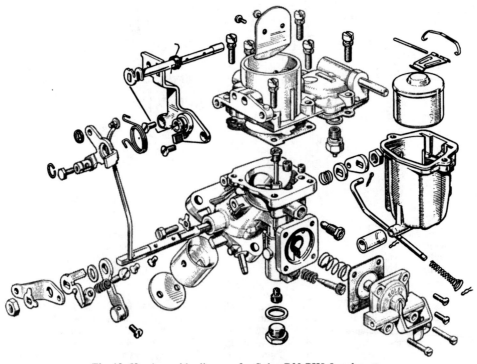

Fig. 12:58   Assembly diagram for Solex B30 PIH-5 carburetter

mechanism and that the outer end of the hot air pipe (which connects from the thermostat to the exhaust manifold) has not become disconnected.

Tuning Solex carburetters

The idling speed of the engine is controlled by the idling screw mounted on the abutment plate of the throttle lever (see figs. 12:49 and 12:51, etc.). Adjust the tickover speed to approximately 700 rev/min, then gradually open the volume screw (see fig. 12:49), turning

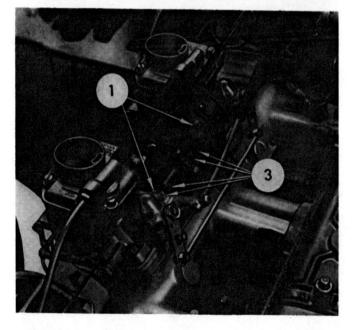

Fig. 12:59 Lh and rh views of Solex B32 PIH carburetter fitted to the Triumph Vitesse up to engine number HB6798. (1) Throttle stop screws. (2) Mixture control screws. (3) Linkage clamp bolts.

On later engines, to improve hot start conditions, the accelerator pump has been discontinued and blanking plates fitted. On early cars, the pump jets should be removed and blanked jets, part number 512087, fitted, the pump operating rods disconnected and removed, and the operating cams removed from the diaphragm covers. From engine number HB858-HE, new jet setting were used and may be adopted for earlier carburetters (the manufacturers should be consulted); see fig. 12:57 for diagrammatic view of jets, etc.

anticlockwise, until the engine begins to 'hunt' then turn it slightly in a clockwise direction half a turn at a time until the engine ceases to 'hunt'. If the engine speed is then too high reduce the speed slightly (to about 500 rev/min). If the engine begins to 'hunt' again, the volume screw must be screwed in slightly to restore a smooth idle speed.

Some Solex carburetters are fitted with a diaphragm-type accelerator pump. A Solex P32

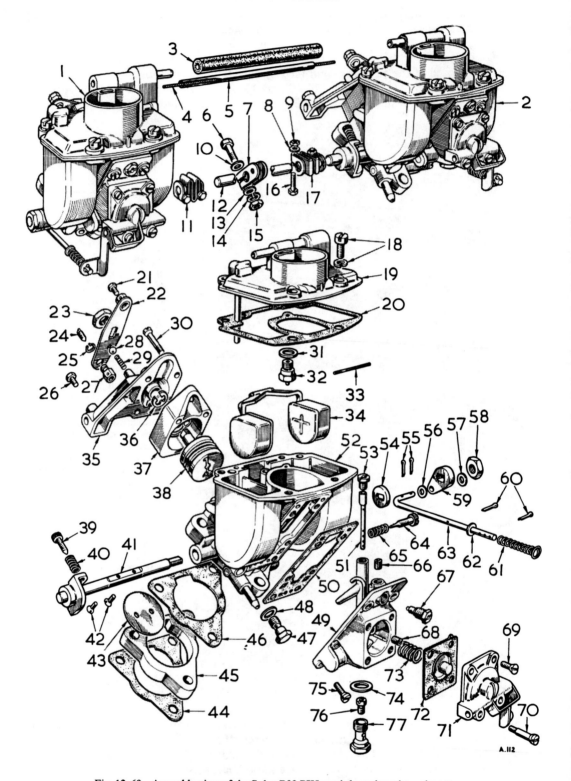

Fig. 12:60   Assembly view of the Solex B32 PIH semi-downdraught carburetter

KEY TO COMPONENTS OF SOLEX B32 PIH CARBURETTER

1.	Rear carburetter	27.	Nipple	53.	Air correction jet
2.	Front carburetter	28.	Ball	54.	Distance piece
3.	Fuel hose	29.	Spring	55.	Split pins
4.	Choke cable—inner	30.	Bolt	56.	Plain washer
5.	Choke cable—outer	31.	Fibre washer	57.	Plain washer
6.	Pinch bolt	32.	Needle valve	58.	Nut
7.	Accelerator level	33.	Pivot pin	59.	Lever
8.	Plain washer	34.	Float assembly	60.	Split pins
9.	Nut	35.	Starter cover	61.	Spring
10.	Plain washer	36.	Circlip	62.	Plain washer
11.	Coupling assembly	37.	Starter body	63.	Push rod
12.	Coupling rod	38.	Disc valve	64.	Idling mixture adjusting screw
13.	Spring washer	39.	Stop screw	65.	Spring
14.	Plain washer	40.	Spring	66.	Idling mixture air bleed jet
15.	Nut	41.	Throttle spindle	67.	Idling mixture fuel jet
16.	Pinch bolt	42.	Screws	68.	Pump jet
17.	Spring coupling	43.	Throttle disc	69.	Screw
18.	Screw and spring washer	44.	Gasket	70.	Screw
19.	Top cover	45.	Insulation gasket	71.	Pump cover plate assembly
20.	Gasket	46.	Gasket	72.	Pump diaphragm
21.	Pinch screw	47.	Starter jet	73.	Spring
22.	Lever	48.	Washer	74.	Fibre washer
23.	Nut	49.	Jet block assembly	75.	Screw
24.	Pinch screw	50.	Gasket	76.	Main jet
25.	Circlip	51.	Emulsion tube	77.	Main jet carrier
26.	Screw	52.	Carburetter body		

PBL2, similar to those fitted to the Jaguar 2·4, is depicted in fig. 12:52 and shows the carburetter pump, power system and jets. The accelerator pump is spring loaded so that, when the throttle is closed, the pump cavity is filled with petrol. The diaphragm is coupled to the throttle spindle by a rod, which conveys the movement of the throttle to the diaphragm and forces fuel through the pump jet into the main air stream.

In the ZIC type Solex (see fig. 12:53), designed to overcome cold-start conditions in cold climates where icing is a problem, the emulsion tube has been moved from the centre of the venturi to the side and the carburetter is fitted with a modified form of the Solex bi-starter device. Other Solex carburetters are illustrated in figs. 12:54 to 12:61.

Weber carburetters

Although there are Weber carburetters of downdraught, horizontal and updraught types, the double venturi twin-choke horizontal Weber carburetter type DCOE (see fig. 12:62), an assembly diagram of which will be found on page 298 (fig. 12:63), is now one of the world's foremost carburetters for high-performance motor cars. The double venturi provides an added velocity boost to the centre 'core' of the air stream and consequently a higher metering depression is available than with a single venturi. The use of this principle is not of course confined to Weber carburetters, for the double-venturi boost is also used on some Strombergs, the W type Zeniths; several American carburetters, in fact use triple venturi systems.

Access to the jets (see fig. 12:63) of the DCOE type for cleaning is obtained by taking off the cover and removing the jet holders with a screwdriver. The jet position of the IMPE type Weber is indicated in fig. 12:64.

The carburetter in fig. 12:65 is the 28/36 DCD fitted to the GT Cortina and the position of the various jets is seen while fig. 12:66 indicates the float setting and fig. 12:67 shows the general assembly. This carburetter has an idling-jet, main-jet, and piston-type accelerator-pump system. The barrels and throttle plates differ in size, the primary being the smaller. Throttle plate opening is at different rates, the primary commencing to open first followed by the secondary, both plates reaching full throttle simultaneously. Each barrel incorporates a main and an auxiliary venturi. The idling system operates in the primary barrel only, although progression is provided for both. Also the accelerator pump only discharges into the primary barrel. Cold starting is by a piston valve type device in the rear of the carburetter.

To adjust the carburetter, screw in the throttle-stop screw to obtain a fast idling speed (engine at normal operating temperature) then screw the volume control in or out so that the engine runs evenly. Readjust the throttle stop screw if the engine is running too fast and follow with a further adjustment with the volume screw.

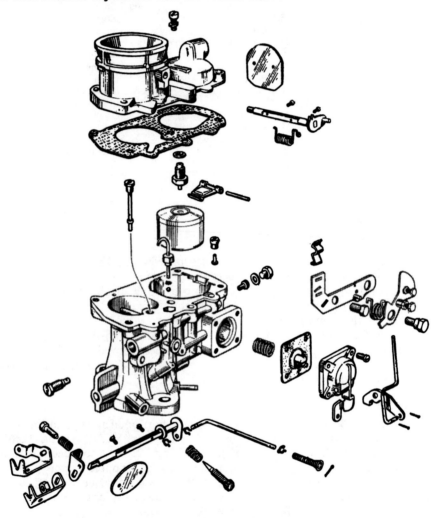

Fig. 12:61 Assembly diagram of Solex B33 PSE1-2 carburetter as fitted to the
Singer and Hillman Minx 1964–65

To clean the Weber 28/36 DCD carburetter (as fitted to the GT Cortina), remove the air cleaner cover and element. Bend back the ears of the tab washers and remove the four nuts, two tab washers, flat washers and rubber insulators with sleeves which retain the air cleaner to the carburetter. Remove the air cleaner and the rubber gasket then detach the fuel supply pipe and the distributor vacuum pipe and disconnect the choke control. Unscrew the six retaining screws and spring washers and remove the carburetter cover. Lift the gasket from the cover. Push out the float fulcrum pin from the cover and remove the floats and needle valve. Next, unscrew the needle valve housing and remove the sealing washer, and then remove the accelerator pump from the carburetter body by pulling out the inverted 'U'-shaped control rod. This removes the split retainer, spring and piston. The accelerator pump delivery valve and inlet valve can now be removed; the delivery valve from between and on top of the barrels, the inlet from the base of the float chamber.

Lift the accelerator pump jet from the body. Unscrew the starting air-correction jet and the starting jet assembly which is partly countersunk in the top of the carburetter, between the barrels and float chamber. Pull the starting jet from the starting air-correction jet. Unscrew the primary air-correction jet and emulsion tube assembly, followed by the secondary air-correction jet and emulsion tube assembly from the top of the carburetter between the barrels and float chamber. Remove the circular idling jet holders from the side of the body, one each side, just below the top of the body. Unscrew the hexagon-headed main jet holders from the outside of the body, one each side above the mounting flange. Unscrew the main jets. Blow out the jets, the accelerator pump and the float chamber. Wash the floats, float chamber and jets in clean petrol.

Fig. 12:62 Weber type 40 DCOE-2 horizontal twin choke carburetter

To reassemble the carburetter, screw each main jet into its hexagonal holder and tighten. Check the condition of the jet-holder copper gaskets and screw each holder in position above the mounting flange. Note that the primary main jet has the lower number (140 on the Cortina) and the secondary main jet the higher number (155 on the Cortina), although the jet holders are identical (see fig. 12:67). The secondary barrel is of a larger diameter than the primary.

Next inspect the idling jet holder's sealing rings, push a jet into each circular-headed holder and fit one in each side of the carburetter below the top of the body. The two jet holders are identical but the primary idling jet has the lower number (50 on the Cortina) and the secondary idling jet the higher (70 on the Cortina). When fitting an air-correction jet into each emulsion tube, note also that the emulsion tubes are identical but the primary air-correction jet has the higher number (230 on the Cortina) and the secondary air-correction jet the lower number (180 on the Cortina). Fit the assemblies on top of the carburetter

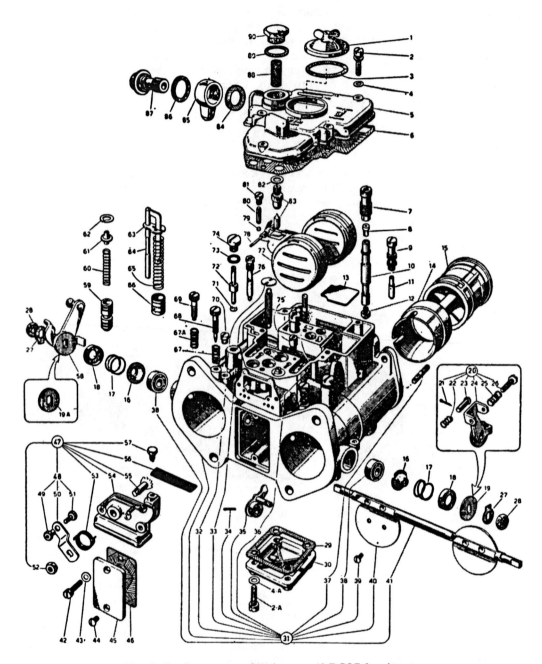

Fig. 12:63   Components of Weber type 40 DCOE-2 carburetter

between the barrels and float chamber. Push the starting petrol jet into the starter air jet, open end first, and screw the assembly into the top of the carburetter between the float chamber and barrels. The hole is counterbored.

When fitting the accelerator pump jet and delivery valve, place a fibre washer in the recess between the two barrels, followed by the pump jet which has a lug on one side that fits a cut-away in one side of the recess. Place another fibre washer on top of the pump jet,

KEY TO COMPONENTS OF WEBER 40 DCOE-2 CARBURETTER

Key No.	Part Name	Quantity
1.	**Jets inspection cover**	1
2.	Cover fixing screw	5
2A.	Fixing screw	4
3.	Gasket	1
4.	Washer for screw	5
4A.	Washer for screw	4
5.	**Carburetter cover**	1
6.	Gasket for carburetter cover	1
7.	Emulsioning tube holder	2
8.	Air corrector jet	2
9.	Idling jet holder	2
10.	Emulsioning tube	2
11.	Idling jet	2
12.	Main jet	2
13.	Plate	1
14.	**Choke**	2
15.	**Auxiliary venturi**	2
16.	Dust cover	2
17.	Spring	2
18.	Small lid for spring retainer	2
19.	Distance washer (rear carburetter)	1
19A.	Distance washer (front carburetter)	1
20.	**Throttle control lever, complete** (front carburetter) including:	1
21.	Split pin	1
22.	Spring	1
23.	Pin	1
24.	**Throttle control lever**	1
25.	Spring	1
26.	Screw	
27.	Lock washer	2
28.	Fixing nut	2
29.	Gasket	1
30.	Bowl bottom small lid	1
31.	**Carburetter body** including:	1
32.	Plate for spring	1
33.	Shaft return spring	1
34.	Spring pin	1
35.	**Pump control lever**	1
36.	Stud bolt	1
37.	Stud bolt	2
38.	Ball bearing	2
39.	Throttles fixing screw	4
40.	Throttle	2
41.	Throttle shaft	1
42.	Screw securing support	2
43.	Washer for screw	2
44.	Fixing screw	2
45.	Plate	1
46.	Gasket	1

Key No.	Part Name	Quantity
47.	**Starting control** including:	
48.	Starting control lever complete with:	1
49.	Nut for screw	1
50.	Starting control lever	1
51.	Screw securing wire	1
52.	Lever fixing nut	1
53.	Lever return spring	1
54.	**Sheath support cover**	1
55.	**Starting shaft**	1
56.	Strainer	1
57.	Screw securing sheath	1
58.	**Throttle control lever** (rear carburetter)	1
59.	Starting valve	2
60.	Spring for valve	2
61.	Spring guide and retainer	2
62.	Circlip	2
63.	Spring retainer plate	1
64.	**Pump control rod**	1
65.	Spring for plunger	1
66.	**Pump plunger**	1
67.	Spring for idling mixture adjustment screw	2
67.	Spring for throttles adjustment screw (rear carburetter)	1
68.	Idling mixture adjustment screw	2
69.	Throttles adjustment screw (rear carburetter)	1
70.	Progression holes inspection screw	2
71.	Pump jet gasket	2
72.	Pump jet	2
73.	Gasket	2
74.	Screw plug	2
75.	**Inlet valve**	1
76.	Starting jet	2
77.	**Float**	1
78.	Pivot	1
79.	Valve ball	2
80.	Stuffing ball	2
81.	Screw for stuffing ball	2
82.	Gasket for needle valve seat	1
83.	**Needle valve seat**	1
84.	Gasket for fuel filter casing	1
85.	Fuel filter casing	1
86.	Gasket for fuel filter casing	1
87.	Fuel filter bolt	1
88.	Strainer	1
89.	Gasket for strainer inspection plug	1
90.	Strainer inspection plug	1

shake the delivery valve, to ensure that the ball slides freely, pass it through the pump jet and screw it into the carburetter. Check that the accelerator pump inlet-valve ball can also move freely, and screw the valve into the base of the float chamber. Replace the accelerator pump assembly in the carburetter and secure it with the split retainer. Screw the needle valve seat into the cover, after ensuring that the seat sealing washer fitted between the seat and cover is in good condition. Refit the needle valve and floats. Check the needle valve damping ball for free operation and then place the needle valve in its seat. Slide the float tab under the needle-valve retaining hook and then push the float fulcrum pin through the cover 'legs' and float hinge. Next, place the carburetter-cover gasket on the carburetter and carefully fit the cover, ensuring that the floats are free to move in the body. Secure the cover with the six screws and spring washers, tightening them evenly.

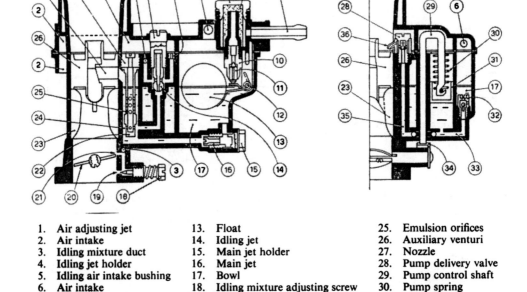

1.	Air adjusting jet	13.	Float	25.	Emulsion orifices
2.	Air intake	14.	Idling jet	26.	Auxiliary venturi
3.	Idling mixture duct	15.	Main jet holder	27.	Nozzle
4.	Idling jet holder	16.	Main jet	28.	Pump delivery valve
5.	Idling air intake bushing	17.	Bowl	29.	Pump control shaft
6.	Air intake	18.	Idling mixture adjusting screw	30.	Pump spring
7.	Strainer inspection plug	19.	Idling hole to intake pipe	31.	Pump plunger
8.	Strainer gauze	20.	Throttle	32.	Pump intake valve
9.	Fuel inlet connection	21.	Progression orifice	33.	Pump exhaust duct
10.	Needle valve	22.	Emulsioning tube	34.	Pump control lever
11.	Needle	23.	Choke	35.	Pump delivery duct
12.	Pivot	24.	Emulsioning tube well	36.	Pump jet

Fig. 12:64   Section through Weber downdraught carburetter type 32 IMPE

Reconnect the choke control, securing the outer cable with the clamp screw on the starting device arm and the inner cable in the operating lever with the clamp screw. Reconnect the fuel and the distributor vacuum pipes to the carburetter. Refit the air cleaner. Place the rubber gasket on top of the carburetter and fit the air filter body. Place a rubber insulator around each mounting stud and then pass a sleeve through each insulator. Place a flat washer and a new double-type tab washer on each of the studs and secure with the four nuts, tightening carefully and locking the nuts with the tab washers.

Fit the air cleaner element and the top cover.

1. Secondary air corrector jet
2. Accelerator pump delivery valve
3. Accelerator pump jet
4. Primary emulsion tube
5. Primary main jet holder

6. Secondary emulsion tube
7. Starting jet and air corrector jet
8. Accelerator pump inlet valve
9. Primary air corrector jet
10. Primary idling jet holder

Fig. 12:65   The Weber 28/36 DCD 23 carburetter as fitted to the GT Cortina

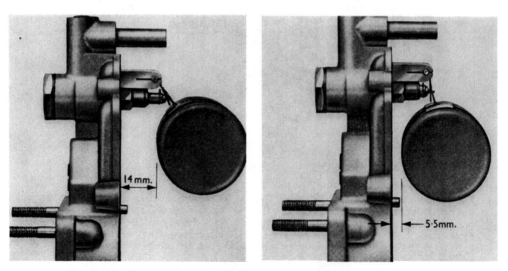

Fig. 12:66   The float setting on the Weber carburetter as fitted to the Cortina
engine

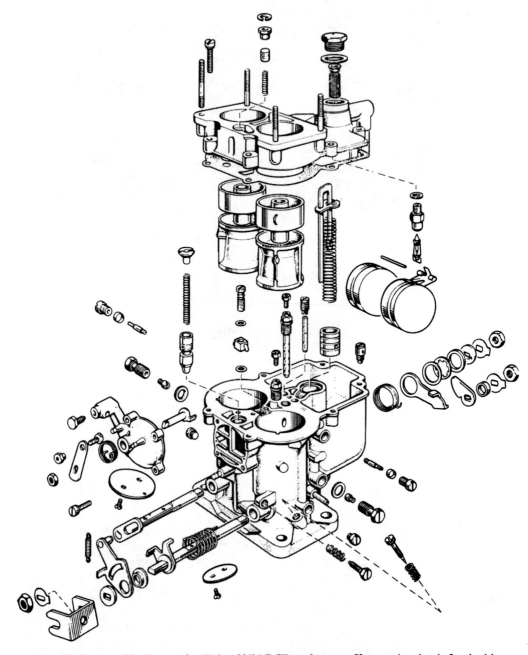

Fig. 12:67   Assembly diagram for Weber 28/36 DCD carburetter. If an engine that is fitted with a progressive type carburetter, such as this DCD on the Cortina GT, starts to use more fuel than normal, is slow to return to normal idle speed, and it is not due to an air leakage or sticking throttle cable, it may be due to the second stage throttle not closing completely. If the secondary linkage at the side of the carburetter is lubricated, this will often cure the trouble

Fig. 12:68 Automatic choke carburetter as fitted to some Cortina engines. Arrowed: the mixture volume control screw

Ford carburetters

To adjust the slow running on Ford carburetters such as fitted to the standard 1300cc and 1500cc Cortinas, warm the engine to normal operating temperature and then turn the throttle stop clockwise (see fig. 12:68) until the engine is running at about 800 rev/min. Next, screw the volume control screw in or out until the engine runs evenly. Readjust the throttle stop screw until the idling speed is satisfactory (580 to 620 rev/min), making any necessary adjustments to the volume control screw to ensure an evenly running engine. Float setting is given in fig. 12:69.

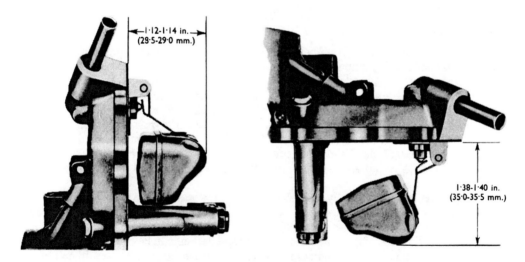

Fig. 12:69 Float settings for the Ford Cortina carburetter

CHAPTER 13

Solving Cooling System Problems

Although the optimum operating temperature of an engine varies to some extent according to its design, too low a temperature may result in poor performance due to incomplete fuel combustion, and will also result in the car heater failing to function satisfactorily, for the heater relies upon the water in the cooling system to provide heat energy.

Until the normal engine operating temperature is reached, the unequal expansion of the different metals used in the construction of the engine, together with the condensation of the unburnt fuel on the cylinder walls, may result in excessive wear taking place. This, indeed, is why an engine should not be subjected to fierce acceleration under heavy load until the normal operating temperature is reached.

When an engine overheats it is because the cooling system is unable to dissipate the heat from the engine rapidly enough. This is due to more heat being generated than the system is designed to dissipate, or alternatively, to some malfunction of the system. Common faults are a slipping fan belt, a blocked (possibly frozen) radiator, a faulty thermostat, faulty radiator cap, a leaking or faulty water pump, or on hot days it may be due to long periods of idling, or long pulls uphill with an overladen vehicle. Too high a running temperature may cause pre-ignition and plug troubles (see page 15).

Cooling system leaks

Internal leaks, resulting in the cooling water contaminating the engine oil, if not discovered, can result in considerable damage to an engine. If cooling water disappears rapidly for no apparent reason, a quick test should be carried out. When the engine is hot, remove the dipstick from the engine and allow a few drops of oil to fall into a small receptacle, such as the screw top from a can. Using a match or, preferably a gas-type cigarette lighter, burn the oil. If the oil spits and crackles, water is present and if the cause cannot be ascertained, the cylinder head should be removed and the cylinder head gasket examined. If the gasket is satisfactory, the cylinder head and block must be pressure-tested for cracks. When the cause of the leak has been found and rectified, the oil must be drained and the engine recharged with fresh oil.

External water leaks are usually fairly easy to find. It is nevertheless surprising how many cars become delayed on journeys due to a top water hose disintegrating. All water hoses are best replaced before they are four years old.

Leakage from the base of the cylinder head studs and gaskets can be cured by removing the offenders and fitting new parts. Temporary relief from leaks can be obtained by use of one of the proprietary leak cures, but they should be drained from the system as soon as possible and if the leak re-occurs it must be cured by rectifying the cause of the leak.

On older cars, leaky core plugs can be a nuisance; they should be removed by driving a small chisel into the centre of the plug. Make sure to obtain a new core plug of suitable size and to clean the area where the plug seats. Fit the new plug, using a liberal amount of sealing compound, by driving home with a hammer and a suitably sized drift.

Thermostats

If overheating is experienced, or if the engine is slow to warm up, the thermostat may be at fault. The thermostat is a valve placed in the cooling system (see fig. 13:1) which opens and closes according to changes in temperature of the cooling fluid.

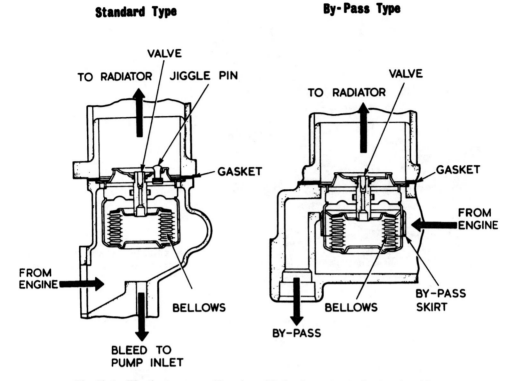

Fig. 13:1   The thermostat and housing with the thermostat in the closed position.
In all cases of overheating, pinking and excessive loss of coolant, the thermostat,
radiator cap and fan belt tension should be among the first things to be checked

Fig. 13:2   Changing a thermostat. The
thermostat is in the closed position

Fig. 13:3 Although a bellows-type thermostat can sometimes be replaced with the wax pellet type, before doing so it is advisable to ascertain that it is recommended practice for the type of engine

If the thermostat ceases to function correctly it must be renewed. Overheating is frequent due to the thermostat not opening at the proper temperature whereas slow warming-up may be due to the thermostat sticking open. The thermostat is usually fitted in a housing on the front of the cylinder head and is quite easy to remove (see fig. 13:2).

A thermostat may be tested by suspending it in a can of water and raising the temperature. Should the thermostat fail to open at a temperature within 5° of that specified (on the thermostat) it should be renewed, for the opening temperature is set by the manufacturer and cannot be altered.

If a second-hand car has been purchased and cooling system troubles develop, it is possible that the wrong thermostat has been fitted, and it should be removed and checked against the manufacturer's recommendation (see also figs. 13:3 and 13:4).

Radiator caps

With modern pressurized cooling systems, the radiator cap contains a pressure release valve designed to operate at a specified pressure (usually marked on the cap).

On sealed systems with overflow tanks, as on the B.M.C. 1100 (see fig. 13:9), the radiator cap does not contain valves and the pressure release cap is fitted to the overflow tank. It is important that the pipe, which connects the radiator to the overflow tank, does not become disconnected or break, as this can cause a coolant shortage which if not noticed may result in overheating and engine seizure. The pipe should be inspected at regular intervals.

Radiator caps are manufactured from corrosion-resistant metal, but the valves sometimes become inoperative, or the sealing washer perished.

Usually a visual inspection reveals the need for replacement of a radiator cap but if the valves are thought to be faulty, when the engine is at full operating temperature, place a thick rag over the radiator cap and *partially* release the cap. Do this carefully, and on no account remove the cap completely, for it is easy to get scalded hands.

As the cap is released, if it was seating correctly and maintaining pressure, a definite hiss will be heard as the pressure is released. This will not confirm whether the valve is sticking, however, allowing too much pressure to build up, but if the valves are not corroded, a thorough cleaning is normally sufficient to restore satisfactory operation.

If a new cap is fitted, it is important to ascertain that it is of the correct pressure and type. The wrong type of cap can cause a complete loss of cooling water and resultant engine seizure.

Curing heater failure

The failure of a heater to supply an adequate heat output may be due to a faulty heater fan, a blocked heater pipe or a faulty valve unit, to incorrectly adjusted heater controls or, if the cooling system has been drained and refilled recently, to an air lock in the heater unit.

By systematic procedure, and elimination of the possible faults, it is fairly simple to locate heater troubles. If the heater motor is not functioning, check the electric supply to the motor. If current is available at the motor, the heater motor is probably faulty and it will have to be removed and overhauled, or a replacement fitted.

If the heater motor is satisfactory, check the flow of water from the heater outlet pipe when the engine is hot (see fig. 13:10). Water should flow immediately, but if a delay occurs before the flow starts, it may be due to an air lock. Reconnect the pipe quickly and tighten the hose clip. If necessary, replenish the radiator with coolant. If water does not flow when the pipe is disconnected, check the heater controls to ensure they are set correctly.

The valve on the coolant outlet from the engine (inlet to the heater) should be in the full-flow position. If the control is set correctly with the valve in the full-flow position, and a flow from the heater outlet pipe cannot be obtained, check the system for a blockage. Detach the inlet and outlet pipes and apply air pressure or a garden hose at the inlet pipe and check that a discharge is obtained. If a blockage is detected, examine the rubber connecting pipes. If these are not blocked, the fault must be in the heater radiator matrix and the heater must

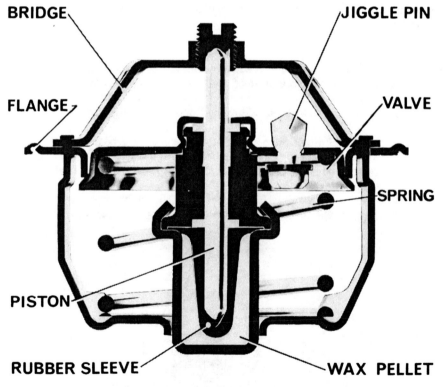

Fig. 13:4   A cross section of a typical wax pellet type thermostat. Although this type is now widely used, it has not completely superseded the bellows type. It is not advisable to substitute a wax pellet type in place of a bellows thermostat unless it is recommended

Fig. 13:5   Cortina water pump

be removed from the car, the matrix removed and checked and if necessary a service replacement fitted.

If the heater matrix and pipes are not blocked, the valve which controls the coolant flow from the engine to heater may be at fault. Most complaints of inadequate heater operation prove due to a faulty thermostat, a faulty regulating valve or improperly adjusted controls or an air lock.

Descaling the system

Over a period of time, rust and scale tend to accumulate in the radiator and the engine water jacket. The scale restricts the circulation of water and the engine tends to become overheated. If the water in the radiator is rusty and muddy in appearance, rust is present and the entire system should be drained and then refilled and descaled, using a proprietary water-jacket cleanser.

After the descaling solution has been used, the system should be flushed, preferably by the reverse-flush method, using a high pressure hose. If high water pressure is not available, a normal water hose can be tried, and should remove any loose deposits freed by the descaling solution.

A defective head gasket can allow gases to leak into the cooling system. If this is suspected, remove the fan belt, the thermostat and top water hose. Top up the water jacket and accelerate

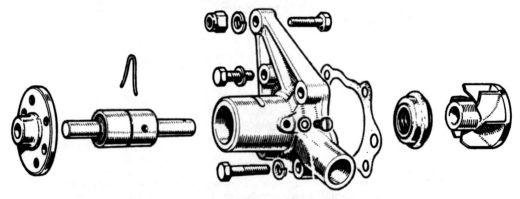

Fig. 13:6   Water pump fitted to B.M.C. 'B' type engines

the engine rapidly several times. If the water hose joints are all tight and bubbles appear, or if the water level rises appreciably, gases are probably penetrating the cooling system, and a new cylinder-head gasket should be fitted.

Water pumps

If the water pump is thought to be faulty in operation, although there is no accurate way of testing it whilst the pump is fitted to the car, a quick check can be obtained when the system is up to operating temperature, by squeezing the radiator top hose and accelerating the engine. If the pump is functioning pressure will be felt as the engine speeds up.

When a water pump begins to leak it should receive attention. Most water pumps are extremely easy to remove and refit and reconditioned exchange pumps are readily available at most of the main agents for the make of car in question.

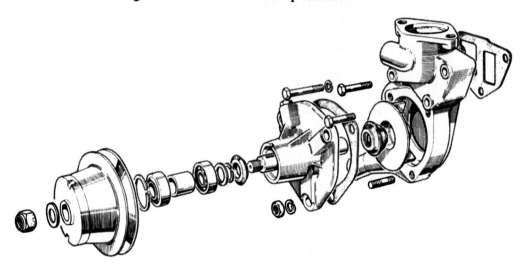

Fig. 13:7   Assembly details of Triumph water pump

If the old water pump is to be stripped and overhauled, some form of hand press will be required to press the impeller and fan blade hub from the main spindle. The parts required for the overhaul are: a set of bearings and spindle (on most pumps this is a complete unit), a spring-retaining clip or clips, a gland seal, gaskets and, for some pumps, hmp grease to pack the bearings. Assembly diagrams of water pumps will be found in figs. 13:5 to 13:8.

Water pumps must be reassembled so that the clearance between the impeller blades and the pump housing is between 0·020 and 0·030 in. It is also essential to line up any lubricating hole in the bearing spindle assembly with that in the water pump body.

On B.M.C. cars, the fan blade hub should be pressed on until the end of the spindle is flush with the face of the hub.

Water pump squeak

In the event of water pump squeak, such as is rather noticeable on some Series III Hillman Minxes, treat the coolant with Shell Donx C fluid. One-eighth of a pint should be sufficient to cure the trouble although if the squeak persists this may be slightly increased.

Frost precautions

Cooling systems are usually equipped with separate drain taps for the engine and radiator and during frosty weather if a car is drained and left overnight the coolant must be drained from both taps. It must be remembered, however, that even if the cooling system is drained,

Fig. 13:8 Typical water pump and fan components (Austin Westminster B.M.C. 'C' type engine). Service replacement pumps for most British cars are readily available

the heater is not and, in very cold weather, damage can be caused to the heater matrix.

On sealed cooling systems, the expansion tank can only be drained with difficulty and if not protected it may be damaged. Rather than attempt to drain such sealed systems it is better to add anti-freeze to the coolant.

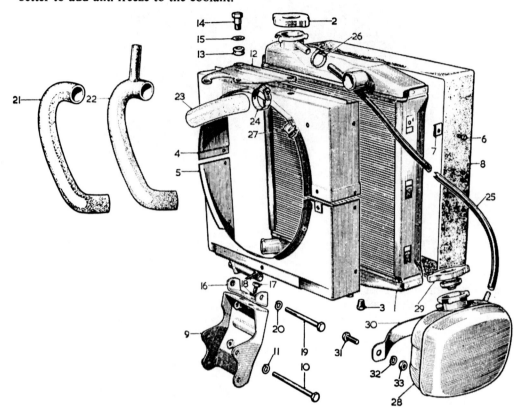

No.	Description	No.	Description	No.	Description
1.	Radiator assembly.	12.	Radiator upper support bracket.	23.	Radiator outlet hose.
2.	Filler cap (radiator).	13.	Support bracket grommet.	24.	Hose clip.
3.	Drain plug.	14.	Screw.	25.	Radiator to expansion tank hose.
4.	Top cowl.	15.	Plain washer.	26.	Hose clip.
5.	Bottom cowl.	16.	Radiator lower support bracket.	27.	Radiator cowl hose clip.
6.	Screw.	17.	Screw.	28.	Expansion tank.
7.	Nut.	18.	Shakeproof washer.	29.	Expansion tank cap.
8.	Cowl surround (rubber).	19.	Bolt.	30.	Expansion tank strap.
9.	Radiator mounting bracket.	20.	Spring washer.	31.	Screw.
10.	Bolt.	21.	Radiator hose to pump.	32.	Spring washer.
11.	Spring washer.	22.	Radiator hose to pump (heater).	33.	Nut.

Fig. 13:9   The cooling system components of the B.M.C. 1100 and 1300

Anti-freeze solutions

Before adding anti-freeze to cooling systems it is advisable to drain and flush the system. When this has been done, close the drain taps, pour in the correct quantity of anti-freeze for

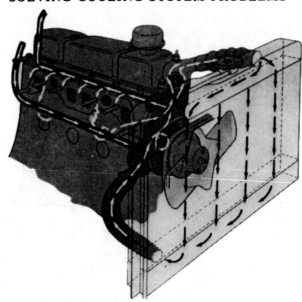

Fig. 13:10   A typical cooling system. Note the heater take off and the direction of flow

the degree of protection required (see accompanying table) and fill the radiator with water to the bottom of the filler neck.

On B.M.C. sealed cooling systems (see fig. 13:9), remove the expansion-tank filler cap and pour in ¼ pint of anti-freeze. Make sure that the caps of the radiator and the tank are correctly fitted.

If a heater is fitted, ensure that the heater control is in the hot position and run the engine briskly until the normal operating temperature is reached, then carefully remove the filler cap and top up with water and replace the cap.

Use only an ethylene glycol type of anti-freeze conforming to British Standard specification B.S. 3151 or B.S. 3152.

Solution (%)	Commences freezing		Frozen solid	
	°C.	°F.	°C.	°F.
25	−13	9	−26	−15
33⅓	−19	−2	−36	−33
50	−36	−33	−48	−53

CHAPTER 14

Emission Control Equipment

Worldwide interest has been focused recently on the dangers to the environment from pollution of many different kinds and research continues on means of controlling the exhaust emissions from motor cars. It can only be a matter of time before emission control equipment becomes compulsory in all civilized countries; already in the United States of America, where certain climatic conditions have accentuated the problem, the Federal regulations (Federal Register Vol. 33 No. 108 Part II) now limit the permitted emission of hydrocarbons and carbon monoxide from the exhaust systems of new vehicles and these regulations may soon be tightened and extended to limit the emission of oxides of nitrogen and also the amount of gasoline evaporating from the fuel system.

This chapter explains briefly the methods of control in use at present, the types of equipment, the maintenance work they require and the diagnosis and remedy of common faults.

All British cars imported into the USA are fitted with one of the following systems of which there are three main types
  (1) Manifold-Air-Oxidation
  (2) Clean-Air System (also known as Cleaner-Air Package or Improved-Combustion System)
  (3) Zenith Duplex System
The following table indicates the system used on specific models (1969–70)

MANIFOLD-AIR-OXIDATION

British Leyland	
B.M.C. division	MGB
	MGC
	MG Midget/Sprite
	Austin America
	Austin/Morris 1275 S
Ford	Cortina GT
Vauxhall (Canadian exports)	Victor 2000
	Viva

CLEAN-AIR SYSTEM

British Leyland	
Rover division	Rover 2000 TC
Standard-Triumph division	TR 250
	GT 6
	Spitfire
Ford	Cortina 1600
Rootes	Sunbeam Arrow
	Sunbeam Alpine GT

ZENITH DUPLEX

British Leyland	
Jaguar division	'E' type
	XJ 6

MANIFOLD-AIR-OXIDATION

In this system, air from an air compressor (belt driven from the crankshaft) is pumped via a check valve (which protects the pump from any backflow of exhaust) through a manifold system into the exhaust port of each cylinder just outside the exhaust valve.

Here the extra air allows the partially combusted hydrocarbons and carbon monoxide in the exhaust gas to continue to more complete combustion in the exhaust manifold before being finally emitted from the tailpipe. When the throttle is closed and the engine is over-running, the pressure in the inlet manifold is very low and fuel which is normally present on the manifold walls is suddenly stripped off, resulting in an over-rich mixture being fed to the

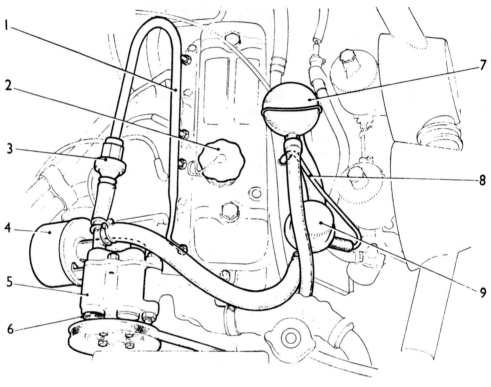

1 Air manifold.	4 Emission air cleaner.	7 Crankcase emission valve.
2 Filtered oil filler cap.	5 Air pump.	8 Vacuum sensing tube.
3 Check valve.	6 Relief valve.	9 Gulp valve.

(From 1969/70 models onwards the crankcase is vented straight to the constant depression area of the carburetter so that item 7 is no longer fitted)

Fig. 14:1   A typical engine emission control system layout

combustion chambers and thus producing exceptionally high emissions. In an attempt to prevent this happening, a second air line runs from the air compressor via a 'gulp' valve into the inlet manifold. The gulp valve is actuated, via another smaller tube connected to the inlet manifold, by the sudden drop in manifold pressure. This depression acts on a diaphragm in the gulp valve causing the valve to open and allow air from the pump to be forced into the inlet manifold where it corrects the over-rich mixture. A bleed hole in the gulp valve diaphragm allows the pressures to equalize fairly rapidly, shutting off the supply of extra air. In some systems a restrictor is fitted in the line between the air compressor and the gulp valve to prevent surging when the valve is operating.

Carburetters fitted to vehicles equipped with emission control are manufactured to a special control specification and are tuned to give optimum engine performance with maximum emission control. A poppet-type limit valve is incorporated in the carburetter throttle butterfly which limits the inlet manifold depression, ensuring that under conditions of high inlet manifold depression the air/fuel mixture entering the cylinders is at a combustible ratio.

Fault diagnosis on the Manifold-Air-Oxidation System

Note: It is assumed that the ignition and fuel system is in good order and that the tappets are correctly set.

Backfire in exhaust system

Possible causes	Cure
Leak in exhaust system	Locate and rectify leak
Leak in hoses or connections to gulp valve or vacuum sensing pipe	Locate and rectify leak
Faulty gulp valve	Test valve, renew if faulty
Leak in intake system	Locate and rectify leak
High inlet manifold depression on over-run—faulty carburetter limit valve (SU carburetters)	Fit new throttle disc and limit valve assembly

Hesitation to accelerate after sudden throttle closure

Leaks in hoses or connections to the gulp valve or vacuum sensing pipe	Locate and rectify leak
Faulty gulp valve	Test valve, renew if faulty
Leak in intake system	Locate and rectify leak

Engine surges (erratic operation at varying throttle openings)

Leaks in hoses or connections to gulp valve or vacuum sensing pipe	Locate and rectify leak
Faulty gulp valve	Test valve, renew if faulty

Erratic idling or stalling

Leaks in hoses or connections to gulp valve or vacuum sensing pipe	Locate and rectify leak
Faulty gulp valve	Test valve, renew if faulty
Carburetter limit valve not seating correctly	Renew throttle disc and limit valve assembly

Heat damaged hose (air pump to check valve)

Faulty check valve	Test valve, renew if faulty. Fit new hose
Air pump not functioning	Test pump, renew if faulty. Fit new hose

Noisy air pump

Incorrect belt tension	Adjust belt tension
Pulley loose, damaged or misaligned	Tighten loose pulley, renew if damaged, realign pump
Pump failing or seized	Test pump, renew if faulty

Excessive exhaust system temperature

Incorrect ignition setting	Check ignition timing, reset if necessary
Air injector missing	Remove air manifold and check injectors
Air pump relief valve inoperative	Test pump, renew if faulty

Air pump (testing)

Before testing the air pump ensure that the drive belt is not slack; it should be tensioned just like a generator belt. Disconnect the gulp valve air supply hose at the gulp valve and plug the hose, then disconnect the air manifold supply hose at the check valve and connect a pressure gauge to the hose (see fig. 14:2).

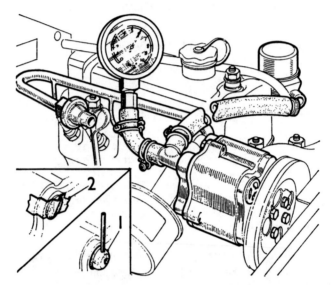

1 Relief valve test tool   2 Tape used to duct air
Fig. 14:2   Testing the air pump

Run the engine at the test speed specified by the manufacturer (Austin America, Mk IV Sprite and Midget Mk III 1200 rev/min; MGB/MGC 1000 rev/min) and read the gauge. If less than $2 \cdot 75$ lb/in^2 is registered, remove the air cleaner and fit a new element. Reassemble and repeat the test. Should the reading still be unsatisfactory or if no air cleaner is fitted, temporarily blank off the relief valve and repeat the test. If the reading is then correct renew the relief valve. If the gauge reading is still not satisfactory remove the pump and overhaul it or fit a new pump.

To test the relief valve, use a test tool if available or improvise a temporary air duct over the face of the valve (see fig. 14:2) then start the engine and gradually increase the speed until an air flow from the relief valve is detected. At this point a gauge reading of $4 \cdot 5$ to $6 \cdot 5$ lb/in^2 should be registered. If the valve fails to operate correctly, remove the pump and renew the valve.

The general assembly of the air pump is given in fig. 14:3. If the pump is overhauled, the following points should be noted:

(1) The bearing should be packed with Esso Andok 260 or other approved lubricant
(2) On Lucas pumps, the slots which carry the carbon and springs are the deeper ones. The carbons should be fitted with the chamfered edge to the inside
(3) The underside of the heads of the rotor-bearing end-plate screws must be treated with 'Locktite' before being tightened

If the relief valve has to be renewed, the air pump must be removed from the car, the pulley taken off and the valve removed from the pump body by using a $\frac{1}{2}$ in. diameter soft-metal drift against the relief valve and carefully driving the valve from the pump body. When a new relief valve is being fitted make sure to:

(1)   Use a new copper seating washer

(2)   Use a tool to the dimension given in fig. 14:4 to drive the valve into the pump

The check valve

Before the valve can be tested it must be removed from the injector manifold by disconnecting the air supply hose and unscrewing the valve. To test the valve, blow through it in turn from each side (on no account use a high-pressure air line). Air should pass only when blown from the air-connection side of the valve. If air can pass from the manifold-connection side, the valve is faulty and should be renewed.

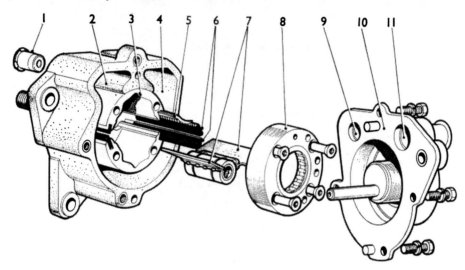

1 Relief valve.	5 Spring.	9 Outlet port.
2 Inlet chamber.	6 Carbons.	10 Port-end cover.
3 Rotor.	7 Vane assemblies	11 Inlet port.
4 Outlet chamber.	8 Rotor bearing end plate.	

Fig. 14:3   The air pump (four-cylinder engines)

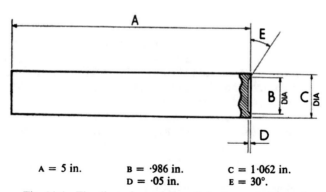

A = 5 in.	B = ·986 in.	C = 1·062 in.
	D = ·05 in.	E = 30°.

Fig. 14:4   The dimensions of the relief valve replacing tool

The injector manifold

To test the air manifold and injectors, detach the manifold and slacken the air supply hose clip at the check valve connection, then turn the manifold until the injector connections are accessible and retighten the hose clip.

Run the engine at idle speed and check for a flow of air from each manifold connection tube. If the air flow from any tube is restricted, remove the manifold and with an air blast clear the obstruction. Next, with the engine running at idling speed, check that exhaust gases blow from each of the injectors.

Note: On B.M.C. engines the injectors may be loose in the cylinder head and are sometimes displaced when carrying out this test.

If an injector is found to be restricted, turn the engine until the exhaust valve below the injector is closed. Using a hand-drill, pass a ⅛ in. drill through the bore of the injector, taking care that the drill does not contact the exhaust valve, or it may break the drill. On no account should a power drill be used. On completion of the cleaning, a blast of air will clear the carbon dust from the exhaust port.

Reassembly is a reversal of the dismantling procedure.

Fig. 14:5   The vacuum gauge connected for testing the gulp valve

The gulp valve

To check that the gulp valve is in order, disconnect the air supply hose (from the air pump connection) and connect a vacuum gauge with a 'T' connection to the gulp-valve air hose (see fig. 14:5). Next, start the engine and allow it to idle. Temporarily close the open end of the tee piece with the thumb and check that a zero reading is maintained for about 15 seconds. If a vacuum is registered, renew the valve.

Next, with the tee connection sealed again, operate the throttle rapidly from closed to open. The gauge should register a vacuum. Repeat the opening and closing of the throttle several times, temporarily lifting the thumb before each operation of the throttle to destroy the vacuum. If a vacuum is not registered on the gauge renew the gulp valve.

Limit valve (manifold depression, SU carburetters)

To test the limit valve, disconnect the gulp-valve sensing pipe from the inlet manifold and connect a vacuum gauge to the sensing-pipe connection on the inlet manifold. Warm the engine at idle until normal operating temperature is reached and then increase the engine speed to 3000 rev/min. On releasing the throttle, the vacuum gauge reading should immediately rise to 20·5–22 in.Hg. If the reading is outside this range, the carburetter must be removed and the throttle butterfly and limit valve renewed.

Tuning Data for Manifold-Air-Oxidation Systems

AUSTIN AMERICA
(engine type 12H 157 and 12H 185)

Idle speed	850 rev/min
Ignition timing (static)	t d c
Ignition timing (stroboscopic)	3° b t d c at 1000 rev/min (vacuum pipe disconnected)

Carburetter
 single SU HS4
 Needle DZ
 Piston spring, red
 Initial jet adjustment 13 flats from bridge

SPRITE MK IV / MG MIDGET MK III

Idle speed	1000 rev/min
Ignition timing (static)	4° b t d c
Ignition timing (stroboscopic)	10° b t d c (vacuum pipe disconnected)

Carburetters
 Twin SU HS2
 Needle AN
 Piston spring, blue
 Initial jet adjustment 11 flats from bridge

MGB

Idle speed (manual transmission)	900 rev/min
Ignition timing (static)	10° b t d c
Ignition timing (stroboscopic)	20° b t d c at 1000 rev/min (vacuum pipe disconnected)

Carburetters
 Twin SU HS4
 Needle FX
 Piston spring, red
 Initial adjustment 14 flats from bridge

MGC

Idle speed (manual transmission)	850 rev/min
Ignition timing (static)	t d c
Ignition timing (stroboscopic)	4° b t d c at 1000 rev/min (vacuum pipe disconnected)

Carburetters
 Twin SU HS6
 Needle KM
 Piston spring, yellow
 Initial jet adjustment 8 flats from bridge

FORD CORTINA GT

Idle speed (manual transmission)	800 rev/min
Ford close-tolerance carburetter	
Ignition timing	4° b t d c (static)

THE CLEAN-AIR SYSTEM

This method of controlling exhaust emissions consists basically of modifications to the engine to improve the efficiency of combustion.

Different ignition time settings and distributor advance curves are employed together

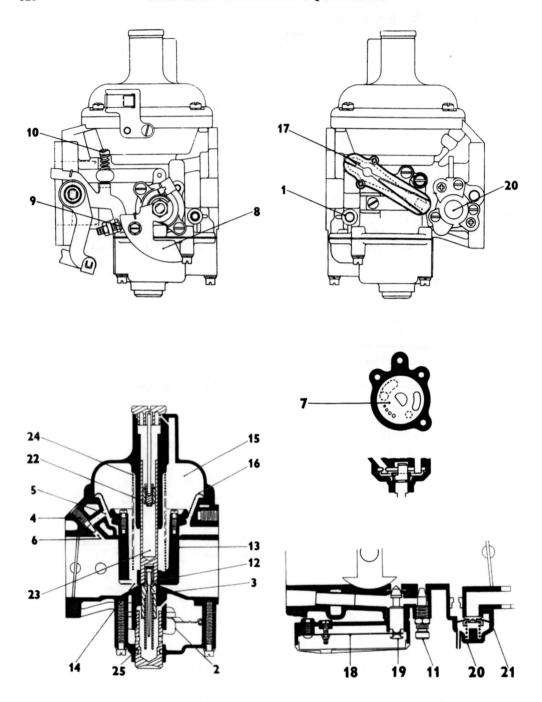

Fig. 14:6   The Zenith Stromberg series CDSE and CD2SE emission control carburetters

1 Petrol inlet
2 Twin floats
3 Jet orifice
4 Leak balancing screw
5 Duct
6 Duct
7 Starting disc
8 Starter lever cam
9 Fast-idle stop screw
10 Throttle stop screw
11 Idle trimming screw
12 Biased jet needle
13 Air valve
14 Duct
15 Depression chamber
16 Diaphragm
17 Temperature compensator
18 Bi-metallic blade
19 Tapered plug
20 Throttle by-pass valve
21 Diaphragm (by-pass valve)
22 Damper
23 Hollow guide rod
24 Coil spring (air valve)
25 Rubber 'O' ring

with reductions in compression ratio and valve overlap. In addition closer tolerance carburetters, which meter the air/fuel mixture more consistently, allow these engines to be operated at maximum combustion efficiency at all times.

The most popular carburetters for Clean-Air systems are the constant vacuum Stromberg CDSE and SU, although some manufacturers do use the fixed-choke type.

The Zenith Stromberg CDSE carburetter

One close-tolerance carburetter, currently being used to control exhaust emissions, is the Stromberg CDSE; this model differs from the normal CD.

All carburetters of the variable choke type require a certain mechanical clearance around the air slide and bore. This clearance results in a proportion of the engine's air supply by-passing the bridge or metering section of the carburetter; because of manufacturing tolerances this leakage is not consistent and can result in small changes in air/fuel ratio which on normal CDs would be corrected by adjustment to the fuel jet. Emission carburetters however are fitted with a leak adjuster which allows every unit to be pre-set by the manufacturers thus eliminating the need for further adjustment when fitted to a vehicle. This adjuster is incorporated in a drilling (see fig. 14:6) taken from the atmospherically vented region beneath the diaphragm to meet another drilling which runs into the carburetter mixing chamber, downstream of the air valve. Once the adjustment has been set by the manufacturers, access is sealed with a plug which should not be removed.

Another feature of the CDSE carburetter is the idle trimming screw. When an engine is very new the mechanical stiffness has a pronounced effect on the mixture required at idle. A stiff engine needs a slightly leaner idle mixture than one that is fully run-in. To compensate for this difference, the trimming screw (see fig. 14:6) is initially set to allow extra air into the mixing chamber and if the idle quality deteriorates during the running-in period, the screw should be slowly turned in a clockwise direction until a smooth idle is just restored. It should be noted that to the ear there may be no detectable difference between the fully closed and open positions; adjustment should therefore be made to achieve the best driveability. Do not overtighten this adjusting screw.

If a satisfactory idle cannot be obtained, check the manifold and carburetter retaining nuts in case there are any air leaks.

The jet needle on the CDSE carburetter is biased so that the needle is permanently in contact with the back of the jet, this ensures a consistent flow from a given needle profile. If the jet/needle relationship is disturbed and/or the needle is changed, the emission level will be increased.

Emission testing has shown the need for a temperature compensator (see fig. 14:6) to cater for minor air/fuel-ratio changes brought about by heat transfer to the carburetter and by variations in underbonnet air temperature. These are significant only in the context of the precision demanded by exhaust emission requirements. The method employed is an air flow channel which permits some of the air passing through the carburetter to by-pass the venturi or bridge section. The air valve drops to a lower position, thus giving a smaller fuel-flow annulus. By adjusting the quantity of air by-passed, the degree of temperature compensation may be varied to suit a particular application: for this purpose a bi-metallic blade is used to regulate the movement of a tapered 'plug' or valve.

The compensator assembly is pre-set and should not be adjusted. If it fails to function, the assembly should be renewed, care being taken to obtain the correct pre-adjusted regulator assembly.

If malfunction of the regulator is suspected, carefully check that the tapered plug is free to move. If it is 'free', the compensator is suspect and the assembly should be renewed.

During deceleration, when manifold depression exceeds 22–23 in.Hg, hydrocarbon and emission can be extremely high. To prevent any rise in excess of the critical figure, which varies to some extent with differing engines, a diaphragm-operated throttle by-pass valve is incorporated in the CDSE carburetters fitted to certain engines. Manifold depression acting on the diaphragm causing the by-pass valve to open when a pre-determined depression is reached allowing a small quantity of mixture to bleed into the engine to maintain combustion.

The valve is pre-set and is fairly trouble-free, but in the event of diaphragm failure or other malfunction, a new complete unit should be fitted.

In the normal servicing of CDSE carburetters, three adjustments can be made: (1) Idle speed, by rotation of the throttle stop screw. (2) Idle emission, by rotation of the idle trim screw. (3) Fast idle, by the fast idle screw.

Fault diagnosis on the Clean-Air system

As the Clean-Air system is basically a 'tune' method of emission control, faults (apart from those mentioned earlier) are largely confined to normal 'run of the mill' troubles.

In the event of erratic idling etc., the carburetter faults set out below under the appropriate heading should be investigated prior to renewing any emission control parts. It is assumed that the tappets, ignition timing and ignition system are in order.

Erratic or poor idling

Incorrect fuel level, caused by incorrect float level and/or worn or dirty needle valve: Check float height and wash in clean spirit. Renew needle valve and seating if worn.

Air valve sticking: Check free movement of spring-loaded metering needle, clean the air-valve rod and guide, and lubricate with SAE 20 oil.

Metering needle incorrectly fitted: Check that the needle is free to move against its biasing spring and that the housing has not been distorted by overtightening the retaining screws.

Obstructed air valve and/or float chamber ventilation holes: Make sure that the air-cleaner case and gaskets are properly aligned and not causing an obstruction of air flow.

Diaphragm incorrectly positioned or damaged (Stromberg): Check the location with the depression chamber cover removed. The two depression holes at the base of the air valve should be in line with and towards the throttle spindle. Renew the diaphragm if damaged and make sure when refitting the depression chamber cover that the damper ventilation boss faces toward the air intake.

On multi-carburetter installations check that the carburetters are correctly synchronized.

Leakage from vacuum pipe connection or chafed pipe: Remove any faulty connections, renew chafed pipe.

Temperature compensator (Stromberg) not operating correctly: When the engine and carburetters are cold, remove the cover from the temperature compensator assembly and ensure that the tapered valve is seated. Check the operation by carefully lifting the valve from its seat; when released, the valve should return freely. If mechanical damage has been caused, or if the valve is not free to move, renew the compensator assembly.

Check the throttle spindle and end seals; fit new parts as required.

Check the induction manifold joints. Re-face the flanges if necessary and fit new gaskets, tightening the nuts evenly and securely.

Hesitation or flat spot

Hesitation, or flat spots, may be caused by any of the faults listed under the preceding heading, but in addition the following should be checked.

Damper missing, faulty, or inoperative due to lack of oil.

Air valve spring missing or incorrect spring fitted. (Identification of the air valve spring is by the spring wire diameter and the number of coils. The springs are colour coded.)

Heavy petrol consumption

Although most of the faults listed under the two previous headings can contribute to heavy petrol consumption on Stromberg CDSE carburetters, the bottom of the carburetter should be inspected to make sure there is not a leakage from the sealing plug 'O' ring which is situated between the jet cover and the float chamber spigot boss.

The following is a list of the equipment fitted to various makes of car, with details of ignition setting to be used on the emission-controlled engines.

Emission Control Equipment	*Ignition timing*
Rover 2000 TC Two SU HS8 carburetters, fitted with biased needles. The throttle butterflies are fitted with poppet valves which limit the deceleration manifold depression. Temperature compensators are fitted to the jets. The inlet manifold incorporates a toothed fuel-deflector plate to prevent fuel streaming on the manifold walls.	4° at dc (static)
Triumph TR 250 Two Stromberg CDSE carburetters, fitted with biased needles, manifold depression limiters and temperature compensation. The throttle linkage is arranged to operate in a slotted quadrant so that initial throttle movement causes the ignition to advance and increases the engine speed from 850 rev/min at idle to 1100 rev/min. Further movement actuates the throttle butterfly. The throttle then operates normally. The distributor is fitted with both an advance and a retard diaphragm (giving maximum retard at idle).	4° at dc (static)
Triumph Spitfire Two SU HS2 carburetters, fitted with biased needles and incorporating throttle-butterfly poppet valves to limit the deceleration manifold depression. A compression ratio of 8·5:1 and a camshaft with which the valve overlap has been reduced from 50° to 20°. Modified distributor advance curve.	6° at dc (static) (2° at dc at idle— 800–850 rev/min)
Triumph GT6 Two Stromberg CDSE carburetters, fitted with biased needles, manifold depression limiters and temperature compensation.	6° bt dc (static) (4° at dc at idle— 800–850 rev/min)
Sunbeam Arrow A Stromberg CDSE carburetter as fitted to the TR250 detailed earlier.	5° bt dc at 1000 rev/min
Ford Cortina 1600 Fitted with Ford close-tolerance emission carburetter and modified distributor advance curve.	4° bt dc (static)

ZENITH DUPLEX SYSTEM

Studies of the effect of mixture strength on exhaust emissions have indicated that the carbon monoxide content falls to a minimum as the amount of fuel is decreased, whilst hydrocarbon emissions also fall to a minimum at an air/fuel ratio approximately stoichiometric but then increase rapidly with further weakening. Hence it can be seen that minimum emission levels of both components are only obtained over a narrow band of mixture strength.

In multi-cylinder engines, cylinder-to-cylinder variations in mixture distribution make the range of acceptable ratios even narrower so that any improvement in distribution can be used to good effect.

The Zenith Duplex emission control system was developed to meet these requirements. Basically the system (see fig. 14:7) consists of either a single or more usually twin Zenith–Stromberg CD carburetters to which are added auxiliary throttles mounted vertically below the main throttles. During idling and part-throttle conditions, the air/fuel mixture metered by the carburetter passes the auxiliary throttles into a pipe leading to the swirl chamber. This chamber is mounted on the exhaust manifold and is therefore quickly heated when the engine is started. The heat, together with the internal design of the chamber, ensures a thoroughly mixed and vaporized mixture which is then fed to the main induction manifold. The main throttles are accurately fitted to seal off the main carburetter bore completely, until the auxiliary throttle reaches just over two-thirds of its travel, when a mechanical linkage picks up the main throttle. Both throttles reach the fully open position simultaneously and mixture flow at full throttle is unimpaired.

This system is currently fitted to Jaguar cars being exported to the United States; on these engines an additional problem is created by the 'cross flow' layout of inlet and exhaust ports and in order that the idle-mixture swirl chamber may be heated by the exhaust gas, the secondary induction tract passes over the engine to the swirl chamber and returns over the engine to the inlet manifold.

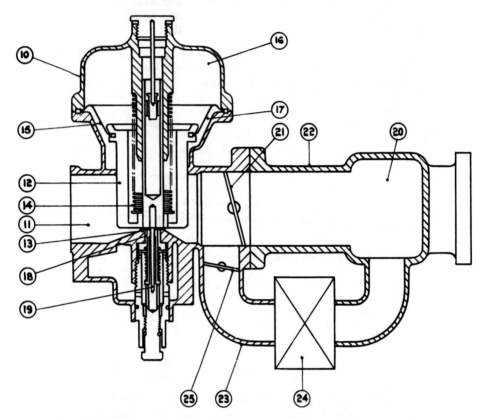

Fig. 14:7 Arrangement of Zenith-Stromberg Duplex carburetter. 10 = Dashpot cover. 11 = Air intake. 12 = Air piston. 13 = Jet housing. 14 = Piston spring. 15/17 = Diaphragm. 16 = Dashpot. 18 = Jet. 19 = Jet needle. 20 = Manifold. 21 = Main throttle. 22 = Main throughway. 23 = Auxiliary duct. 24 = Heated swirl chamber. 25 = Auxiliary throttle.

The ignition timing used on the export Jaguar engines is t d c at idling speed; a retard capsule changes this to 10° a t d c during deceleration with the throttle closed and maximum manifold depression.

APPENDIX

APPENDIX 1

CHAMPION SPARK PLUG RECOMMENDATIONS

A.C.

1948–57 2 litre (gap 0·020 in.)	L-7
1953–58 Petite (3-wheeler)	
18 mm (gap 0·020 in.)	K-9
14 mm (gap 0·020 in.)	L-5
1954–63 Ace & Aceca	
2 litre AC eng. (gap 0·020 in.)	L-7
2·6 litre Ford eng. Std (gap 0·032 in.)	N-8
2·6 litre Ford eng. Tuned	N-5
Greyhound (gap 0·020 in.)	L-7
Cobra 4727cc, 289 (gap 0·035 in.)	F-11Y
7010cc, 428 (gap 0·035 in.)	F-9Y

Alvis

1946–58	L-10
1959 TD21, ½ in. reach	L-10
¾ in. reach	N-5
1960 onwards TD21	N-5
TE21	N-4
Other models	
18 mm (gap 0·020 in.)	UK-10
14 mm	L-10
TF21	N-9Y

Aston Martin

1950–52 DB2 (Std eng.) (gap 0·020 in.)	Z-10
1950–53 DB2 (Vantage eng.) (gap 0·020 in.)	Z-10
1954–57 DB2-4 (gap 0·020 in.)	Z-10
1957–59 DB Mk 3, DB2-4 Mk 3	L-7
1959 onwards DB4	N-9Y or N-5
DB5, DB6, DBS, Volante	N-9Y
DB6 Mk II. F.I, DB5 V-8	N-9Y

Austin

A30, A35, A40 Sports, A90 Atlantic	N-5
Westminster A90 & A95	UN-12Y or N-8
A70, A125, A135	N-8
To 1958, A40, A50, A55, Metropolitan	N-8
1959 onwards A40, A55, Seven, Metropolitan, Mini, Mini-Cooper, 1100, 1800	N-5
Mini-Cooper 'S', all models 1300, 1800 Mk II, Mk IIS, 3 litre, Maxi	N-9Y
A99, A105	N-5 or UN-12Y
A60 (petrol)	N-5
A60 (diesel)	AG32
A110	UN-12Y
Austin-Healey 100, 100M	N-5
Austin-Healey 100S	N-3
Austin-Healey 100/Six	N-5 or UN-12Y
Austin-Healey 3000	UN-12Y
Sprite 948cc, 1098cc	N-5
1275cc	N-9Y

Bond

250cc 4T Twin (gap 0·020 in.)	L-81
Equipe GT (1147cc)	L-7 or L-87Y
GT4S, 2 litre GT	N-9Y
875	N-9Y

Bristol

401 to 406 (gap 0·020 in)	Z-10
407, 408, 409, 410 (gap 0·035 in.)	UJ-10Y

Ford

V-8 30 hp (18 mm)	7
Popular, Squire, Escort sv	L-10
Anglia & Prefect sv	L-10
Anglia & Prefect ohv	N-9Y
Consul, Consul 375, Zephyr Mks I & II, Zodiac Mks I & II (gap 0·032 in.)	UN-12Y
Zephyr 4, Zephyr 6, Zodiac Mk III	UN-12Y
Classic, Corsair, 1961–3 Capri, Cortina	N-9Y
Cortina Lotus	N-6Y
Cortina GT, 1961–3 Capri GT, Corsair GT	N-9Y
Corsair V-4, V-4GT	N-9Y
Capri 1969–on	N-9Y
Zephyr V-4, V-6, Zodiac V-6, Mk IV	N-9Y
1968 Escort, Escort GT	N-9Y
Escort Twin-Cam	N-6Y

Healey

2·4 litre	N-5
3 litre	L-10
Nash-Healey	L-7
Others—See Austin	

Hillman

1939–58 sv eng. (gap 0·032 in.)	L-10
1955–59 ohv models except Husky 'B' series chassis numbers	N-8
Husky 'B' series chassis numbers	UN-12Y
875cc Husky	N-9Y
1959–66 Minx, Super Minx (1500, 1600)	N-5
1725	N-9Y
1967 onwards Minx (1700, 1725)	N-9Y
Imp, Super Imp, Hunter, GT	N-9Y
Imp Californian	N-9Y
Avenger	N-9Y

H.R.G.

1½L Meadows eng. (gap 0·016 in.)	K-57R
1100, 1500 ¾ in. reach (gap 0·030 in.)	N-5
1100, 1500 ½ in. reach (gap 0·020 in.)	L-7

Humber

Hawk Mk 1–3 (gap 0·030 in.)	L-10
Snipe Mk 1–3 (gap 0·030 in.)	L-10
Pullman, Mk 1–3 (gap 0·030 in.)	L-10
Hawk, Mk 4, 5 (gap 0·030 in.)	N-8
Hawk, Mk 6, ohv (gap 0·030 in.)	N-8
Snipe, Mk 4 (gap 0·030 in.)	N-8
Pullman, Mk 4 (gap 0·030 in.)	N-8
1957 onwards Hawk series 1, 1A, 2, 3, 4	UN-12Y
1959 onwards Super Snipe 2·6L, 3L	N-9Y
Sceptre 1600cc	N-5
1725cc	N-9Y
Imperial	N-9Y

Jaguar

1949–50 Mk V 2½L	N-8
XK 120, XK 140, 8:1 cr	N-8 or UN-12Y
XK120, Mk VII, Mk VIIM, Mk VIII, Mk IX, XK140, 2·4 litre, XK150, 7·1 cr	L-7
2·4 litre Mk 1 & 2, 240, 8:1 cr	N-5

325

Mk VII, Mk VIIM, Mk VIII, Mk IX,
　　XK150s, XK150, except 7:1 cr　UN-12Y or N-5
XK140, 8:1 cr ('C' type head) N-5
XK140, 9:1 cr ('D' type head) (gap
　　0·020 in.) N-3
XK120C, 'D' type, 'C' type Sports Car N-3
XK 'SS' N-5*
Mk I 3·4, 3·8 ....UN-12Y or N-5
Mk II 3·4, Mk II 3·8, Mk 10 3·8, 'E'
　　type 3·8, 340, 'S' type UN-12Y
4·2 'E' type, 2+2 ....N-11Y
4·2 Mk 10, 420, 420G N-11Y
XJ6 2·8 ....N-9Y
XJ6 4·2 ....N-11Y

　　* For competitions use N-3

Jensen
4 litre 14 mm. short reach L-7
4 litre Interceptor N-8
Model 541 ....N-8
541 R & S Series* N-8
R51 ....N-5
C-V8, V-8FF Mks I–III (gap 0·035 in.) J-11Y
Mk IV (gap 0·035 in.) J-13Y
V-8 4475cc (gap 0·035 in.) N-9Y
Director, V8-Interceptor (gap 0·035 in.) ..J-11Y

　　* N-5 for hard driving

Lagonda V-12 ....L-7
1962 onwards
　　Rapide ....N-9Y or N-5

Lotus
Seven (1172cc), Sports 45 ....L-10
Seven Two F, Eleven Sports L-10
Seven (997cc, or 948cc), Seven A, Seven
　　Two A, Super Seven N-4
Le Mans 75, 85, Club, Club 75 ....N-3
Eleven le Mans, Fifteen, Seventeen N-3
Elite, Elite Super 95 ....N-3
Elan 1500, Elan 1600, SE, Elan+2 ....N-6Y
Europa N-4

Marcos
Mini GT850 ....N-5
1500, 1600 3 litre.......................N-9Y

Metropolitan
To 1958 ....N-8
1959 onwards N-5

M.G.
1946 onwards TC, Y L-7
TD to Eng. No. 22734 ....L-7
TD from Eng. No. 22735 ....N-5
TF N-5
TD Mk II to Eng. No. 17028 ....L-5
TD Mk II Eng. No. 17029 onwards N-3
YB to Eng. No. 17993 ....L-10
YB Eng. No. 17994 onwards N-8
Magnette 1954–56 ....N-8
　　1956 onwards N-5
MGA 1·4, 1·5 (ohv) N-5
MGA 1·6 litre (twin cam) N-4

MGB, GT, MGC N-9Y
1100 ....N-5
1300 ....N-9Y
Midget 948cc, 1098cc N-5
　　1275cc N-9Y

Mini
Mark II Saloon, Clubman N-5
1275 GT, Cooper S N-9Y

Morgan
Triumph TR eng. L-87Y
4/4 Aquaplane Head N-5
4/4 Series 3, 4, Ford ohv eng. N-9Y
4/4 Series V Ford 1500cc, 1600cc
　　Normal N-9Y
　　GT version N-9Y
Plus-Four-Plus, Plus-8 ....L-87Y

Morris
Minor (sv) Series MM L-10
Minor (ohv Series II) Traveller 1000 ......N-5
Oxford (Series MO) L-10
Cowley 1200 ....N-8
Isis ....N-8 or UN-12Y
Six (Series MS) L-10
To 1958—Oxford ohv models, Cowley
　　1500 ....N-8
1959 onwards Oxford, Cowley 1500, 1100,
　　1800 Mk I, Minor 1000, Mini-Minor,
　　Mini-Cooper N-5
Mini-Cooper 'S' All models, 1300, 1800 Mk II
　　Mk IIS N-9Y

Princess
3L, Mk I ....UN-12Y or N-8
3L, Mk II UN-12Y
4 litre limousine N-8
1100 ....N-5
4 litre R UN-12Y
1300 ....N-9Y

Reliant
Regal (gap 0·020 in.) L-10
Regal 3/25 ....N-5
Sabre (Ford eng.) (4 cyl.) (gap 0·032 in.) N-8
Sabre (Ford eng.) (6 cyl.) N-5
Sabre GT N-4
Rebel N-5
Scimitar GT, GTE N-9Y

Riley
1939–55 1½ litre L-7
1946 on 2½ litre Pathfinder N-5
Two-Point-Six ....N-5 or UN-12Y
One-Point-Five, 4/68, 4/72, Elf, Kestrel
　　1098cc N-5
Kestrel 1300 ....N-9Y

Rolls-Royce
to 1935 7
1935–39 all models except Phantom 3 .....LB-8
Phantom 3, Silver Dawn N-8
To 1955 Silver Wraith ....N-8 or UN-12Y

Silver Wraith Series E6·6:1, Silver
    Cloud 6·6:1 ....RN-8 or UN-12Y
Silver Wraith 8:1, Silver Cloud
    8:1 ....N-5 or UN-12Y
Phantom V (V-8), Silver Cloud Series II,
    III (V-8), Silver Shadow N-14Y

Rover
    60, 75, 90 (gap 0·030 in.) N-8
    80 ....N-8
    95, 100, 105R, 105S (gap 0·030 in.) N-5
    110 (gap 0·030 in.) N-4
    3 litre 7·5:1, 3 litre 8·75:1 (gap 0·030 in.) N-5
    3·5 litre V-8 ....L-87Y
    Three Thousand Five L-87Y
    2000 single carb. N-9Y
        twin carb. N-6Y, N-7Y

Singer
    Sports J-6
    Le Mans (gap 0·016 in.) L-57R
    1936–40 ....L-10
    1946–51 except 1500 ....L-7
    1500 (single carb.) N-8
    1500 (twin carb.) (gap 0·032 in.) N-5
    Hunter (single carb.) N-8
    Hunter (twin carb.) N-5
    SM Roadster (gap 0·030 in.)
        (single carb.) N-8
        (twin carb.) N-5
    Hunter 75 ....L-7
    Gazelle Mk I & II N-8
    Gazelle Mk III, Series IIIA, IIIB, IIIC, V N-5
        Series VI N-9Y
    Vogue 1600cc N-5
        1725cc N-9Y
    Chamois N-9Y
    Chamois Sport N-9Y
    1967 onwards New Gazelle (1500, 1725) ...N-9Y

Standard
    1945–48 8 h p N-5
    1946–47 12 h p, 14 h p (gap 0·035 in.) N-8
    Ensign L-7
    1953 onwards 'Eight', 'Ten' N-5
    Sportsman L-7
    Companion, Pennant N-5
    1948–58 Vanguard L-10
    1959 Vanguard, Vignale IV L-7
    1960 onwards Vanguard, Vignale
        Luxury VI N-5

Sunbeam
    Rapier I (gap 0·030 in.) N-8
    Rapier Series II, III, IIIA, IV, V-1600cc
        engs. N-5
    Rapier V-1725cc, H120 ....N-9Y
    Alpine-1600cc N-5
    Alpine-1725cc N-9Y
    Tiger V-8 (gap 0·035 in.) F-11Y
    Imp Sport, Stiletto N-9Y

Triumph
    Model 1800, 2000 Renown L-10
    TR I–IV L-87Y
    Spitfire Mk 1 & 2
        Normal L-87Y
        Stage 2 Tuned L-5
    Spitfire Mk 3 ....N-9Y
    Mayflower (gap 0·022 in.) N-8
    Herald, 948cc, Herald 'S', 948cc N-5
    Herald 1200, 12/50 ....L-87Y
    1962 onwards, Vitesse 6 cyl. N-9Y
    1300, 1300TC, 13/60, 2000 6 cyl., 2000 Mk II,
        GT6, TR5, TR6, 2·5 PI N-9Y

Vauxhall
    1939–59 all models (gap 0·030 in.) J-8
    Velox, Cresta, Friary, 2262cc (gap
        0·030 in.) J-8
        2651cc (gap 0·030 in.) N-5
    Velox, Cresta, Viscount, Ventora 3294cc
        (gap 0·030 in.) N-9Y
    Victor Series F (gap 0·030 in.) J-8
    Victor FB, FC (101) (gap 0·030 in.) J-6
    Victor ohc 1600, 2000, Viva GT (gap
        0·030 in.) L-5
    VX4/90, Viva, Viva SL (gap 0·030 in.) N-9Y
    Viva 90, Viva SL90 (gap 0·030 in.) N-6Y
    2000 SL, VX4/90 (FD) (gap 0·030 in.) L-5

Wolseley
    1936–52 14 mm L-10
    1953 onwards 6/80 ....L-10
    1953 onwards 4/44 ....N8
    1954–57 6/90 (gap 0·020 in.) ....UN-12Y or N-8
    1958 onwards 6/90 ....UN-12Y or N-8
    To 1958, 1500, 15/50 ....N-8
    1959 onwards 1500, 15/50, 15/60, 16/60,
        Hornet 1100, 18/50 ....N-5
    18/85 (eng. No. Prefix 18H) N-9Y
    6/99 ....N-5 or UN-12Y
    6/110 ....UN-12Y
    1300 ....N9Y

SPARK PLUG EQUIVALENTS

This chart is reproduced by courtesy of Champion Spark Plugs. Manufacturers' code reference denotes the nearest equivalent.

SIZE		CHAMPION	A.C.	AUTOLITE	BOSCH	LODGE	N.G.K.	K.L.G.
14 mm ¾ in. reach	colder	N-5	44N 44XL	AG3	W175T2	HLN HL14 HLNP	B·6E	FE70 FE75
		N-4	C42N C41N	AG2	W225T2	2HLN	B·7E	FE80
		N-3	43XL	AG901	W260T2	3HLN	B·7E	FE100
		N-60						
		N-57		AG701				
14 mm ¾ in. reach P.C.N.	colder	N-9Y	C42N	AG32	W200T30	HLNY	BP7E	FE65P
		N-66Y		AG22		2HLNY		
		N-65Y		AG12		2HLNY		
		N-6Y or N-64Y			W260T28	3HLNY		FE125P
		N-63Y				4HLNY		FE135P
		N-60Y				5HLNY		FE155P
14 mm ¾ in. reach	colder	N-62R		AG23	W240T17	RL47	B·8EN	FE220
		N-60R		AG903	W270T17	RL49	B·9EN	FE260
		N-57R		AG603	W310T17	RL50	B·10EN	FE280
		N-54R		AG403	W400T17	RL51	B·11EN	FE310
		N-52R		AG203 AG103	W440T17	RL53	B·12EN	FE340
14 mm ¾ in. reach	colder	J-6J	M44C M44	A3X	W225T3	HAN HANP	B·6	FS·75H
		J-4J	42 M42K	AT2 A21XM A2X A21X	W240T3	2HAN	B·7 B·76	FS100H
		J-2J	M41K M41G	AT1	W260T3	3HAN	B·77C	FS·100
		●J-79		A901 AT-1				
14 mm ¾ in. reach P.C.N.	colder	UJ-10Y J-10Y	C44S 43S	AT42 A32	W145T7	CANY	BP·6	FS·55P
		J-63Y	42S	A22				
		J-61Y J-86Y	C421 C42	A12				
		J-83Y						
14 mm ¾ in. reach	colder	J-62R		A23	W240T16	R·47 with gasket	B·8N B·776C	
		J-60R		A903	W270T16	R·49 with gasket	B·9N	
		J-57R	41	A603	W290T16 W310T16	R·50 with gasket	B·10N	
		J-54R		A403 A203	W340T16 W400T16	R·51 with gasket	B·11N	
14 mm ½ in. reach	colder	L-7	44F	AE4	W175T1	HF2H HN N14	B6H	F·75
		▲L-85	46FF	AE4 AE3	W225T1	HF2H	B6H B7HZ	
		▲L-81	M42F	AE2	W240P11S	HH14	B7H	DF80
		L-5	42F 41F	AE2	W225T1 W240T1	2HN	B7H	F·100
		L-4J		AE2	W175T1	2HN	B7H	F·100
14 mm ½ in. reach P.C.N.	colder	L-87Y		AE32	W175T1 W175T7	CNY		F55P
		L-82Y		AE22	W225T7			F65P
		L-66Y		AE32				
		L-64Y		AE22				
		L-61Y						
14 mm ½ in. reach	colder	L-62R		AE23	W175T1	R·47	B8HN	F220 F250
		L-60R		AE903	W225T1	R·49	B9HN	F260
		L-57R		AE603	W240T16 W270T16	R·50	B10HN	F280
		L-54R		AE403 AE203	W310T16 W340T16	R·51	B11HN	F320 F340
10 mm 0·700 in. reach	colder	G-63						
		G-61						
		G-59R		PG603		10RL·49	C·10EN	TE220
		G-56R		PG403		10RL·50	C·11EN	TE240
		G-54R		PG203		10RL·51	C·12EN	TE260
12 mm 0·492 in. reach	colder	P-7						
12 mm ¾ in. reach	colder	R-63					D80	
		R-61					D8E	
		R-59R					D9E	
		R-56R					D10E	
		R-54R					D12E	
		R-52R					D13E	
18 mm tapered seat	colder	F-10	C83T C84T	BTF3 BTF31	MA175T1	HTN	A-7F	TNT50
		F-82		BTF1		HTN81		
18 mm tapered seat P.C.N.	colder	F-9Y	83TS	BF-32		HTNY		MT65P
		F-83Y		BF-22				
		F-62Y		BF-12				
		F-60Y						
18 mm tapered seat	colder	F-62R		BF-703				
		F-60R		BF-601				
		F-57R		BF-603				
		F-54R		BF-403				
				BF-203				

APPENDIX 2

The following guide is for use in the absence of the manufacturers' recommendations. The figures are based upon normal new averages. Maximum permissible wear should be regarded as approximately 50% of the maximum clearance figure.

Cars up to 1300cc *inches*

Big end bearing/crankshaft journal diametrical clearance	0·001 –0·0025
Main bearing/crankshaft journal diametrical clearance	0·001 –0·0025
Crankshaft end-float	0·002 –0·004
Piston ring side clearance	0·0015–0·002
Exhaust valve stem/guide clearance	0·002 –0·003
Inlet valve stem/guide clearance	0·0015–0·002
Camshaft end-float	0·002 –0·003

Cars from 1300cc–3000cc

Big end bearing/crankshaft journal diametrical clearance	0·0025–0·004
Main bearing/crankshaft journal diametrical clearance	0·0015–0·003
Crankshaft end-float	0·003 –0·008
Camshaft end-float	0·003 –0·006

Clutch adjustment
Hydraulically operated
Free play between slave cylinder piston and rod ¼ in.
Mini and 1100, 0·025 in. between clutch arm and stop

Mechanically operated
Free play at pedal 1 in.

Tyres
Minimum tread depth
Average 1 mm.

Tyre pressures

Tyre pressures increase as tyres warm up. Pressures also increase more in hot weather than in cold and also increase as a result of speed.

The pressure in warm tyres should not be reduced to the standard pressure for cold tyres. Such 'bleeding, increases a tyre's deflection and causes the temperature to rise even higher, moreover the tyres will be under-inflated when cooled.

High speed causes increased temperatures due to a faster deflection and recovery rate. In addition, resistance to tread abrasion decreases with increased tyre temperature. Tread wear may be twice as rapid at 50 mile/h (80 km/h) as at 30 mile/h (50 km/h). Always use manufacturers' recommended pressures.

Cylinder Arrangement and Firing Order of British Passenger Cars

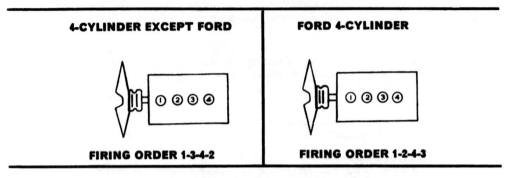

4-CYLINDER EXCEPT FORD

FIRING ORDER 1-3-4-2

FORD 4-CYLINDER

FIRING ORDER 1-2-4-3

**6-CYLINDER EXCEPT
ROLLS-ROYCE, BENTLEY & JAGUAR**

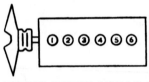

FIRING ORDER 1-5-3-6-2-4

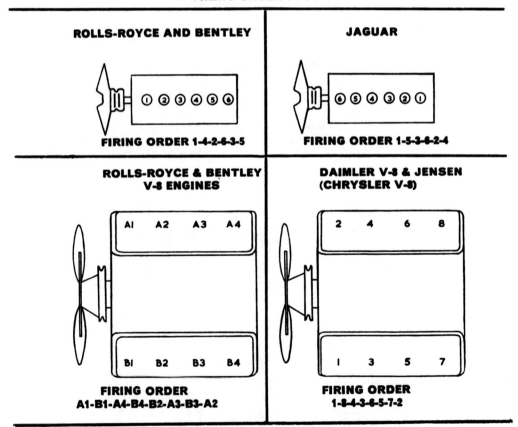

ROLLS-ROYCE AND BENTLEY

FIRING ORDER 1-4-2-6-3-5

JAGUAR

FIRING ORDER 1-5-3-6-2-4

**ROLLS-ROYCE & BENTLEY
V-8 ENGINES**

A1   A2   A3   A4

B1   B2   B3   B4

FIRING ORDER
A1-B1-A4-B4-B2-A3-B3-A2

**DAIMLER V-8 & JENSEN
(CHRYSLER V-8)**

2   4   6   8

1   3   5   7

FIRING ORDER
1-8-4-3-6-5-7-2

APPENDIX 4

Car and model	Ignition timing	Tappet clearance: Inlet (in.)	Tappet clearance: Exhaust (in.)	Contact-breaker gap (in.)	Spark-plug gap (in.)	Oil pressure (lb/in.²)	Toe in (in.)	Camber	Castor	King pin inc.
A99 AUSTIN WESTMINSTER (1959–61)	tdc	0·012	0·012	0·014–0·016	0·024–0·026	55	⅛	1°	1¼°	7½°
A110 AUSTIN WESTMINSTER Mk I and Mk II	5° btdc	0·012	0·012	0·015	0·025	55	⅛	1° Mk II ½°	1¼°	7½°
AUSTIN MINI (1959–67)	tdc or 7° btdc									
AUSTIN MINI SUPER (1961–64)	According to distributor									
AUSTIN MINI COUNTRYMAN (1960–67)		0·012	0·012	0·014–0·016	0·025	60	(toe out) 1/16	1°	3°[1]	9½°
AUSTIN MINI AUTOMATIC (1965–67)	tdc	0·012	0·012	0·014–0·016	0·024–0·026	60	(toe out) 1/16	1°–3°	3°	9½°
AUSTIN MINI COOPER (1961–67)	5° btdc	0·012	0·012	0·014–0·016	0·025	70 (30 idling)	(toe out) 1/16	1°	3°	9½°
AUSTIN MINI COOPER S (1963–64)	3° btdc	0·012	0·012	0·014–0·016	0·024–0·026	60	(toe out) 1/16	1°–3°	3°	9½°
AUSTIN MINI COOPER 970 (1964–65)	12° btdc	0·012	0·012	0·014–0·016	0·024–0·026	60	(toe out) 1/16	1°–3°	3°	9½°
AUSTIN MINI COOPER 1275S (1964–65)	2° btdc	0·015	0·015	0·014–0·016	0·024–0·026	60	(toe out) 1/16	1°–3°	3°	9½°
AUSTIN 1100 (1963–67)	3° btdc	0·012	0·012	0·014–0·016	0·024–0·026	60	(toe out) 1/16	½°	6°	10°
AUSTIN 1100 AUTOMATIC (1965–68)	7° btdc	0·012	0·012	0·014–0·016	0·024–0·026	60	(toe out) 1/16	½°	6°	10°
AUSTIN 1800 (1964–68)	14° btdc	0·015	0·015	0·014–0·016	0·024–0·026	30–50	(toe out) 1/16	1½°	¼°	12°
AUSTIN HEALEY 100S (1955–56)	6° btdc	0·015	0·015	0·014–0·016	0·024–0·026	50–55	1/16–⅛	1°	1¾°	6½°
AUSTIN HEALEY BN4 (1956–57)	6° btdc	0·015	0·015	0·014–0·016	0·024–0·026	50–55	1/16–⅛	1°	1¾°	6½°
AUSTIN HEALEY BN6 (1957–59)	6° btdc	0·012	0·012	0·014–0·016	0·024–0·026	55–60	1/16–⅛	1°	2°	6½°
AUSTIN HEALEY SPRITE Mk I (1958–61)	6° btdc	0·012	0·012	0·014–0·016	0·024–0·026	30–60	0–⅛	1°	3°	6½°
AUSTIN HEALEY SPRITE Mk II (1961–62)	6° btdc	0·012	0·012	0·014–0·016	0·024–0·026	30–60	0–⅛	1°	3°	6½°
AUSTIN HEALEY SPRITE Mk II (1962–Feb. 64)										
AUSTIN HEALEY SPRITE Mk III (March 1964–66)	5° btdc	0·012	0·012	0·015	0·025	30–60	0–⅛	1°	3°	6½°
AUSTIN HEALEY 3000 Mk I (1959–61)	5° btdc	0·012	0·012	0·014–0·016	0·025	50	1/16–⅛	1°	2°	6½°
AUSTIN HEALEY 3000 Mk II (1961–62)	12° btdc	0·012	0·012	0·014–0·016	0·025	50	1/16–⅛	1°	2°	6½°
AUSTIN HEALEY 3000 SPORTS CONV. (1962–Feb. 1964)	12° btdc	0·012	0·012	0·014–0·016	0·025	50	1/16–⅛	1°	2°	6½°
AUSTIN HEALEY 3000 Mk III (1964–67)	10° btdc	0·012	0·012	0·014–0·016	0·024–0·026	50	1/16–⅛	1°	2°	6½°
AUSTIN A40 (1950–57)	4° btdc 12° btdc	0·015	0·015	0·014–0·016	0·020	55–60	0–⅛	½°	1½°	7°
AUSTIN A35 (1956–59)	6½° btdc 10° btdc	0·012	0·012	0·014–0·016	0·025	50–55	1/16–⅛	1°	3°	6½°
AUSTIN A50 (1955–57)	4° btdc 10° btdc	0·015	0·015	0·014–0·016	0·020	55–60	0–⅛	½°	1½°	7°
AUSTIN A55 Mk I (1957–58)	— 10° btdc	0·015	0·015	0·014–0·016	0·018	50	0–⅛	1°	1½°	7°
AUSTIN A95–A105 (1957–59)	5° btdc	0·012	0·012	0·014–0·016	0·025	55–60	0–⅛	¾°–1°	1¾°	6½°
AUSTIN PRINCESS IV (1956–58)	tdc	0·012	0·012	0·014–0·016	0·025	50	1/16–⅛	¾°	0°	5°
AUSTIN A135 (1957)	tdc	0·012	0·012	0·010–0·012	0·025	50–55	⅛	¾°	¼°	5°
AUSTIN A40 Mk I (1958–61)	5° btdc[2]	0·012	0·012	0·014–0·016	0·025	60	1/16–⅛	0°	3½°	6½°
AUSTIN A40 Mk II (Oct. 1961–Oct. 1962)	5° btdc[2]	0·012	0·012	0·014–0·016	0·025	60	1/16–⅛	0°	3½°	6½°
AUSTIN A40 Mk II (Oct. 1962–64)	3° btdc	0·012	0·012	0·014–0·016	0·025	60	1/16–⅛	0°	3½°	7°
AUSTIN A55 Mk II (1959–61)	5° btdc[3]	0·015	0·015	0·014–0·016	0·025	50	1/16–⅛	¾°–1°	1½°	6½°
AUSTIN METROPOLITAN (1956–59)	5° btdc[3]	0·015	0·015	0·014–0·016	0·025	50	0–1/16	½°–1½°	2°–3°	6½°
AUSTIN A60 CAMBRIDGE (1961–67)	5° btdc[4]	0·015	0·015	0·014–0·016	0·025	50	1/16–⅛	¾°–1°	¼°–1°	6½°
BOND EQUIPE GT 4S (1963–64)	13° btdc	0·009	0·009	0.015	0·025	Warning light	1/16 to parallel	2° pos. laden	4° pos. laden	6¾° laden
BOND EQUIPE (Triumph Spitfire engine) (1965–67)	17° btdc[5]	0·009	0·009	0·015	0·025	Warning light	1/16 to parallel	2° pos. laden	4° pos. laden	6¾° laden
BOND (Hillman Imp Ic engine) (1966–67)	3° btdc	0·004–0·006	0·006–0·008	0·015	0·025	50		Single front wheel		
FORD 105E (Jan. 1966–67)	10° btdc	0·010	0·017	0·025	0·020–0·024	35–40	0·12–0·19	0°30′–2°	1°30′–3°	4°45′–6°15′
FORD POPULAR (1953–59)	5° btdc	0·0115–0·0135	0·015–0·017	0·014–0·016	0·020–0·022	30	1/16–⅛	2°	7°50′	7°
FORD POPULAR (1959–62)[7]	5° btdc	0·0115–0·0135	0·015–0·0135	0·014–0·016	0·025	30	1/16–⅛	0°30′–2°15′	1°–3°	3°30′–5°

[1] 1¼° up to chassis 11670
[2] High compression engine. Low compression engine 2° btdc
[3] High compression engine. Low compression engine tdc
[4] Low compression 6°
[5] Triumph Spitfire Mk II engine
[6] 0° to 1¼° pos. after Sept. 1956
[7] Also Anglia, Prefect, Escort and Squire (1954–59)

Car and model	Ignition timing	Tappet clearance Inlet (in.)	Tappet clearance Exhaust (in.)	Contact-breaker gap (in.)	Spark-plug gap (in.)	Oil pressure (lb/in.²)	Toe in (in.)	Camber	Castor	King pin inc.
FORD ANGLIA 105E (1959–Dec. 65)	10° btdc[8]	0·010	0·017	0·015	0·023–0·028	35–40	0·12–0·19	0°30′–2°	1°30′–3°	4°45′–6°15′
FORD PREFECT 107E (1959–62)	10° btdc	0·010	0·017	0·015	0·023–0·028	35–40	0·06–0·12	$\frac{3}{4}$°–2$\frac{1}{4}$°	1°–2$\frac{1}{2}$°	3$\frac{1}{2}$°–5°
FORD ANGLIA SUPER 123E (1962–67)	6° btdc	0·010	0·017	0·014–0·016	0·023–0·028	35–40	$\frac{1}{8}$–$\frac{3}{16}$	0°30′–2°	1°30′–3°	4°45′–6°15′
FORD CONSUL 204E Mk II (1956–62)	8° btdc	0·014	0·014	0·014–0·016	0·032	50–60	$\frac{1}{16}$–$\frac{1}{8}$	$\frac{1}{2}$°–2$\frac{1}{4}$°	0°–1$\frac{1}{4}$°[9]	3$\frac{1}{2}$°–4$\frac{1}{2}$°
FORD ZEPHYR 206E (1956–62)	8° btdc	0·014	0·014	0·014–0·016	0·032	50–60	$\frac{1}{16}$–$\frac{1}{8}$	$\frac{1}{2}$°–2$\frac{1}{4}$°	−$\frac{1}{4}$°–+$\frac{3}{4}$°[6]	3$\frac{1}{2}$°–4$\frac{1}{2}$°
FORD ZODIAC Mk II (1956–62)	8° btdc	0·014	0·014	0·014–0·016	0·032	50–60	$\frac{1}{16}$–$\frac{1}{8}$	$\frac{1}{2}$°–2$\frac{1}{2}$°	−$\frac{1}{4}$°–+$\frac{3}{4}$°	3$\frac{1}{2}$°–4$\frac{1}{2}$°
FORD CORSAIR 120E (1963–65)	8° btdc[10]	0·010	0·017	0·014–0·016	0·023–0·027	35–40	$\frac{1}{16}$–$\frac{1}{8}$	0°40′–2°10′	1$\frac{1}{4}$°–2$\frac{3}{4}$°	4°25′–5°56′
FORD ZEPHYR 6 213E	8° btdc	0·014	0·014	0·014–0·016	0·023–0·028	50–60	$\frac{1}{16}$–$\frac{3}{16}$	1°37′–2°37′	−19′–0°41′	5°40′–6°40′
FORD ZODIAC Mk III 213E (1962–66)	8° btdc[11]	0·014	0·014	0·014–0·016	0·023–0·028	50–60	$\frac{1}{16}$–$\frac{3}{16}$	1°37′–2°37′	−19′–0°41′	5°40′–6°40′
FORD ZEPHYR 4 211E (1962–Apr. 66)	8° btdc[11]	0·014	0·014	0·014–0·016	0·023–0·028	50–60	$\frac{1}{16}$–$\frac{3}{16}$	1°37′–2°37′	−19′–0°41′	5°40′–6°40′
FORD CLASSIC 315 and CAPRI (1961–62)	6° btdc	0·010	0·017	0·014–0·016	0·023–0·028	35–40	$\frac{1}{8}$–$\frac{3}{16}$	$\frac{3}{4}$°–2$\frac{1}{2}$°	1°–2$\frac{1}{2}$°	6$\frac{1}{4}$°–7$\frac{3}{4}$°
FORD CLASSIC 315 (1962–63)	8° btdc	0·010	0·017	0·014–0·016	0·023–0·028	35–40	$\frac{1}{8}$–$\frac{3}{16}$	$\frac{3}{4}$°–2$\frac{1}{2}$°	1°–2$\frac{1}{2}$°	6$\frac{1}{4}$°–7$\frac{3}{4}$°
FORD CAPRI 116E (1962–64)	8° btdc	0·010	0·017	0·014–0·016	0·023–0·028	35–40	$\frac{1}{8}$–$\frac{3}{16}$	$\frac{3}{4}$°–2$\frac{1}{2}$°	1°–2$\frac{1}{2}$°	6$\frac{1}{4}$°–7$\frac{3}{4}$°
FORD CORTINA 113E (1962–66)	6° btdc	0·010	0·017	0·014–0·016	0·020–0·024	35–40	0·06–0·12	0°40′–2°10′	0°33′–2°03′	4°26′–5°56′
FORD CORTINA 118E (1963–66)	8° btdc	0·010	0·017	0·014–0·016	0·020–0·024	35–40	0·06–0·12	0°40′–2°10′	0°33′–2°03′	4°26′–5°56′
FORD CORTINA GT 118E (1963–66)	10° btdc	0·014	0·021	0·014–0·016	0·023–0·027	35–40	0·06–0·12	0°40′–2°10′	0°33′–2°03′	4°26′–5°56′
FORD CORSAIR GT 120E (1963–65)	10° btdc	0·014	0·021	0·014–0·016	0·023–0·027	35–40	0·06–0·12	0°40′–2°10′	0°33′–2°03′	4°26′–5°56′
FORD CAPRI GT 116E (1963–64)	10° btdc	0·014	0·021	0·014–0·016	0·023–0·027	35–40	0·06–0·12	0°50′–2°10′	0°33′–2°03′	4°26′–5°56′
FORD LOTUS CORTINA 125E (1963–66)	14° btdc	0·005	0·006	0·014–0·016	0·023–0·028	35–40	$\frac{1}{16}$–$\frac{1}{8}$	0°40′	0°36′	5°54′
FORD CORSAIR V4 DE LUXE and GT (1966)	8° btdc[11]	0·010	0·018	0·014–0·016	0·023–0·028	50	$\frac{1}{8}$–$\frac{3}{16}$	1°11′–2°41′	1°45′–3°15′	3°53′–5°23′
FORD ZEPHYR Mk IV (V4 eng. 1966)	8° btdc[11]	0·010	0·018	0·014–0·016	0·023–0·028	45–50	0·26–0·32	1°03′–2°03′	−2′–−1°2′	5°57′–6°57′
FORD ZEPHYR 6, ZODIAC Mk 4 (V6 eng., May 1966)	12° btdc[12]	0·010	0·018	0·014–0·016	0·023–0·028	45–50	0·26–0·32	1°03′–2°03′	−2′–−1°2′	5°57′–6°57,
HILLMAN HUNTER (1967–)	6°–10° btdc	0·012	0·014	0·015	0·025	40	$\frac{1}{8}$±$\frac{1}{16}$	0°±$\frac{3}{4}$′[13]	$\frac{1}{4}$° neg.±$\frac{1}{2}$°[13]	11$\frac{3}{4}$°±$\frac{3}{4}$°[13]
HILLMAN HUSKY (1954–55)	4°–6° btdc 7°–9° btdc	0·012	0·014	0·014–0·016	0·025	30–40	$\frac{1}{8}$	$\frac{3}{4}$°–15′	3$\frac{3}{4}$°	8$\frac{1}{4}$°±15′
HILLMAN HUSKY (1959[14]–65)	6°–8° 6°–8°[15]	0·012	0·014	0·016	0·025	30–50	$\frac{1}{8}$–$\frac{3}{16}$	$\frac{3}{4}$°±$\frac{1}{2}$°	1°20′±$\frac{1}{2}$°	5$\frac{1}{4}$°±$\frac{1}{4}$°
HILLMAN MINX (1956–57) Series I	8°–10° btdc	0·012	0·014	0·014–0·016	0·028–0·032	50	$\frac{1}{8}$–$\frac{3}{16}$	$\frac{3}{4}$°	3°40′	5$\frac{1}{4}$°
HILLMAN MINX (1958) Series II	4°–6° btdc	0·012	0·014	0·016	0·025	30–50 at 50 mile/h	$\frac{3}{16}$	0°45′±15′	+3°33′[13]±$\frac{1}{2}$°[19]	5°15′±15′
HILLMAN MINX (1958–61 Series III[16]	6°–8° btdc[15]	0·012	0·014	0·016	0·025	30–50	$\frac{3}{16}$[13]	$\frac{3}{4}$°±$\frac{1}{4}$°[13]	$\frac{1}{2}$°[13]±$\frac{1}{2}$°[18]	5$\frac{1}{4}$°±$\frac{1}{4}$°[13]
HILLMAN MINX (1961–63) Series IIIC	6°–8° btdc[15]	0·012	0·014	0·015	0·025	40–50	$\frac{1}{8}$[13]	$\frac{3}{4}$°±$\frac{1}{4}$°[13]	$\frac{1}{2}$°[13]±$\frac{1}{4}$°[18]	5$\frac{1}{4}$°±$\frac{1}{4}$°[13]
HILLMAN MINX (1963–Sept. 65) Series V	6°–8° btdc[15]	0·012	0·014	0·015	0·025	40–50	$\frac{1}{8}$[13]	$\frac{3}{4}$°[13]±$\frac{1}{4}$°	$\frac{1}{2}$°[13]±$\frac{1}{4}$°[18]	5$\frac{1}{4}$°±$\frac{1}{4}$°[13]
HILLMAN SUPER MINX (1961–62) Mk I	9°–11° btdc	0·012	0·014	0·015	0·025	40–50	$\frac{1}{8}$[13]	$\frac{1}{2}$°±$\frac{1}{4}$°[13]	$\frac{3}{4}$°±$\frac{1}{2}$°[13]	5$\frac{1}{4}$°±$\frac{1}{4}$°[13]
HILLMAN SUPER MINX (1962–Sept. 64) Mk II[17]	9°–11° btdc	0·012	0·014	0·015	0·025	40–50	$\frac{1}{8}$[13]	$\frac{1}{2}$±°$\frac{1}{4}$°[13]	$\frac{3}{4}$°±$\frac{1}{2}$°[13]	5$\frac{1}{4}$°±$\frac{1}{4}$°[13]
HILLMAN SUPER MINX Series IV and Mk VI MINX (Sept. 1965–66)	6°–10° btdc	0·012	0·014	0·015	0·025	40	$\frac{1}{8}$[13]	$\frac{1}{2}$°±$\frac{1}{4}$°[13]	$\frac{3}{4}$°±$\frac{1}{2}$°[13]	5$\frac{1}{4}$°±$\frac{1}{4}$°[13]
HILLMAN IMP Mk I Mk II (1963–66)	3° btdc	0·006	0·008	0·015	0·025	50	$\frac{3}{16}$±$\frac{1}{8}$	7°±1°[13]	10°±1°[13]	3°±1°
HUMBER HAWK Mk VI (1954–57)	tdc	0·007	0·009	0·016	0·028–0·032	45–50	0·125	$\frac{3}{4}$° pos.±$\frac{1}{4}$°[20]	0	8$\frac{1}{4}$°
HUMBER HAWK Series I and Series IA[21]	tdc to 2° atdc	0·007	0·009	0·016	0·025	30–50	$\frac{1}{8}$	$\frac{3}{4}$°±$\frac{1}{4}$°[20]	$\frac{1}{2}$° pos.[24]	8°15′±15′
HUMBER HAWK Series II, III and IV[21]	tdc to 2° atdc	0·007	0·009	0·016	0·025	40–50	$\frac{1}{8}$	$\frac{3}{4}$°±$\frac{1}{4}$°[20]	$\frac{1}{2}$° neg.[22]	8$\frac{1}{4}$°±$\frac{1}{4}$°
HUMBER SUPER SNIPE Mk IV (1952–56)	5°–7° btdc	0·010	0·012	0·016	0·025–0·032	50–65	$\frac{1}{8}$	$\frac{3}{4}$°	1°5′	5°15′
HUMBER SUPER SNIPE Series I (1958–59)	5°–7° btdc	0·014	0·014	0·015	0·025	45–50	$\frac{1}{8}$	$\frac{3}{4}$°±$\frac{1}{4}$°[20]	$\frac{1}{2}$° neg.[22]	8°15′±15′

[8] 8° btdc from Sept. 1963
[9] −$\frac{1}{4}$° to +$\frac{3}{4}$° prior to Sept. 1956
[10] 6° btdc from April 1965
[11] Low compression engine 4° btdc
[12] Zephyr and Zodiac V6; low compression engine 8°
[13] Laden onto gap gauges
[14] Series I 1959; Series II 1960–63; Series III 1963–65
[15] Low compression engine 9°–11° btdc premium fuel
[16] Series III 1958–59; Series III A 1959–61; Series III B 1961
[17] Also Mk III, Sept. 1964, Sept. 1965
[18] Estate cars 1°40′
[19] Saloon and Coupé to chassis A181559. Special models to chassis A1867144. From chassis 1867145 the castor angle is 1°45′±$\frac{1}{2}$°
[20] Laden with 360 lb weight 34$\frac{1}{2}$ in. to rear of front-wheel centres
[21] Series I 1958–59; Series IA 1959–60; Series II 1960–62; Series III 1962–64; Series IV 1962–66
[22] 2° neg. with car laden onto gap gauges

Car and model	Ignition timing	Tappet clearance: Inlet (in.)	Tappet clearance: Exhaust (in.)	Contact-breaker gap (in.)	Spark-plug gap (in.)	Oil pressure (lb/in.²)	Toe in (in.)	Camber	Castor	King pin inc.
HUMBER SUPER SNIPE[23] Series II (1959–60)	5°–7° btdc	0·014	0·014	0·015	0·025	40–50 at 50 miles/h	$\frac{1}{8}$	$\frac{3}{4}$° ± $\frac{1}{4}$°[20]	$\frac{1}{2}$ neg.[22]	8°15′ ± 15′
HUMBER SUPER SNIPE Series IV (1962–Sept. 64)	5°–7° btdc	0·014	0·014	0·016	0·025	45–50	$\frac{1}{8}$	$\frac{3}{4}$° ± $\frac{1}{4}$°[20]	$\frac{1}{2}$ neg.[22]	8°15′ ± 15′
HUMBER SUPER SNIPE Series V (1964–66)	5°–7° btdc	0·014	0·014	0·016	0·025	45–50	$\frac{1}{8}$	$\frac{3}{4}$° ± $\frac{1}{4}$°[20]	$\frac{1}{2}$ neg.[22]	8°15′ ± 15′
HUMBER SCEPTRE (1963–Aug. 65) Series I	7°–9° btdc (Solex carb.) / 9°–11° btdc	0·012	0·014	0·015	0·025	55–60	$\frac{1}{8}$	$\frac{1}{2}$° ± $\frac{1}{4}$°	$\frac{3}{4}$° ± $\frac{1}{2}$°	5$\frac{1}{4}$° ± $\frac{1}{4}$°
HUMBER SCEPTRE (1965–66) Series II	6°–10° btdc	0·012	0·014	0·015	0·025	40	$\frac{1}{8}$	$\frac{1}{2}$° ± $\frac{1}{4}$°	$\frac{3}{4}$° ± $\frac{1}{4}$°	5$\frac{1}{4}$° ± $\frac{1}{4}$°
JAGUAR XK140 (1955)	7:1 cr 8° btdc / 8:1 cr 10° btdc	0·004	0·006	0·014–0·016	0·022	40	0–$\frac{1}{8}$	1$\frac{1}{2}$°–1° pos.	2$\frac{1}{2}$°–3° pos.	5°
JAGUAR Mk VII 'M' type (1955)	7:1 cr 5° btdc / 8:1 cr 0° tdc	0·004	0·006	0·014–0·016	0·022	40	$\frac{1}{8}$–$\frac{3}{16}$	1° ± $\frac{1}{4}$° pos.	0° ± $\frac{1}{4}$° pos.	8°
JAGUAR 2·4 Mk I (1956)	6° btdc	0·004	0·006	0·014–0·016	0·030	40	0–$\frac{1}{16}$	$\frac{1}{2}$°–1°	−$\frac{1}{2}$°– −1°	6$\frac{3}{4}$°
JAGUAR 2·4 Mk I (1957–59)	7:1 cr 4° btdc / 6° tdc	0·004	0·006	0·014–0·016	0·030	40	0–	$\frac{1}{2}$°–1° pos	−$\frac{1}{2}$°– −1°	6$\frac{3}{4}$°
JAGUAR 2·4 Mk II (Oct. 1959–66)	7:1 cr 6° btdc / 8:1 cr 8° btdc	0·004	0·006	0·014–0·016	0·025	40	0–$\frac{1}{16}$	$\frac{1}{2}$°–1° pos.	0° ± $\frac{1}{4}$°	3$\frac{1}{2}$°
JAGUAR Mk VIII (1957–58)	6° btdc	0·004	0·006	0·014–0·016	0·022	40	$\frac{1}{8}$–$\frac{3}{16}$	1° ± $\frac{1}{4}$°	0° ± $\frac{1}{4}$°	8°
JAGUAR Mk IX (1959–61)	8:1 cr 4° btdc / 9:1 cr 5° btdc	0·004	0·006	0·014–0·016	0·025	40	$\frac{1}{8}$–$\frac{3}{16}$	1° ± $\frac{1}{4}$°	0° ± $\frac{1}{4}$°	8°
JAGUAR 3·4 Mk I (1957–59)	8:1 cr 2° btdc / 7:1, 9:1 cr tdc	0·004	0·006	0·014–0·016	0·025	40	0–$\frac{1}{16}$	$\frac{1}{2}$°–1° pos.	$\frac{1}{2}$°–1° neg.	6$\frac{3}{4}$°
JAGUAR 3·4 Mk II (Oct. 1959–66)	Paper air-cleaner: 7:1 cr tdc[26] / 8:1 cr 7° btdc	0·004	0·006	0·014–0·016	0·025	40	0–$\frac{1}{16}$	$\frac{1}{2}$°–1° pos.	0° ± $\frac{1}{4}$°	3$\frac{1}{2}$°
JAGUAR 3·4 S (1963–66)	Paper air-cleaner: 7:1 cr tdc[26] / 8:1 cr 7° btdc	0·004	0·006	0·014–0·016	0·025	40	0–$\frac{1}{16}$	$\frac{1}{2}$°–1° pos.	0° ± $\frac{1}{4}$°	3$\frac{1}{2}$°
JAGUAR 3·8 Mk II (1959–66)	Paper air-cleaner: 7:1 cr tdc[25] / 8:1 cr 7° btdc[26]	0·004	0·006	0·014–0·016	0·025	40	0–$\frac{1}{16}$	$\frac{1}{2}$°–1° pos.	0° ± $\frac{1}{4}$°	3$\frac{1}{2}$°
JAGUAR 3·8 S (1963–66)	AS 3·8 Mk II			0·014–0·016	0·025	40	0–$\frac{1}{16}$	$\frac{1}{2}$°–1° pos.	0° ± $\frac{1}{4}$°	3$\frac{1}{2}$°
JAGUAR XK150 (1957–59)	8:1 cr 6° btdc[27] / 7:1 cr 4° btdc[27]	0·004	0·006	0·014–0·016	0·025	40	0–$\frac{1}{8}$	$\frac{1}{2}$°–1° pos.	1$\frac{1}{2}$°–2° pos.	5°
JAGUAR XK150S (1959–61)	8:1 cr 9° btdc / 9:1 cr 9° btdc	0·004	0·006	0·014–0·016	0·025	40	0–$\frac{1}{8}$	$\frac{1}{2}$°–1° pos.	1$\frac{1}{2}$°–2° pos.	5°
JAGUAR E type (1961–64)	9:1 cr 10° btdc / 8:1 cr 9° btdc	0·004[28]	0·006[28]	0·014–0·016	0·025	40	$\frac{1}{16}$–$\frac{1}{8}$	$\frac{1}{4}$° ± $\frac{1}{4}$°	1$\frac{3}{4}$° ± $\frac{1}{4}$°	4°
JAGUAR Mk X (1961–64)	9:1 cr 10° btdc / 8:1 cr 9° btdc	0·004	0·006	0·014–0·016	0·025	40	$\frac{1}{16}$–$\frac{1}{8}$	$\frac{1}{2}$° ± $\frac{1}{4}$°	0° ± $\frac{1}{4}$°	3$\frac{3}{4}$°
JAGUAR 4·2 E type (1965–	9:1 cr 10° btdc / 8:1 cr 9° btdc	0·004	0·006	0·014–0·016	0·025	40	$\frac{1}{16}$–$\frac{1}{8}$	$\frac{1}{4}$° ± $\frac{1}{4}$°	1$\frac{3}{4}$° ± $\frac{1}{4}$°	4°
JAGUAR 4·2 Mk X (1965–)	9:1 cr 10° btdc / 8:1 cr 9° btdc	0·004	0·006	0·014–0·016	0·025	40	$\frac{1}{16}$–$\frac{1}{8}$	$\frac{1}{2}$° ± $\frac{1}{4}$°	0° ± $\frac{1}{4}$°	3$\frac{3}{4}$°
LOTUS ELITE (std. 1959–54)	2°–3° btdc	0·006	0·008	0·014	0·018	55	$\frac{1}{16}$–$\frac{1}{8}$	1°	7°	—
LOTUS ELAN 1600 Series I (1963–64)	23D dist. 14° btdc / 25D dist. 7° btdc	0·005	0·006	0·014–0·016	0·023–0·028	40	$\frac{1}{8}$	0°–$\frac{1}{2}$°	3°–[29]	9°
LOTUS ELAN 1600 Series II (1965–66)	23D dist. 14° btdc / 25D dist. 7° btdc	0·005	0·006	0·014–0·016	0·023–0·028	40	$\frac{1}{8}$	0°–$\frac{1}{2}$°	3°–[29]	9°
LOTUS COUPE (1966)	7° btdc	0·005	0·006	0·014–0·016	0·023–0·028	40	$\frac{1}{8}$	0°–$\frac{1}{2}$°	3°–[29]	9°
M.G. MIDGET Mk III (1966–67)	tdc	0·012	0·012	0·014–0·016	0·024–0·026	60	0–$\frac{1}{8}$	$\frac{3}{4}$°	3°	6$\frac{3}{4}$°
M.G. MAGNETTE Series ZA (1953–56)	8° btdc	0·015	0·016	0·014–0·016	0·019–0·021	50–60	Nil	1°	3° pos.	6°
M.G. MAGNETTE Series ZB (1956–58)	4° btdc	0·015	0·016	0·014–0·016	0·019–0·021	50	Nil	1°	3°	6°

23 Super Snipe Series III 1959–60
24 To chassis number A5706749, $\frac{1}{2}$° neg. from chassis number A5706750
25 Oil-bath air cleaner, 7:1 cr tdc; 8:1 cr 4° btdc; 9:1 cr 10° btdc. Paper-element air cleaner, 9:1 cr 5° btdc
26 Oil-bath air cleaner, 7:1 cr tdc; 9:1 cr tdc; 8:1 cr 2° btdc. Paper-element air cleaner, 9:1 cr 5° btdc
27 5° btdc for 9:1 cr
28 Racing clearance: Inlet 0·006 in.; exhaust 0·010 in.
29 Up to car number 26/3075, caster angle was 7°

Car and model	Ignition timing	Tappet clearance Inlet (in.)	Tappet clearance Exhaust (in.)	Contact breaker gap (in.)	Spark-plug gap (in.)	Oil pressure (lb/in.²)	Toe in (in.)	Camber	Castor	King pin inc.
M.G. MAGNETTE Series III (1959–61)	8° btdc	0·015	0·015	0·014–0·016	0·024–0·026	75	1/16–1/8	1/2°–1°	1 1/2°	6 1/4°
M.G. MAGNETTE Series IV (1961–6)	4° btdc	0·015	0·015	0·014–0·016	0·024–0·026	70	1/16–1/8	1/2°–1°	1/2°–1°	6 1/4°
M.G.A twin cam (1959)	tdc	0·016	0·016	0·014–0·016	0·024–0·026	50–60	Parallel	1° pos.–1° neg.[30]	4°	10 1/2°[30]
M.G.A (1955–59)	7° btdc	0·017	0·017	0·014–0·016	0·019–0·026	50	P	1° pos.–1° neg.[30]	4°	10 1/2°[30]
M.G.A Mk I Series 1600 (1959–61)	6° btdc (5° late cars)	0·015	0·015	0·014–0·016	0·019–0·026	50	P	1° neg.[30]	4°	10 1/2°[30]
M.G.A Mk II Series 1600 (1961–62)	lc 10° btdc	0·015	0·015	0·014–0·016	0·024–0·026	75	P	1° neg.[30]	4°	10 1/2°[30]
M.G.B 1962–M.G.B GT 1966	hc 8° btdc / lc 10° btdc	0·015	0·015	0·014–0·016	0·024–0·026	70	P–3/32	1° neg.[30]	7°	8°
M.G. MIDGET 948cc (1961–62)	lc 1° btdc	0·012	0·012	0·014–0·016	0·024–0·026	30–60	0–1/16	1°	3°	6 1/4°
M.G. MIDGET Mk I (1962–Feb. 64)	4° btdc	0·012	0·012	0·014–0·016	0·024–0·026	60	1/16–1/8	1°	3°	6 1/4°
M.G. MIDGET Mk II (1964–66)	5° btdc	0·012	0·012	0·014–0·016	0·024–0·026	60	1/16–1/8	1°	3°	6 1/4°
M.G. 1100 (1962–68)	5° btdc	0·010	0·012	0·012	0·032	40–60	(toe out) 1/16–1/8	2°	6°	10°
MORGAN PLUS IV (Vang. eng.) (1951–58)	4° btdc	0·010	0·012	0·012	0·032	40–60	1/8	2°	4°	2°
MORGAN PLUS IV (2 str. TR.2 eng.) (1954)	5° btdc	0·013	0·013	0·015	0·032	40–50	1/8	2°	4°	2°
MORGAN 4/4 Series II Ford eng. (1956–60)[31]	5° btdc	0·010	0·017	0·015	0·023–0·028	30–50	1/16–1/8	2°	4°	2°
MORGAN PLUS 4 TR3 eng. (1957–62)	6° btdc	0·010	0·010	0·014–0·016	0·023–0·028	35–40	1/8	2°	4°	2°
MORGAN 4/4 Series IV Ford eng. (1961–63)	8° btdc	0·008	0·018	0·016	0·023–0·028	35–40	1/8–3/16	2°	4°	2°
MORGAN 4/4 Series V Ford eng. (1963–67)	2° btdc	0·011	0·010	0·014–0·016	0·024–0·026	50–75	1/16–1/8	2°	4°	2°
MORGAN PLUS 4 (1963–68)	4° btdc	0·015	0·015	0·014–0·016	0·019–0·021	30–40	1/16–1/8	1°	3°	7 1/2°
MORRIS MINOR Series II (1952–56)	2° btdc	0·011	0·011	0·014–0·016	0·019–0·021	50	1/8	1°	3° pos.	8 1/2°
MORRIS OXFORD Series II (1954–56)	4° btdc	0·012	0·012	0·014–0·016	0·019–0·021	40	1/16	1°	3° pos.	8 1/2°
MORRIS COWLEY (1954–56)	3°–4° btdc	0·015	0·015	0·014–0·016	0·025	55	1/4	1°	3 1/2°	7°
MORRIS ISIS Series II (1957–58)	hc tdc	0·012	0·012	0·014–0·016	0·025	40–60	3/32	1°	3°	7°
MORRIS MINOR 1000 (1956–62)	lc 4° btdc	0·012	0·012	0·014–0·016	0·024–0·026	60	3/32	1°	3°	7 1/2°
MORRIS MINOR 1000 (1962–)	3° btdc	0·012	0·012	0·014–0·016	0·024–0·026	60	3/32	1°	3°	7 1/2°
MORRIS OXFORD Series III Cowley 1500 (1956–58)	5° btdc	0·015	0·015	0·014–0·016	0·024–0·026	40–50	1/16–1/8	1 1/2°–1°	3°	8 1/2°
MORRIS OXFORD Series V (1959–61)	hc 5° btdc	0·014	0·014	0·014–0·016	0·025	40–50	1/16–1/8	1 1/2°–1°	1 1/2°	6 1/4°
MORRIS OXFORD Series VI (1961–67)	lc 5° btdc	0·015	0·015	0·014–0·016	0·024–0·026	50	1/16–1/8	1/2°–1°	1/2°–1°	6 1/4°
MORRIS MINI MINOR (1959–67)	See AUSTIN MINI									
MORRIS MINI MINOR SUPER (1961–64)	See AUSTIN MINI									
MORRIS MINI MINOR TRAVELLER (1961–)	See AUSTIN MINI COUNTRYMAN									
MORRIS MINI MINOR AUTOMATIC (1965–)	See AUSTIN MINI AUTOMATIC									
MORRIS MINI COOPER 998cc	See AUSTIN MINI COOPER									
MORRIS MINI COOPER 970 S type (1964–65)	See AUSTIN 970 S type									
MORRIS MINI COOPER 1275 S type (1964–)	See AUSTIN MINI COOPER 1275 S type									
MORRIS MINI COOPER 997cc	See AUSTIN MINI COOPER									
MORRIS 1100 AUTOMATIC	See AUSTIN 1100 AUTOMATIC									
MORRIS 1100 (1962–68)	See AUSTIN 1100									
MORRIS 1800 (1966–)	See AUSTIN 1800									
MORRIS MINI COOPER 1071	3° btdc									
RILEY PATHFINDER (1953–57)	4°–8° btdc	0·012	0·012	0·015	0·025	60	(toe out) 1/16	1°–3°	3°	9 1/2°
RILEY 1·5 (1957–65)	tdc 6° btdc (1959)	0·011	0·011	0·015	0·025	40	Nil	1°	3°	6°
RILEY 2·6 (1957–59)	4°–5° btdc	0·015	0·015	0·014–0·016	0·025	50	P	1°	3°	6°
RILEY 4/68 (1959–61)	8° btdc	0·015	0·015	0·014–0·016	0·024–0·026	75	P	1 1/2°–1°	1 1/2°–1°	6 1/4°
RILEY 4/72 (1961–66)	4° btdc	0·012	0·012	0·014–0·016	0·024–0·026	70	1/16–1/8	1/2°–1°	1 1/2°	6 1/4°
RILEY ELF (1961–Apr. 63)	tdc	0·015	0·015	0·014–0·016	0·024–0·026	50	(toe out) 1/16–1/8	1/2°–1°	3°	9 1/2°
RILEY ELF Mk II (Apr. 1963–)	5° btdc	0·012	0·012	0·014–0·016	0·024–0·026	50	(toe out) 1/16–1/8	1°–3°	1°–3°	9 1/2°
RILEY KESTREL (1965–67)	5° btdc	0·012	0·012	0·014–0·016	0·024–0·026	60	(toe out) 1/16–1/8	1°–3°	6°	10°
ROVER 60 (1954–59)	10° btdc	0·008	0·017	0·014–0·016	0·029–0·032	55–65 (at 30 mile/h)	0–1/8	1°–3°[32]	0–2° neg.[32]	2 1/4°–4 1/2°[32]

[30] On full bump
[31] Series III 1960–61 fitted with Ford 105E 996-6cc engine
[32] In datum position

Car and model	Ignition timing	Tappet clearance Inlet (in.)	Tappet clearance Exhaust (in.)	Contact-breaker gap (in.)	Spark-plug gap (in.)	Oil pressure (lb/in.²)	Toe in (in.)	Camber	Castor	King pin inc.
ROVER 75 (1955–59)	10° btdc	0·008	0·012	0·014–0·016	0·029–0·032	55–65	0–⅛	1°–3°	0°–2° neg.	2½°–4½°
ROVER 90 (1954–59)	10° btdc	0·008	0·012	0·014–0·016	0·029–0·032	55–65	0–⅛	1°–3°	0°–2° neg.	2½°–4½°
ROVER 105 R and S (1957–59)	3 °btdc (1959 10°)	0·008	0·012	0·015	0·029–0·032	55–65	0–⅛	1°–3°	[33]	2½°–4½°
ROVER 80 (1960–62)	RF 3° btdc PF 6° btdc	0·010	0·010			55–65 at 30 mile/h	0–⅛	1°–3°	0°–2° neg.	2½°–4½°
ROVER 100 (1960–62)	10° btdc	0·006	0·010	0·014–0·016	0·029–0·032	55–65 at 30 mile/h	0–⅛	1°–3°	0°–2° neg.	2½°–4½°
ROVER 3 LITRE										
Mk I (1959–60)	3° btdc	0·006	0·010	0·014–0·016	0·029–0·032	55–65 at 30 mile/h	0–1/16	2°–1½°	1°	4°
Mk IA (1961–62)	3° btdc	0·006	0·010	0·014–0·016	0·029–0·032	55–65 at 30 mile/h	0–1/16	2°–1½°	1°	4°
Mk II (1963–65)	3° btdc	0·006	0·010	0·014–0·016	0·029–0·032	55–65 at 30 mile/h	0–1/16	2°–1½°	1¼° pos.	4½°
Mk III (1966)	3° btdc	0·006	0·010	0·014–0·016	0·029–0·032	55–65 at 30 mile/h	0–1/16	2°–1½°	¼° pos.	4½°
ROVER 95 Mk I 110 Mk I (1963)	6° btdc	0·006	0·010	0·014–0·016	0·029–0·032	55–65 at 30 mile/h	0–⅛	1°–3°	0°–2° neg.	2½°–4½°
ROVER 2000 (1964–66)	4° btdc	0·008	0·013	0·014–0·016	0·023–0·027[34]	50–60 at 30 mile/h	0–3/32	0°±1°	½°±½°	8°
ROVER 2000 TC (1966)	10:1 cr 6° btdc	0·008	0·013	0·014–0·016	0·023–0·027	50–60	0–3/32	0°	¾°	8°
SINGER GAZELLE Mk I (1956–57)	10° btdc[35]	0·018	0·020	0·014–0·016	0·025	30	⅛	¾°±¼°	3°40′	5½°±¼°
SINGER GAZELLE Mk II (1957–58)	5° btdc Series III 12°–14°	0·018	0·020	0·014–0·016	0·025	30	0·125–0·250	0°45′±15′[36]	1°45′[36]	5°15′±15′[36]
SINGER Mk IIA and Mk III (1958–59)	9°–11° btdc	0·012	0·014	0·015	0·025	55	0·128–0·250	0°45′±15′[36]	1°45′[36 37]	5°15′±15′
SINGER Mk IIIA and Mk IIIB (1959–61)	6°–8° btdc	0·012	0·014	0·014–0·016	0·024–0·026	40–50	3/16	0°45′±15′[36]	1°45′[36 37]	5°15′±15′
SINGER GAZELLE Mk IIIC (1961–63)	6°–8° btdc	0·012	0·014	0·014–0·016	0·024–0·026	40–50	⅛[36]	0°30′±15′[36]	½°[36]±½°[38]	5°15′±15′[36]
SINGER GAZELLE Mk V (1963–65)	6°–8° btdc	0·012	0·014	0·014–0·016	0·024–0·026	40–50	⅛[36]	0°30′±15′[36]	½°[36]±½°[38]	5°15′±15′[36]
SINGER GAZELLE Mk VI (1965–66)	6°–10° btdc	0·012	0·014	0·014–0·016	0·024–0·026	40	⅛[36]	½°±¼°[36]	½°±¼°[36]	5¼°±¼°[36]
SINGER VOGUE Mk I (1961–62)	9°–11° btdc	0·012	0·014	0·014–0·016	0·024–0·026	40–50	⅛[36]	0°–30′[36]	0°45′±15′[36]	5°15′±15′[36]
SINGER VOGUE Mk II (1962–64)	9°–11° btdc	0·012	0·014	0·014–0·016	0·024–0·026	40–50	⅛[36]	0°–30′[36]	0°45′±15′[36]	5°15′±15′[36]
SINGER VOGUE Mk III (1965)	7°–11° btdc	0·012	0·014	0·015	0·024–0·026	55–65 at 30 mile/h	⅛	½°±¼°[36]	¾°±¼°[36]	5¼°±¼°[36]
SINGER VOGUE Mk IV (1965–66)	6°–10° btdc	0·012	0·014	0·015	0·024–0·026	40	⅛	½°±¼°[36]	½°±¼°[36]	5¼°±¼°[36]
SINGER VOGUE Mk V (1966–67)	6°–10° btdc	0·012	0·014	0·015	0·024–0·026	40	⅛±1/16	0°±¾°[36]	¼° neg.±½°[36]	11¾°±¾°[36]
SINGER CHAMOIS Mk I (Oct. 1964–Sept. 65)	3° btdc	0·004	0·006	0·015	0·025	50	3/16±⅛	7°±1°	10°±	3°±1°
SINGER CHAMOIS Mk II (Sept. 1965–67)	3° btdc	0·006	0·013	0·015	0·025	50	3/16±⅛	7°±1°	10°±	3°±1°
SINGER CHAMOIS SPORT (1967)	3° btdc	0·006	0·012	0·015	0·025	50	3/16±⅛	7°±1°	10°±	3°±1°
STANDARD VANGUARD Mk III (1955–59)	12° btdc	0·010	0·012	0·015	0·035	70	(toe out) P–1/16	2° pos.[36]	—	5½°[36]
STANDARD VANGUARD VIGNALE (1959–61)	12° btdc	0·010	0·010	0·015	0·025	70	(toe out) P–1/16	2° pos.[36]	1½° neg.[36]	5½°[36]
STANDARD 8 (1953–59)	10° btdc	0·010	0·010	0·015	0·032	40–60	(toe out) 0–1/16	1°55′[36]	1°45′[36]	7°5′[36]
STANDARD 10 (1953–61)	10° btdc	0·010	0·010	0·015	0·032	40–60	(toe out) 0–1/16	1°55′[36]	1°45′[36]	7°5′[36]
STANDARD PENNANT (1957–60)	10° btdc	0·010	0·010	0·015	0·032	40–60	(toe out) 0–1/16	1°55′[36]	1°45′[36]	7°5′[36]
STANDARD ENSIGN (1958–60)	3° btdc	0·010	0·010	0·015	0·032[45]	70	P–1/16			5½°
STANDARD SPORTSMAN (1957–58)	9° btdc	0·010	0·010	0·015	0·025	70	P–1/16			5½°
STANDARD 6 (1960–63)	15° btdc	0·010	0·010	0·014–0·016	0·024–0·026	45–55	(toe out) P–1/16	2°[36]	0°–1°[36]	5½°
ENSIGN DE LUXE (1962–63)	15° btdc	0·010	0·010	0·014–0·016	0·024–0·026	70	(toe out) P–1/16	2°[36]	0°–1°[36]	5½°
SUNBEAM ALPINE Mks I, II and III (1954–56)	tdc	0·009	0·011	0·016	0·028–0·032	35–45	0·125	0°45′±15′[36]	3°[36]	8°15′±15′
SUNBEAM RAPIER Series I (1956)	5°–7° btdc hc lc	0·012	0·014	0·016	0·028–0·032	30–50	⅛	¾°	3°40′	5¼°
SUNBEAM RAPIER Series I (1956–57)	10°–12° btdc 8° btdc	0·012	0·014	0·016	0·028–0·032	30–50	⅛–3/16	¾°	3°40′	5¼°
SUNBEAM RAPIER Series II (1958–59)	7°–9° btdc	0·012	0·014	0·016	0·025	30–50 at 50 mile/h	⅛–3/16	0°45′±¼°[36]	1°45′±¼°[36]	5°15′±¼°[36]

[33] 105R models to chassis 615700257 1° neg.; from chassis 615700258 1¾° neg.
[34] Champion N9Y plugs. Lodge HLMY 0·025–0·028 in.
[35] Distributor DKY4A, distributor D2A4—5°
[36] Laden

Car and model	Ignition timing	Tappet clearance: Inlet (in.)	Tappet clearance: Exhaust (in.)	Contact breaker gap (in.)	Spark-plug gap (in.)	Oil pressure (lb/in.²)	Toe in (in.)	Camber	Castor	King pin inc.
SUNBEAM RAPIER Series III (1959–60)	5°–7° btdc	0·012	0·014	0·016	0·025	30–50 at 50 mile/h	1/8–3/16	0°45′±¼°[36]	1°45′±¼°[36]	5°15′±¼°[36]
SUNBEAM RAPIER Series IIIA (1961–63)	5°–7° btdc	0·012	0·014	0·014–0·016	0·024–0·026	40–50 at 50 mile/h	3/16	0°45′±15′[36]	1¾°±½°[36]	5°15′±15′[36]
SUNBEAM ALPINE Series I (1959–60)	5° btdc	0·012	0·014	0·016	0·025	30–50 at 50 mile/h	1/8[36]	¾°±¼°[36]	4°41′[41]	5¼°±¼°[36]
SUNBEAM ALPINE Series II (1961–63)	5°–7° btdc	0·012	0·014	0·014–0·016	0·024–0·026	40–50 at 50 mile/h	1/8	¾°±¼°[36]	4°41′[41]	5¼°±¼°[36]
SUNBEAM ALPINE Series III and GT (early 1963)	9°–11°	0·012	0·014	0·014–0·016	0·024–0·026	40	1/8	½°±¼°[36]	4°41′[42]	5¼°±¼°[36]
SUNBEAM ALPINE SPORTS and GT										
Series III (late 1963)	9°–11° btdc	0·012	0·014	0·015	0·025	40–50	1/8	½°±¼°[36]	4°41′[42]	5¼°±¼°
Series IV (1964–65)	9°–11° btdc	0·012	0·014	0·015	0·025	40–50	1/8	½°±¼°[36]	4°41′[42]	5¼°±¼°
SUNBEAM RAPIER Series IV (1963–65)	7°–9° btdc	0·012	0·014	0·015	0·025	40–50	1/8	½°±¼°[36]	½°±¼°[43]	5¼°±¼°
SUNBEAM RAPIER Series IV (1965–66)	6°–10° btdc	0·012	0·014	0·015	0·025	40	1/8	½°±¼°[43]	½°±¼°[43]	5¼°±¼°[43]
SUNBEAM ALPINE SPORTS and GT Series V (1965–66)	6°–10° btdc	0·012	0·014	0·015	0·025	40 (2000 rev./min)	1/8	½°±¼°[43]	3°50′±¼°[43]	5¼°±¼°[43]
SUNBEAM TIGER 260 (1965–66) (American Ford engine)	6° btdc	(Exhaust and Inlet)[44] 0·082–0·152		0·015	0·032–0·036	35–55	1/8	½°±¼°[43]	3°50′±¼°[43]	5½°±¼°[43]
SUNBEAM IMP SPORT (1967)	3° btdc	0·006	0·012	0·015	0·025	50	3/16±1/8	7°±1°[43]	10°±1°[43]	3°±1°[43]
TRIUMPH GT6 (1966–67)	4° btdc	0·010	0·010	0·015	0·025	60	1/16–1/8	2¾°	3½°	6°
TRIUMPH VITESSE 2 litre (1966–67)	13° btdc	0·010	0·010	0·015	0·025	60	1/16–1/8	3¼°	2¾°	5½°
TRIUMPH TR2 (1954–55)	4° btdc	0·010	0·012	0·014–0·016	0·032	40–60	0–1/8	2°[36]	0°	7°
TRIUMPH TR3 (1955–56)	4° btdc	0·010[49]	0·012[49]	0·015	0·032	70[48]	0–1/8	2°	0°	7°
TRIUMPH TR3 (1957–61)	4° btdc	0·010	0·010	0·015	0·025	70	0–1/8	2°	0°	7°
TRIUMPH TR4 (1961–65)	4° btdc	0·010	0·010	0·014–0·016	0·025	65–70	0–1/16	2½°	2¾°	6½°
TRIUMPH TR4A (1965–67)	4° btdc	0·010	0·010	0·014–0·016	0·025	70	1/16–1/8	½°	2¾°	8½°
TRIUMPH HERALD (1959–61)	10°–12° btdc	0·010	0·010	0·014–0·016	0·025	50	1/16–1/8	3¼°	2¾°	5½°
TRIUMPH HERALD Sports Saloon (1961–63)	10°–12°	0·010	0·010	0·014–0·016	0·025	50	1/16–1/8	3¼°	2¾°	5½°
TRIUMPH HERALD 1200 (1961–67)	15° btdc[46]	0·010	0·010	0·014–0·016	0·025[50]	65–70	1/16–1/8	3¼°	2¾°	5½°
TRIUMPH HERALD 12/50 Saloon (1963–67)	13° btdc	0·010	0·010	0·014–0·016	0·025	60	1/16–1/8	3¼°	2¾°	5½°
TRIUMPH VITESSE (1962–66)	10° btdc[47]	0·010	0·010	0·014–0·016	0·025	70	1/16–1/8	3¼°	2¾°	5½°
TRIUMPH SPITFIRE Mk I (1962–65)	14° btdc	0·010	0·010	0·014–0·016	0·025	70	1/16–1/8	2°	3°	6¾°
TRIUMPH SPITFIRE Mk II (1965–67)	17° btdc	0·010	0·010	0·014–0·016	0·025	70	1/16–1/8	2°	3°	6¾°
TRIUMPH 2000 Saloon (1964–67)	8° btdc	0·010	0·010	0·015	0·025	60	1/16–1/8	2½°	2°	11°
TRIUMPH 2000 Estate (1966–67)	8° btdc	0·010	0·010	0·015	0·025	60	1/16–1/8	2½°	2°	11°
TRIUMPH 1300 (1966–67)	9° btdc	0·010	0·010	0·015	0·025	70–75	(toe out) 0–1/16	0°	2°	7°
VAUXHALL WYVERN (1956–67)	9° btdc	0·012	0·012	0·014–0·016	0·028–0·030	25–35	1/16–1/8	1°–2½°	¾°–1¾°	2¼°–3¾°
VAUXHALL VELOX and CRESTA (1956–57)	9° btdc	0·012	0·012	0·014–0·016	0·028–0·030	25–35	1/16–1/8	1°–2½°	¾°–1¾°	2½°–2¾
VAUXHALL VICTOR (F) Series I and Series II (1957–61)	9° btdc	0·013	0·013	0·019–0·021	0·028–0·030	25–35	1/8	0°38′[36]	1°30′[36]	4°
VAUXHALL VICTOR (FB) (1961–63)	9° btdc	0·013	0·013	0·019–0·021	0·028–0·030	25–35	1/8–5/32	¼°–1¼°	1°–2½°	5°–6°
VAUXHALL VX 4/90 (1962–63)	9° btdc	0·013	0·013	0·019–0·021	0·028–0·032	25–35	1/8–5/32	¼°–1¼°	1°–2½°	5°–6°
VAUXHALL VX 4/90 (1964–65)	9° btdc	0·013	0·013	0·019–0·021	0·028–0·032	25–35	0·18	¾°[36]	1½°[36]	5½°

37 Estate car 2°58′
38 Estate car 1°40′ laden
39 If Lodge CN plugs fitted 0·025 in.
40 If alloy rocker pedestals fitted 0·010 in.
41 Laden on to gap gauges 3°10′
42 Laden on to gap gauges 3°50′
43 On gap gauges
44 At valve stem tip, with hydraulic lifter collapsed
45 1959 0·025 in.
46 Low compression engine 9° btdc
47 From late 1965 4° btdc
48 70 lb/in² at 70°C at 2000 rev/min
49 For high speed motoring increase to 0·013
50 Low compression engine 0·030 in.

Car and model	Ignition timing	Tappet clearance		Contact-breaker gap (in.)	Spark-plug gap (in.)	Oil pressure (lb/in.²)	Toe in (in.)	Camber	Castor	King pin inc.
		Inlet (in.)	Exhaust (in.)							
VAUXHALL VICTOR (1964–65)	9° btdc	0·013	0·013	0·019–0·021	0·028–0·032	25–35	0·18	¼°[36]	1¼°[36]	5¼°
VAUXHALL VELOX and CRESTA (PA) (1958–60)	9° btdc[51]	0·013	0·013	0·019–0·021	0·028–0·030	25–35	0·125	0°38'[36]	2°[36]	4°
VAUXHALL VELOX and CRESTA SPA (1961–62)	9° btdc[52]	0·013	0·013	0·019–0·021	0·028–0·032	25–35	⅛	0°38'[36]	2°[36]	4°

[51] 3°–4° btdc low compression engines
[52] tdc from 1961

APPENDIX 5

Apart from the motor manufacturers, there are many firms which are able to supply high quality spare parts and kits of parts for motor cars. For instance, Quinton Hazell supply water pumps, and kits for suspension, steering and king pin overhaul for most British and Continental cars. The Armstrong shockabsorber company as well as supplying shockabsorbers. supply suspension struts, and suspension-strut kits for Ford, Peugeot, Porsche, BMW, Hillman, Humber, Singer, Triumph and numerous other cars.

ENGINE OVERHAUL KITS

Supplied by Associated Engineering Co.

AES Kit No.	Model	Bore Size (Standard)
B1	B.M.C. Mini 848cc 8·3:1 cr	2·478 in.
B3	Morris 1000, Austin A35, A40 Farina, 948cc 8·3:1 cr	2·478 in.
B6	Morris Oxford IV & V, Austin A55 (A50 hc), 1498cc 8·2:1 cr	2⅞ in.
B8	B.M.C. 1100, 1098cc 8·5:1 cr	2·542 in.
F22	Ford 100E Anglia, Prefect & Popular, 1172cc (sv)	2⅛ in.
F23	Ford Anglia 105E, Prefect 107E, 997cc	3 5⁄16 in.
F29	Ford Cortina 113E, 1198cc	3 1⁄16 in.
F30	Ford Cortina 116E, 1499cc	3 3⁄16 in.
F31	Ford Zephyr 4 Mk III, 1703cc	3¼ in.
H42	Hillman Minx VIII, 1390cc	3 in.
H44	Hillman Super Minx 1600, 1600cc	3·210 in.
H45	Hillman Imp, 875cc	2·676 in.
S54	Triumph Herald 1200, 1147cc	2·7279 in.
V65	Vauxhall Victor, 1507cc	3⅛ in.
V66	Vauxhall Viva, 1057cc	2·925 in.
V67	Vauxhall Victor 101, 1594cc 9·1 cr	3·214 in.

Supplied by G.M.A. Reconsets

AUSTIN

A30 Saloon, Countryman, 5 cwt. Van	803cc	1952/54
A35 Saloon, Estate, 5 cwt. Van	948cc	1955/59
Mini Seven Saloon, Countryman, Van	848cc	1959/65
A40 Farina Mk I	948cc	1959/61
A40 Farina Mk II, Austin 1100	1098cc	1963 on
A40 Cambridge	1200cc	1954/56
A40 Somerset, Countryman, Van	1200cc	1951/54
A50 A55 Cambridge, Farina	1489cc	1955/60
A60 Cambridge	1600cc	1961 on
A60 Cambridge 1800	1795cc	1965/67

FORD

Popular, Anglia, Prefect, Van (93a eng.)	1172cc	1950/53
Anglia, Prefect, Escort, Squire, Popular (1961), Van (All 100E eng.)	1172cc	1954/59

Note 93a and 100E eng., exchange rods available. Main shells included in kit.

105E, Anglia, Prefect, 106E, 107E	997cc	1959/61
Anglia Super, Consul Classic, Cortina, Corsair	997cc	1962/64
Cortina, Anglia, Super 113E	1198cc	1963/66
Consul Mk I	1507cc	1952/56
Consul Mk II, Zephyr '4' Mk III	1703cc	1956/65
Zephyr, Zodiac Mk I, 6 cyl.	2553cc	1951/56
Zephyr, Zodiac Mk II, Mk III	2553cc	1956/65

HEALEY

Sprite, Mk I and II	948cc	1958/61
Sprite, Mk III	1098cc	1962/64

HILLMAN

Minx Series ohv VIII, Husky I, II and III	1390cc	1954/58
Minx Series IIIA, IIIB, III	1494cc	1959/61
Minx 1600, Series IIIC and V Super Minx	1592cc	1961/65

M.G.

Midget	948cc	1961/62
Midget Mk II	1098cc	1962/65
M.G. 1100	1098cc	1962/65
Magnette ZA, ZB, Mk III (B.M.C. 'B') Farina	1489cc	1954/61
Magnette Mk IV, M.G.A. 1600	1622cc	1961/65

MORRIS

Minor Series II, Countryman, ¼-ton Van	803cc	1952/56
Morris Minor 1000, Estate, Van	948cc	1956/61
Mini Minor, Traveller, Van	848cc	1959/65
Minor 1000, ADO 16 Saloon, Estate	1098cc	1962/65
Morris 1100	1098cc	1963 on
Cowley	1200cc	1954/56
Oxford Series, II, III, IV, V, Cowley 1500 Traveller	1489cc	1954/61
Oxford Series, VI	1622cc	1961/65
1800	1795cc	1966/67

RILEY

Elf (B.M.C. 'A' type eng.)	848cc	1962/63
Elf Mk II	998cc	1963 on
One-Point-Five, 4/68	1489cc	1957/61
4/72	1622cc	1961/65

SINGER

Gazelle Series II, IIA, III, IIIA, IIIB	1494cc	1961/65
Gazelle Series IIIC, Vogue Mk II and III	1592cc	1961 on

STANDARD

Eight, ohv	803cc	1954/59
Ten, Pennant, Companion, Estate, Van	948cc	1954/61
Vanguard, Series I and II	2088cc	1952/55
Vanguard Phase III, Sportsman, Vignale	2088cc	1956/61

SUNBEAM

Rapier	1390cc	1956/58
Rapier Series II, III, IIIA, Alpine I	1494cc	1958/61
Rapier Series IIIA (from Ch. B3062665) IV	1592cc	1962/65

TRIUMPH

Herald Saloon, 'S'	948cc	1959/61
Herald Coupe, 'S'	948cc	1959/63
Herald 1200, 5 cwt. Courier Van	1147cc	1961/63
Vitesse Six	1596cc	1962/65
12/50, Spitfire Sports	1147cc	1962/65
TR3, TR4	1991cc	1953/61

VAUXHALL

Wyvern and Bedford eng.	1507cc	1952/57
Viva	1057cc	1963/65
Victor Mks I, II and III (Bedford Victor eng.)	1507cc	1958/62
Victor VX4/90, FXR	1594cc	1962/65
Velox, Cresta, 6 cyl. PA. II, III	2275cc	1952/59
Cresta, Velox IV	2651cc	1961/65

WOLSELEY

Hornet ohv B.M.C. 'A' eng.	848cc	1962
Hornet	998cc	1964 on
15/50, 15/60, 1500	1489cc	1957/64
16/60 Farina	1622cc	1962/63

A high performance kit is available for the following cars:

948cc ENGINES

Austin A35.	Austin ¼-ton Van.
Austin A35 Estate.	Morris Minor 1000.
Austin A40 Farina.	Morris Minor Estate.
Austin Healey Sprite Mk 1.	

Addresses

Associated Engineering Sales Ltd, 123 Mortlake High St, London SW14.
G.M.A. Reconsets Ltd, 119 Uxbridge Rd, London W12.
Quinton Hazell Ltd, Hazell House, Blackdown, Nr Leamington Spa, Warwicks

APPENDIX 6

SU Carburetter Jet Needle Recommendations

Car model		Year	Rich	Std	Weak	Spring Colour
A.C.						
16·56-h.p.	6-cyl.	1952	I	DW	D4	Blue
ALVIS						
	TA21/TC21 3-litre	1950/55	L	ES		Red
	TB21 3-litre Sports	1952		UA		Red
	TC21/100	1954		CE		Red
	TD21	1959/63		TA		Red
	TD21	1963/64		KA		Red
3-litre	6-cyl. TF21 Series IV	1965/66		SC	KA	Red
ASTON MARTIN						
2·5-litre	6-cyl. DB2.................................	1950/53		GB		
2·5-litre	6-cyl. Vantage	1951/54		RL		Yellow
2·5-litre	6-cyl. DB2/4 (Mk I)	1953/54		RJ		Yellow
3-litre	6-cyl. DB2/4 (Mk I)	1954/55		SV		Yellow
3-litre	DB2/4 (Mk II).............................	1955/56		SV		Yellow
3-litre	6-cyl. DB (Mk III)	1957/59		SV		Yellow
3-litre	DB Special (Mk III)	1958/59		SV		Red
3·7-litre	6-cyl. DB4.................................	1958		UJ		Red/Blue
3·7-litre	6-cyl. DB4.................................	1960		UJ		Red/Blue
3·7-litre	6-cyl. DB4	1961/62		UN		Red/Blue
3·7-litre	6-cyl. DB4 Special	1961/62		UP		Blue/Black
3·7-litre	6-cyl. DB5 (ptfe bushes)	1962/64		UX		Red/Green
4-litre	6-cyl. DB6.................................	1965/66		UX		Red/Green
AUSTIN						
	4-cyl. Austin-Healey 100	1953/56	QA	QW	AT	Yellow
	4-cyl. Austin-Healey Le Mans	1954/56	OA6	OA7	OA8	Red
	4-cyl. Austin-Healey 100S	1955	KW	KWI	SA	Red
	6-cyl. Austin-Healey	1957	4	AJ	MI	Red
	6-cyl. Princess DM4	1957/61	RD	CV	SQ	Red
2·6-litre	6-cyl. Austin 105	1957	3	V2	AC	Red
	4-cyl. Sprite	1959	EB	GG	MOW	
	6-cyl. Austin-Healey BN6 3000 (Mk I)	1959	RD	CV	SQ	Yellow
	6-cyl. A99	1959/61		M5	HA	Yellow
	Seven....................................	1959	M	EB	GG	Red
	Austin-Healey BN7	1959	RD	CV	SQ	Green
	4-cyl. A55	1959	AH2	M	EB	Red
	Austin-Healey BN7	1959	RD	CV	SQ	Green
	Austin-Healey BN7 (RC)	1960	RD	CV	SQ	Green
	Austin-Healey BN7 (Mk II)	1961/62	DK	DJ	DH	Red
	Seven and Super	1961/62	M	EB	GG	Red
997cc	4-cyl. Mini Cooper	1961/62	AH2	GZ	EB	Red
1622cc	A60......................................	1961/64	M	GX	GG	Yellow
	4-cyl. A40 (Mk II)	1961/62	AH2	M	EB	Red
950cc	4-cyl. Healey Sprite (Mk II)	1961/62	V2	V3	GX	Blue
	6-cyl. A110 Westminster hc and lc	1961/63	3	AR	HA	Yellow
1098cc	4-cyl. Healey Sprite (Mk II)	1962/63	M	GY	GG	Blue
1098cc	4-cyl. Healey Sprite (Mk III)	1963/64	H6	AN	GG	Blue
	6-cyl. Healey 3000 (Mk II)................	1962/63	RD	BC	TZ	Green
	4-cyl. A35 Van	1962/63	H6	AN	EB	Red
1098cc	4-cyl. A40 and Austin 1100	1962/67	H6	AN	EB	Red
1070cc	4-cyl. Mini-Cooper 'S'	1963/64	3	H6	EB	Red
	6-cyl. Healey BJ8	1964	UN	UH	UL	Red/Green
1275cc	4-cyl. Mini-Cooper	1964/67	AH2	M	EB	Red
970cc	4-cyl. Mini-Cooper 'S'	1964	H6	AN	EB	Red
998cc	4-cyl. Mini-Cooper	1964/67	M	GY	GG	Blue
1800cc	Austin 1800..............................	1964	SW	TW	C1W	Yellow
850cc	A35 Van	1965/67	M	EB	GG	Red
850cc	Mini Automatic	1965/67	H6	AN	EB	Red
1098cc	1100 Automatic	1965/67	BQ	DL	ED	Red
1800cc	1800.....................................	1966/67	SW	TW	C1W	Yellow
850cc	Mini Automatic	1967/68	H6	AN	EB	Red
1098cc	1100 Automatic (Mk II)	1967/68	BQ	DL	ED	Red
2912cc	A110 Westminster........................	1967	3	AR	HA	Yellow
1098cc	4-cyl. 1100 (Mk II)	1967/68	H6	AN	EB	Red
998cc	4-cyl. Mini (Mk II)	1967/68	M	GX	GG	Red
998cc	4-cyl. Mini (Mk II) Automatic	1967/68	M1	AC	HA	Red
1275cc	4-cyl. 1300..............................	1967/68	BQ	DZ	CF	Red
1275cc	4-cyl. 1300 Automatic	1967/68	BQ	DZ	CF	Red
2912cc	6-cyl. 3-litre	1967/68	TU	C1	C1W	Yellow
1275cc	4-cyl. Austin-Healey Sprite (Mk III)	1967/68	H6	AN	GG	Blue
1275cc	4-cyl. Austin-Healey Sprite (Mk III) (American market only)	1968		AN		Blue
1275cc	American Automatic E.P.A.I.	1968		DZ		Red
CONVERSION SETS						
	M.G.—Elva	1959/61		GS		Red
	B.M.C. 'A' Series—Turner	1959/61		BXI		
	Minor 1000—Speedwell	1959/61		M8		Blue
	B.M.C. 'A' Series—Turner	1959/61		M6		Red
	Sprite—Sebring	1960		GX		Blue
	F2 Cooper Climax S/C	1960				
	B.M.C.—FJ Cooper	1960		AM		Blue
	Alexander Herald	1960/61		M6		Blue
	Sprite	1960		A5		Blue
	Mangoletsi Remix........................	1961/63		M8		

Car model		Year	Needle			Spring Colour
			Rich	Std	Weak	
CONVERSION SETS (cont.)						
	Healey 3000 Competition	1961		UH		Blue/Black
	Mini-Cooper (Thermo jets)	1961/63		MME		Blue
	Sprite—Speedwell	1962/63		AO		Red
	Mini Competition	1962/63		MME		Blue
970cc	4-cyl. Mini-Cooper 'S' Group II	1964/68		CP4		Blue
1070cc	4-cyl. Mini-Cooper 'S' Group II	1964/68		MME		Blue
1275cc	4-cyl. Mini-Cooper 'S' Group II	1964/68		BG		Blue
	Formula III B.M.C.-Cooper	1964/68		SS		Red
1098cc	4-cyl. Morris 1100 (Downton)	1964/68		AM		Blue
COVENTRY CLIMAX						
1100cc		1954/58	R6	BE	6	Blue
	F.W.A. Stage I			BE		Blue
	F.W.A. Stage II			BF		Blue
1220cc	Lotus Elite			BQ		Blue
1220cc	Lotus Elite			BF		Yellow
1500cc	F.P.F.			ZB		
DAIMLER						
2½-litre	Conquest Century	1955/57	SC	TC	TJ	Red
3½-litre	Regency	1956/57		SQ		Red
3½-litre	Regency	1958/59		SQ		Red
4½-litre	DK400 Majestic	1956/57		UB		Red/Green
	SP250 Sports	1959		TS	TR	Yellow
	SP250 Sports	1960/61		TS	TR	Yellow
	Majestic Major and Majestic	1960/62		UL		Red/Green
2½-litre	Saloon	1962/63		CI		Red
	Majestic Major and Majestic (ptfe bushes)	1964		UL		Red/Green
2½-litre	V8 Saloon	1964/68		TZ		Red
4½-litre	V8 Majestic Major	1964/68		UL		Red/Green
FORD (Conversions)						
1172cc	E93A	1949/53	M9	EK	MOW	
1172cc	100E Aquasport	1953/57	MI	A5	HA	Red
1172cc	100E Prefect and Anglia	1953		M6		
1172cc	100E Lotus	1954/60	M5	M6	M7	Red
	Consul—Aquaplane (Series I)	1954/57	4	3	L	Red
	Consul—W.H.M.B. Ltd.	1955/57	H2	QA	QW	Red
	Zephyr—Aquaplane (Series I)	1954/57	4	3	L	Red centre, Yellow front and rear
	Zephyr—W.H.M.B. Ltd.	1955/57	EM	ES	AP	Red
	Zephyr—Raymond Mays (Series I)	1954/56	CN	5	GE	Yellow
	Zephyr—Raymond Mays (Series II)	1957/62	MME	7	AO	Yellow
	Consul—R. Owen (Series II) 4 port head	1958/60		RB		Red
	Consul—R. Owen (Series II) 6 port head	1958/60		RB		Red
	105E FJ	1960/62		AM		Blue
	100E Aquaplane	1960/62		GX		Blue
	105E/107E Aquaplane	1960/62		A5		Blue
	Consul—R. Owen (Series II) 4 port head	1962		RB		Red
	Zephyr—Raymond Mays	1962		AY		Yellow
	E93A—Dellow	1950		RLS		Red
	Consul (Series I)—Dellow	1953		M5		Red
	Zephyr (Series I)—Dellow	1954		M5		Red
	Prefect and Anglia—100E—Dellow	1959	M9	EK	MOW	
30 hp	V8 (Special adaptor)	1950	RO	6		Red
	Consul—Series I	1952		61		Yellow
	Consul—Series I	1953		62		Yellow
	Zephyr—Series I	1953		WX		Yellow
30 hp	V8 Allard and Jensen	1937		6		Brown
	Lotus 105E	1961/62		A5		Blue
	Turner 109E	1961/62		DJ		Red
	Reliant Ford	1962/63		CZ		Red
	Reliant Zephyr	1963/64		DH		Red
HILLMAN						
	Imp	1964		H4		Blue
1600cc	Minx	1964		QA		Red
HILLMAN (Conversions)						
1390cc	Minx	1956/58	CU	CZ	CF	Blue
	Alexander Minx	1959/61		GR		Blue
INNOCENTI						
1098cc	1M3	1963/64	D6	D3	GV	Blue
1098cc	1100	1964	D6	D3	GV	Blue
1098cc	1M4	1964	H6	AN	EB	Red
850cc	4-cyl. Mini-Minor	1965/66	M	EB	GG	Red
JAGUAR						
3½-litre	Mk V	1948/50	FL	FW	62	
	XK120	1949/50		RB		Red
	Mk VII 7:1 and 8:1 cr to Eng. No. B2917	1951/52	RH	SM	SK	Red
	Mk VII 7:1 and 8:1 cr from Eng. No. B2918	1952/54	SM	SR	CIW	Red
	XK120	1951/54	53	RF	RG	Red
	XK120 (remote air cleaner)	1951/54	WO4	WO2	WO3	Red
	XK120 7:1 and 8:1 cr 'C' type	1952	75	VR	VE	Black/Red
	XK120 8:1 cr 'C' type	1952		RG		Red
	XK120 8:1 cr (remote air cleaner)	1952		DG		Red
	XK120 9:1 cr 'C' type	1952		RC		Red
	Mk VII 7:1 and 8:1 cr 'M' type	1954/56	SM	SR	CIW	Red
	Mk VII 7:1 cr ('C' type head), A.C. disc cleaner	1954/56		SL		Red
	Mk VII 8:1 cr ('C' type head), A.C. disc cleaner	1954/56		SJ		Red

Car model	Year	Needle Rich	Needle Std	Needle Weak	Spring Colour
JAGUAR 3½ litre (cont.)					
Mk VII 8:1 and 9:1 cr ('C' type head)	1954/56	75	VR	VE	Black/Red
XK140 7:1 and 8:1 cr.	1954	SA	SJ	LBA	Red
XK140C 7:1 and 8:1 cr ('C' type head)	1954		SR		Red
XK140 7:1 and 8:1 cr ('C' type head), disc cleaners	1954		WO2		Red
XK140C 8:1 and 9:1 cr ('C' type head)	1954	75	VR	VE	Black/Red
XK140C 7:1 and 8:1 cr D/H Coupé and Standard	1955		WO2		Red
XK140C 7:1 and 8:1 cr R.H.D. F/H Coupé	1955		WO2		Red
XK140C 7:1 and 8:1 cr L.H.D. F/H Coupé	1955		WO2		Red
XK140 7:1 and 8:1 cr L.H.D. F/H Coupé	1955	SA	SJ	LBA	Red
XK140 7:1 and 8:1 cr R.H.D. F/H Coupé	1955	SA	SJ	LBA	Red
XK140 7:1 and 8:1 cr Borg-Warner	1955	SA	SJ	LBA	Red
XK140 7:1 and 8:1 cr Borg-Warner R.H.D. F/H Coupé	1956	SA	SJ	LBA	Red
XK140 7:1 and 8:1 cr Borg-Warner L.H.D. D/H Coupé	1956	SA	SJ	LBA	Red
XK140 7:1 and 8:1 cr Borg-Warner R.H.D. D/H Coupé	1956	SA	SJ	LBA	Red
2·4 litre — Mk I 8:1 cr Stage III tune	1956		TO		Red
3·4 litre — Mk I standard trans. or O/D	1957	WO3	TL	SJ	Red
3·4-litre — Mk I standard trans. or O/D (oil-bath cleaner)	1957		SC		Red
3·4-litre — Borg-Warner	1957	WO3	TL	SJ	Red
3·4-litre — Borg-Warner (oil-bath cleaner)	1957		SC		Red
Mk VIII standard trans. or O/D	1956/57	WO3	TL	SJ	Red
Mk VIII Borg-Warner	1956/57	WO3	TL	SJ	Red
Mk VIII standard trans. or O/D	1958	WO3	TL	SJ	Red
Mk VIII Borg-Warner	1958	WO3	TL	SJ	Red
3·4-litre — Standard trans. or O/D	1958	WO3	TL	SJ	Red
3·4-litre — Mk I Borg-Warner	1958	WO3	TL	SJ	Red
3·4-litre — XK150S	1959/62		UE		Blue/Black
3·4-litre — Mk I	1959		WO3	SJ	Red
3·4-litre — Mk I (oil-bath cleaner)	1959		SC		Red
Mk VIII	1959	WO3	TL	SJ	Red
Mk IX	1959		TU		Red
XK150	1959	WO3	TL	SJ	Red
3·4 and 3·8 litre — Mk II	1960/61		SC		Red
Mk IX 7:1 and 8:1 cr	1960/62		TU		Red
3·4-litre — XK150	1960/62	WO3	TL	SJ	Red
3·8-litre — XK150	1960/62		TU		Red
3·8-litre — Mk II 7:1 cr	1960/62		TM		Red
3·8-litre — Mk II 8:1 and 9:1 cr	1960/62		C1		Red
Mk IX 7:1 cr (Cooper paper cleaner)	1960/62		TM		Red
Mk IX 8:1 and 9:1 cr (Cooper paper cleaner)	1960/62		TU		Red
3·4-litre — Mk II 7:1 cr (Cooper paper cleaner)	1960/63		TM		Red
3·4-litre — Mk II 8:1 and 9:1 cr (Cooper paper cleaner)	1960/63		TU		Red
3·8-litre — 'E' type	1961/63		UM		Blue/Black
3·8-litre — Mk X 8:1 and 9:1 cr	1961/63		UM		Blue/Black
3·8- and 3·4-litre — Mk III	1963/64		TL		Red
3·8-litre — Mk X	1963/64		UM		Blue/Black
3·8-litre — 'E' type	1963/64		UM		Blue/Black
3·8- and 4·2-litre — Mk X 8:1 and 9:1 cr	1964		UM		Blue/Black
3·8-litre — 'S' type Mk III 8:1 and 9:1 cr (paper cleaner)	1964		TL		Red
3·8-litre — 'S' type Mk III 8:1 and 9:1 cr (oil-bath cleaner)	1964		C1		Red
3·8-litre — 7:1 cr (Cooper paper cleaner)	1964		TM		Red
3·8- and 4·2-litre — Mk X Automatic and O/D	1964		UM		Blue/Black
3·8- and 4·2-litre — Mk X Standard transmission	1964		UM		Blue/Black
2·4-litre — 6-cyl. '240' Manual	1967/68		TL		Red
2·4-litre — 6-cyl. '240' Automatic	1967/68		TL		Red
3·4-litre — 6-cyl. '340' Manual 7:1 cr and Automatic (A.C. paper cleaner)	1967/68		TM		Red
3·4-litre — 6-cyl. '340' Manual and Automatic 8:1 and 9:1 cr (A.C. paper cleaner)	1967/68		TL		Red
3·4- and 3·8-litre — 6-cyl. 'S' type Manual and Auto. 8:1 and 9:1 cr (A.C. paper cleaner)	1967/68		TL		Red
4·2-litre — 6-cyl. '420' Manual 8:1 and 9:1 cr (A.C. paper cleaner)	1967/68		UM		Blue/Black
4·2-litre — 6-cyl. '420' Automatic 8:1 and 9:1 cr (A.C. paper cleaner)	1967/68		UM		Blue/Black
4·2-litre — 6-cyl. '420G' Manual 8:1 and 9:1 cr (A.C. paper cleaner)	1967/68		UM		Blue/Black
4·2-litre — 6-cyl. '420G' Automatic 8:1 and 9:1 cr (A.C. paper cleaner)	1967/68		UM		Blue/Black
4·2-litre — 6-cyl. 'E' type 8:1 and 9:1 cr	1967/68		UM		Blue/Black
JENSEN					
Jensen—Ford V8	1937	CU	6		Brown
541/541R	1954/60		CS	CF	Red
541S	1960/62		CF		Red
M.G.					
1250cc — TC	1939/50	EM	ES	AP	
1250cc — Y Saloon	1947/50	DK	F1	EF	
1250cc — Y Tourer	1948	EM	ES	AP	
1250cc — TD	1950/53	EM	ES	AP	
1250cc — TC Competition	1949	AQ	LS1		Red
1250cc — TC Competition S/C	1949		RM7		Yellow
1250cc — TD Competition	1951	AQ	LS1		Red
1466cc — TD Competition	1953	GK	CV	BC	Blue
1250cc, 1500cc — TF	1954/55	H1	GJ	GL	Blue
ZA Magnette	1954	M	GM	GO	Red
ZA Magnette	1954	M	GM	GO	Red
1500cc — A	1955/59	CC	GS	4	Red
ZA Magnette	1954/55	M	GM	GO	Red
ZA/ZB Magnette	1956/57		EQ	M5	Red
ZB Magnette	1958		EQ	M5	Red

		Car model	Year	Needle Rich	Needle Std	Needle Weak	Spring Colour
M.G. (cont.)							
1588cc		Twin cam	1958	RH	OA6	OA7	Red
		Magnette III	1959	FT	FU	M9	Red
		Magnette III	1960/61	FT	FU	M9	Red
1588cc		'A' (Mks I and II)	1959/62	RO	6	AO	Red
1622cc		Magnette (Mk IV)	1961/63	FU	HB	FK	Red
948cc		Midget	1961/62	V2	V3	GX	Blue
		1100	1962/68	D6	D3	GV	Blue
1098cc		Midget	1962/63	M	GY	GG	Blue
1800cc		'M.G.B'	1962/63	6	MB	21	Red
		'M.G.B' Competition	1963/64		UVD		Blue/Black
		'M.G.B' and GT	1966	6	5	21	Red
1098cc		Midget Mk II	1964	H6	AN	GG	Blue
1800cc		'M.G.B' and GT GHN4, GHD4	1967/68	5	FX	GZ	Red
1275cc		M.G.	1967	BQ	DZ	CF	Red
1275cc	4-cyl.	M.G. Automatic	1967/68	BQ	DZ	CF	Red
1275cc	4-cyl.	M.G. Midget (Mk III)	1967/68	H6	AN	GG	Blue
2912cc	6-cyl.	'M.G.C'	1967/68	SQ	ST	C1W	Yellow
1275cc	4-cyl.	Sedan (American Market only)	1967/68		DZ		Red
1275cc	4-cyl.	Midget (Mk III) (American market only)	1968		AN		Blue
1800cc	4-cyl.	'M.G.B' (American market only)	1968		FX		Red
2912cc	6-cyl.	'M.G.C' (American market only)	1968		KM		Yellow
1275cc	4-cyl.	Sedan (American market only) Automatic	1968		DZ		Red
MORRIS							
8 hp		Minor MM	1948/50	M9	EK	MOW	
		Oxford MO	1948/49	ES	FP	HB	Red
	6-cyl.	MS	1948/50	AA	81	VS	Red
		Oxford MO	1950/54	3	V2	V3	Red
	6-cyl.	MS	1951/54	AA	81	VS	Red
		Minor MM	1951/52	M9	EK	MOW	
		Minor (Series II) ohv	1953	EB	GG	MOW	
		Minor (Series II) ohv	1954/56	EB	GG	MOW	
		Oxford (Series II) (BP15M)	1954	M6	M8	MO	Green
		Hindustan (India only)	1954		V3		Red
		Cowley (BP12M)	1954/56	M6	M8	MO	Red
	6-cyl.	Isis (C26M)	1954/57	BE	7	AO	Green
		Oxford (Series II)	1955/56	M6	M8	MO	Green
		Oxford (Series III) and Cowley, lc	1957	HA	EB	GG	Green
		Oxford (Series III) and Cowley, hc	1957	AH2	M	EB	Yellow
		Oxford (Series III) and Cowley, lc (rubber fuel line)	1957	HA	EB	GG	Green
		Oxford (Series III) and Cowley, hc (rubber fuel line)	1957	AH2	M	EB	Yellow
		Minor 1000	1957	S	BX1	MO	Red
		Minor 1000 (paper cleaner)	1957	AH2	M	EB	Red
		Minor 1000 (rubber fuel line)	1957	S	BX1	MO	Red
		Minor 1000 (rubber fuel line, paper cleaner)	1957		M		Red
	6-cyl.	Isis (rubber fuel line)	1957	BE	7	AO	Green
		Hindustan (rubber fuel line)	1957		V3		Red
		Minor 1000 (steel levers)	1958	S	BXI	MO	Red
		Minor 1000 (steel levers, paper cleaner)	1958	AH2	M	EB	Red
		Oxford (steel levers), lc	1958	HA	EB	GG	Green
		Oxford (steel levers), hc	1958	AH2	M	EB	Yellow
		Minor 1000	1959	AH2	M	EB	Red
		Oxford (Series V)	1959/60	AH2	M	EB	Red
		Mini-Minor	1959/62	M	EB	GG	Red
		Mini-Minor	1962	M	EB	GG	Red
		Minor 1000	1960/62	AH2	M	EB	Red
997cc		Mini-Cooper	1961/62	AH2	GZ	EB	Red
		Oxford VI	1961/62	M	GX	GG	Yellow
		1100	1962	H6	AN	EB	Red
1098cc		Minor	1962/63	H6	AN	EB	Red
1070cc		Mini-Cooper 'S'	1963	3	H6	EB	Red
		Hindustan Oxford	1963/64		M		Yellow
1275cc		Mini-Cooper	1964	AH2	M	EB	Red
970cc		Mini-Cooper 'S'	1964	H6	AN	EB	Red
998cc		Mini-Cooper	1964	M	GY	GG	Blue
850cc		Mini Automatic	1965/66	H6	AN	EB	Red
1098cc		1100 Automatic	1965/66	BQ	DL	ED	Red
1800cc		1800	1966	SW	TW	C1W	Yellow
850cc		Mini Automatic	1967	H6	AN	EB	Red
1098cc		1100 Automatic	1967	BQ	DL	ED	Red
1098cc	4-cyl.	1100 (Mk II)	1967/68	H6	AN	EB	Red
998cc	4-cyl.	Mini (Mk II)	1967/68	M	GX	GG	Red
998cc	4-cyl.	Mini (Mk II) Automatic	1967/68	M1	AC	HA	Red
1275cc	4-cyl.	1300	1967/68	BQ	DZ	CF	Red
1275cc	4-cyl.	1300 Automatic	1967/68	BQ	DZ	CF	Red
NOTE.—Replacement for AUC 928				M	GY	GG	Red
Replacement for AUC 944				AH2	M	EB	Red
Replacement for AUC 912				M	EB	GG	Red
Twin-carburetter Sets							
		Minor MM and Series II—Derrington	1948/56	M9	EK	MOW	
		Oxford MO, Series II and III—Derrington	1950/57	CJ	HB	MO	Red
		Minor (Series II)—Alexander engine	1953/56	EB	GG	MOW	
950cc		Minor—Power drive	1957	EB	GG	MOW	
RILEY							
16 hp	4-cyl.	Export	1948/51	CY	EE	EM	
16 hp	4-cyl.	RMF	1952/54	CY	EE	EM	
16 hp	4-cyl.	Export	1952	CY	EE	EM	
12 hp	4-cyl.	High performance AUC 416	1950/51		6I		
12 hp	4-cyl.	High performance 1½-litre post-war	1952	WXI	AK	D6	
12 hp	4-cyl.	RMA2	1955		3		Red
16 hp	4-cyl.	Pathfinder	1955/57	CY	EE	EM	

Capacity	Cyl.	Car model	Year	Needle Rich	Needle Std	Needle Weak	Spring Colour
RILEY (*cont.*)							
16 hp	4-cyl.	Pathfinder (rubber fuel line)	1955/57	CY	EE	EM	Yellow
2·6-litre	6-cyl.		1957/59	FR	AJ	V2	Red
		One-Point-Five	1957/64	AR	AD	HA	Red
		One-Point-Five (LHD)	1957/62	AR	AD	HA	Red
		4/68	1959/60	FT	FU	M9	Red
		4/68	1960/61	FT	FU	M9	Red
848cc		Elf	1961/62	M	EB	GG	Red
1622cc		4/72 Saloon	1961/64	FU	HB	FK	Red
998cc		Elf	1963/64	M	GX	GG	Red
12 hp	4-cyl.	Service replacement for post-war cars			AK		
1098cc		Kestrel	1965/66	D6	D3	GV	Blue
1275cc		Kestrel	1967/68	BQ	DZ	CF	Red
ROVER							
2·1-litre		75	1950/54	GE	FV	CR	Yellow
2-litre		60	1954/59		GI	GS	Green
2·2-litre		75	1954/59		SS	SY	Yellow
2·6-litre		90	1954/55		SS	SY	Yellow
2·6-litre		90	1956/59		SZ	KW	Yellow
2·6-litre		105R/105S	1957/59		TM		Yellow
2·6-litre		100	1959		SS		Yellow
2·6-litre		100 and 95	1960/62		SS		Yellow
3-litre		P5	1958/59		UF		Red/Green
3-litre		P5	1960/62		UF		Red/Green
3-litre Coupé		P5	1963		UR		Red/Green
2·6-litre		P5	1963		UG		Red/Green
2·6-litre		P5 (ptfe bushes)	1963		UG		Red/Green
2·6-litre		110 P4	1963		UG		Red/Green
2·4-litre		P5	1963		UT		Red/Green
2-litre		'2000'	1963/64		RN		Green
3-litre			1963/64		UR		Red/Green
3-litre		NADA (Smith's valve)	1963/64		UR		Red/Green
2-litre		'2000' (Smith's valve)	1963/64		RR		Green
		2000 TC	1966		U1		Black/Blue
2-litre		2000	1967/68		RN		Green
2-litre		2000 (American market only)	1967/68		RR		Green
2-litre		2000 TC	1967/68		AAA		Blue/Black
2-litre		2000 TC (American market only)	1967/68		AAA		Blue/Black
3·5-litre	V8 cyl.		1967/68		KL		Yellow
3·5-litre	V8 cyl.	3500	1968		KO		Yellow
LAND ROVER							
2·6-litre		109 FWD, Forward control	1963		SS		Yellow
2·6-litre		Station Wagon	1967		SS		Yellow
2·6-litre		109 WB (L.H.D.)	1967/68		UG		Red/Green
STANDARD							
9 hp	4-cyl.		1955/56		D3		
10 hp	4-cyl.		1955/56		D3		
2-litre		Vanguard Sportsman	1957	SD	TC	TJ	Red
10 hp		Pennant	1958		MOW		
STANDARD-TRIUMPH							
		TR2	1953/55	GER	FV	CR	Red
		TR3	1956/58	RH	SM	SL	Red
		Herald	1959/61	EB	GV	CA	
		TR3, TR3A and TR4	1959/62	RH	SM	SL	Red
		Spitfire Mk II	1962/66	H6	AN	EB	Red
		TR4A	1965/66	SW	TW	C1W	Red
		Spitfire Group II	1966		DB		Blue
1300cc		Spitfire (Mk III)	1967/68		BO		Red
1300cc		1300 TC Saloon	1967/68		BO		Red
1300cc		Spitfire (Mk III) (American market only)	1968		DD		Red
		TR4A (American market only)	1968		QW		Red
STANDARD-TRIUMPH (Conversions)							
		Vitesse	1963/64		MO		Red
VANDEN PLAS							
3-litre		Princess	1960		M5	HA	Yellow
3-litre		Princess 1c	1961/62		3		Yellow
3-litre		Princess hc and 1c	1961/63	3	AR	HA	Yellow
1098cc		Princess 1100	1964	D6	D3	GV	Blue
1275cc		Princess and Automatic	1967/68	BQ	D2	CF	Red
WOLSELEY							
	4-cyl.	4/50	1951/53	EE	EM	I	Red
	6-cyl.	6/80	1951/53	CJ	HB	MO	Red
	4-cyl.	4/44	1953	VS	FJ	I	
	6-cyl.	6/90	1954/57	3	GR	V2	Red
	4-cyl.	4/44	1954/65		BT		Red
	6-cyl.	6/90 hc	1956/57	CM	BV	3	Red
	6-cyl.	6/90 hc	1957	FJ	I	M5	Red
	4-cyl.	15/50 lc	1956/57	HA	EB	GG	Green
	4-cyl.	15/50 hc	1957	AH2	M	EB	Yellow
	4-cyl.	1500 lc (rubber fuel line)	1957/58	HA	EB		Red
	4-cyl.	1500 lc (rubber fuel line)	1957/58	HA	EB	GG	Green
	4-cyl.	15/50 hc (rubber fuel line)	1957	AH2	M	EB	Yellow
	6-cyl.	6/90 lc (rubber fuel line)	1957/59	3	GR	V2	Red
	6-cyl.	6/90 hc (rubber fuel line)	1957/59	FJ	I	M5	Red
	4-cyl.	15/50 Manumatic, lc	1957/58	HA	EB	GG	Green
	4-cyl.	15/50 Manumatic, hc	1957/58	AH2	M	EB	Yellow

Car model		Year	Needle			Spring Colour
			Rich	Std	Weak	
WOLSELEY (*cont.*)						
4-cyl.	15/50 lc	1958/59	HA	EB	GG	Green
4-cyl.	15/50 hc	1958/59	AH2	M	EB	Yellow
4-cyl.	1500	1959/62	AH2	M	EB	Red
4-cyl.	15/60	1959/61	AH2	M	EB	Red
6-cyl.	6/99	1959/61		M5	HA	Yellow
6-cyl.	6/99	1960/61		M5	HA	Yellow
4-cyl.	Hornet	1961/62	M	EB	GG	Red
4-cyl.	16/60	1961/62	M	GX	GG	Yellow
6-cyl.	6/110 hc and lc	1961/63	3	AR	HA	Yellow
6-cyl.	6/110 lc	1961/62	3			Yellow
4-cyl.	1500	1962/64	M	GY	GG	Red
998cc	Hornet	1963	M	GX	GG	Red
	1100	1965/66	D6	D3	GV	Blue
1800cc	18/85 Automatic	1967	SW	TW	C1W	Yellow
2192cc	6/110	1967	3	AR	HA	Yellow
1275cc	Wolseley	1967/68	BQ	DZ	CF	Red
1275cc	Wolseley Automatic	1967/68	BQ	DZ	CF	Red
NOTE.—Replacement for AUC 928			M	GY	GG	Red
Replacement for AUC 929			M	GY	GG	Red
Replacement for AUD 27		1961/64	3	AR	HA	Yellow

SU Fuel Pumps—Identification of Basic Types
(Earlier Models)

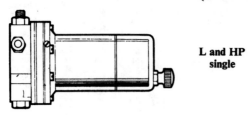

L and HP single

High- or low-pressure. Single: 6-, 12-, or 24-volt. Minimum flow, HP—7 gal. per hour, L—8 gal. per hour. Valves in outlet connection, plain disc, outlet valve in cage. Filter at bottom. Outlet connection at top. Sandwich plate and gasket between diaphragm and body.

Earlier HP coil housing $\frac{3}{16}$ in. longer than L, later models same external length as L.

Large capacity. Single: 12- or 24-volt. Minimum flow, 12½ gal. per hour. Valves inside top cover, outlet valve in cage, earlier valves both plain disc, later inlet plain disc with spring, outlet plastic assembly. Filter inside bottom cover.

LCS

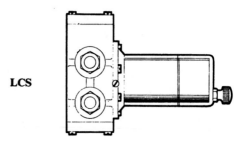

High- or low-pressure. Dual 12-volt, working simultaneously. Minimum flow, HP—16 gal. per hour, L—20 gal. per hour. Valves under top caps, outlet valves in cage plain disc. Filter at bottom.

L and HP dual

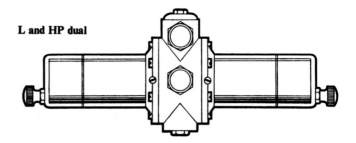

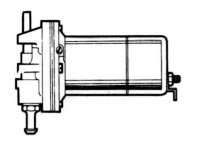

SP

High-pressure. Single: 12-volt. Minimum flow, 7 gal. per hour. Valves inside body, plastic type held by retainer plate and single screw. Filter in inlet connection.

SU Fuel Pumps—Identification of Basic Types
(Later Models)

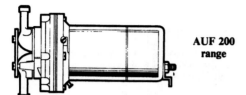

AUF 200 range

High-pressure. Single: 12-volt. Minimum flow, 7 gal. per hour. Valves accessible externally through inlet and outlet nozzles, plastic type all held by circular clamp plate and two screws. Filter under inlet nozzle.

High-pressure. Single- 12-volt. Minimum flow, 15 gal. per hour. Valves inside body, plastic type held by clamp plate and two screws. Filter on inlet valve, plain air bottle on inlet, flow-smoothing device on delivery.

AUF 300 range

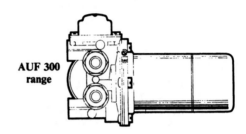

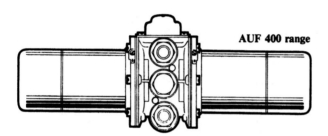

AUF 400 range

High-pressure. Dual working simultaneously. Single inlet, single outlet. 12-volt. Minimum flow from both—30 gal. per hour. Valves inside body, plastic type held by clamp plate and two screws. Filter on inlet valve. Plain air bottle on inlet, flow-smoothing device on delivery.

High-pressure. Dual normally working separately. Dual inlet, single outlet. 12-volt. Minimum flow, 12½ gal. per hour each. Valves inside body, plastic type held by clamp plate and two screws. Filter on inlet valve. Plain air bottle on inlet, diaphragm-type on delivery.

AUF 500 range

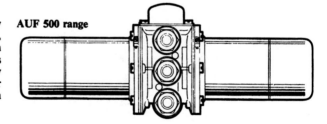

Pump voltage

All S.U. pumps can be identified for voltage by the marking and colour of the end-cover, thus:

6-volt—Brown
12-volt—Black
24-volt—Blue.

Earlier-type pumps

L. HP, LCS, Dual L, and HP and SP pumps are recorded under specification numbers, having the prefix letters AUA or AUB.

All earlier-type pumps in regular production bear this specification number stamped on a metal plate secured under two of the coil housing screws. Any new part fitted during reconditioning must be in accordance with the particular specification.

Externally the earlier HP pump coil housings were $\frac{3}{16}$ in. longer than the L type. All HP pumps are now fitted with the shortened coil housing of the same overall length as that of the L type.

Production of the long coil housing has ceased but it is still used in reconditioning when a coil, similar to that used in short housing pumps, but wound on the longer core, is used. Long and short versions of the pump are interchangeable, but because components differ, and may be required as spares, such pumps carry a prefix to the specification number, thus:

AUA 50 long housing pump, built with short-coil housing, becomes AUA 150.

AUA 52 long housing pump, built with short-coil housing, becomes AUA 152.

As the pumps are functionally identical, in service the non-prefixed pumps may be replaced by prefixed pumps, e.g. AUA 50 by AUA 150 and vice versa.

Later-type pumps

The AUF type of pump has been introduced to provide a simplified range using standardized parts where possible, resulting in simpler servicing. These pumps differ from earlier types mainly in the design of body and valves used.

Pumps in the AUF range are always referred to by their specification number, e.g. AUF 201, and not by the previous type to which they correspond thus:

AUF 200 to 299 corresponds to HP and SP types.

AUF 300 to 399 corresponds to LCS type.

AUF 400 to 499 corresponds to Dual HP type.

AUF 500 to 599 a new double-entry fuel pump.

As with the earlier-type pumps, the AUF range bear the specification number stamped on a metal plate secured under the coil housing screws. Again, any new part fitted during reconditioning must be in accordance with the particular specification. When the specification of a pump in the AUF range is altered by the addition of specially vented end-covers and coil housings, and with diaphragms of new plastic material protected by thin nylon barrier diaphragms etc., then the modified pump is allotted another number, e.g. AUF 200 becomes AUF 201.

Identification of AC Fuel Pump Types

No.	Pump Type	Principal Vehicle Coverage
BD-1	T, TF TG	Austin 1935–48, All; Bedford 1935–44, All bcyl.; Ford 1938–44, All V8; Hillman 1935–48, Minx; Vauxhall 1937–40, All bcyl.
BD-2	U, UE, UF, UG, UH, FE	Austin 1949–60, All exc. A30, A35 A40 Farina and Mini Gipsy Petrol; Ford 1945–56, V8; 1962, Zephyr 4 & 6 Zodiac III K & M Series with Air Brakes; Hillman 1955–66, Minx & Husky; 1966–68, Hunter, Minx 1500; Humber 1945–68, All; Rover 1959–61, Model 80; 1955–67 Land-rover, Petrol; Sunbeam Rapier & Alpine; Singer Gazelle & Vogue; Standard Vanguard & Ensign; Triumph TR's, Vitesse 2000, GT6 & TR5-P1; Vauxhall Models E, PA, PB, PC, FB, FC, FD, FDH, HBR & VX4/90
BD-3	Y	Austin 1952–55, A30; Ford 1954–55, Anglia & Prefect; Hillman 1948–56, Minx (SV). Husky (SV); Standard 1953–55, 8 & 10 h.p.; Vauxhall 1948–53, 4 cyl.
BD-4	YD, YJ	Austin 1955–61, A30, A35, A40; DAF 600 & 750 c.c.; Ford 1956–59, Anglia & Popular; 1959–61, New Anglia; Hillman 1956, Minx (SV) & Imp, 1963–on; Singer 1964–68 Chamois & Sport; Standard 1956–60, 8 & 10 h.p.; Sunbeam 1966–68, Imp Sport, Stiletto; Triumph 1959–68, Herald, 1200, 12/50, 1300, 13/60, Spitfire; Vauxhall 1964–68, Viva & Epic, HA, HB
BD-5	A, B	1929–34, Austin, Bedford, Hillman, Standard Vauxhall; Ford 1962–68, Anglia, Cortina, Corsair 1968, Escort
BD-7	WE (Vacuum)	Bedford Model S, Mar. to Aug., 1952; Ford Consul & Zephyr, to Oct. 1952
BD-8	WE, WG W (Vacuum)	Bedford 1953–56, 6 cyl.; Ford 1952–61, Consul, Zephyr
BD-9	Various	Special Export
BD-10	W, WG (Vacuum)	Holden Model FE, after Eng. L327358; Ford 1959–62, Prefect (107E)
BD-12	CG	Holden Model FB, after initial production
BD-13		Holden Model FB, early production
D-53	WG (Vacuum)	Holden Models 48, 50 & FJ FE to Eng. L327358
D-54	W, WG (Fuel)	Ford 1952–61, Consul, Zephyr, Zodiac; Bedford 1952–56, 6 cyl.
BD-20	CG	Ford Corsair V4 & 2000 Zephyr, Zodiac, Mk IV, Transit Van

Identification of AC Fuel Pump Types

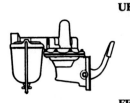

UE Standard type of pump for engines over 1,200 c.c. Detachable valve assemblies and visible side filter. Mounting stud holes arranged diagonally.

FE As UE but mounting holes side by side.

T Early type standard size with separate valves and pump. Mounting stud holes arranged diagonally.

TG As type T but glass filter bowl on top.

TF As type T but mounting holes side by side.

UH Standard type of pump without filter. Mounting stud holes arranged diagonally.

FH As UH but mounting holes side by side.

Y Small size pump for engines up to 1,200 c.c. Mounting stud holes side by side.

YD As type Y but later design detachable valve with assemblies.

YJ As YD but no filter.

U Standard type of pump with metal filter cover. Mounting stud holes arranged diagonally.

UF As U but stud holes side by side.

UG Standard type of pump with glass filter bowl on top. Mounting stud holes arranged diagonally.

FG As UG but mounting holes side by side.

CG Standard size pump, flat link, filter bowl on top, mounting holes side by side.

SU PUMP SERVICE LITERATURE

(Available from the SU Carburetter Company Ltd, Wood Lane, Erdington, Birmingham 24)

General	Sheet No.
Information and identification	AKD 4792 A
List of abridged pump specifications	AKD 4812 C
Pump type/car model reference list	AKD 4813 B
Recommended mounting positions	AKD 4814 A

Description and Fault Diagnosis

L and HP type single pump	AKD 4793 B
LCS type pump	AKD 4794 B
L and HP type dual pump	AKD 4795 B
SP type pump	AKD 4796 B
AUF 200 range pump	AKD 4797 B
AUF 300 range pump	AKD 4798 B
AUF 400 range pump	AKD 4799 B
AUF 500 range pump	AKD 4800 B

Dismantling and Reassembling Instructions

L and HP type single pump	AKD 4801
LCS type pump	AKD 4802
L and HP type dual pump	AKD 4803
SP type pump	AKD 4804
AUF 200 range pump	AKD 4805
Supplement—Plastic Armature Guide Plate	AKD 4805/1
AUF 300 range pump	AKD 4806
AUF 400 range pump	AKD 4807
AUF 500 range pump	AKD 4808

Reconditioning and Testing

SU pump testing instructions (SU test rig)	AKD 4809 A
SU pump reconditioning instructions	AKD 4810 A
SU pump testing instructions (Churchill test rig)	AKD 4811

Index

CPSIA information can be obtained at www.ICGtesting.com
Printed in the USA
LVOW11s0837270813

349699LV00005B/118/P